C0-ANO-855

IONIC EQUILIBRIA IN ANALYTICAL CHEMISTRY

IONIC EQUILIBRIA IN ANALYTICAL CHEMISTRY

HENRY FREISER & QUINTUS FERNANDO
DEPARTMENT OF CHEMISTRY, UNIVERSITY OF ARIZONA

ROBERT E. KRIEGER PUBLISHING COMPANY
MALABAR, FLORIDA

Original Edition 1963
Reprint Edition 1979

Printed and Published by
ROBERT E. KRIEGER PUBLISHING COMPANY, INC.
KRIEGER BUILDING, KRIEGER DRIVE
MALABAR, FLORIDA 32950

Printed in the United States of America

Library of Congress Cataloging in Publication Data

Freiser, Henry, 1920-
Ionic equilibria in analytical chemistry.

Reprint of the edition published by Wiley, New York.
Bibliography: p.
Includes index.
1. Chemistry, Analytic. 2. Ionic equilibrium.
I. Fernando, Wuintus, joint author. II. Title.
[QD75.2.F73 1979] 541.392 79-9746
ISBN 0-88275-955-8

Preface

The plan of this book is to present a pedagogically sound and logical stepwise development of the principles of chemical equilibrium and techniques of calculation. Special effort is made to have the student realize that the material is more than an interesting mathematical exercise. The physical and chemical concepts will be kept in sharp focus and the implications of the work in the several areas of chemistry will be stressed. This book would serve very well as a supplementary text for both qualitative and quantitative analysis, a companion to a laboratory-oriented text or one that stresses mainly stoichiometry. It may also be used as a text for a one-semester course for seniors or first year graduate students. The organization of the book permits the instructor the option of including more advanced material or omitting it without disturbing the continuity.

This book is designed to fill the gap that exists between the relatively superficial treatment of equilibrium principles and calculations presented at the freshman and sophomore level and the relatively complex elaboration of the subject at the senior and graduate level. With proper presentation, a student who has had first year college chemistry is capable of learning chemical equilibrium in a rigorous and penetrating fashion. The material learned at this earlier level will be of considerable use and will not have to be retaught later in a different fashion.

A simple thermodynamic approach to the description of chemical equilibrium is offered as a preferable alternative to the essentially more complex kinetic approach. This approach not only simplifies the derivation of the Mass Action expression but also helps in understanding the basis of the Nernst equation.

In Chapter III the activity concept is developed in a manner that permits practical application to ionic equilibria. The activity corrections are confined to the calculation of the concentration equilibrium constant, K, which is then used without further involvement of activities. A simple method of obtaining the appropriate K value at different ionic strengths which involves a single consulting of tabulated corrections is developed.

Chapter III may be omitted at the instructor's option. Students will still be able to appreciate the effects of deviation from ideality by noticing that the K values given in examples and problems (calculated by us from method shown in Chapter III) change with solution conditions. Since the extent of such changes is often quite marked, the importance of activity corrections in reasonably accurate calculations cannot fail to be realized.

In the past, many of those interested in rigorous ionic equilibrium calculations have been preoccupied with elaborate mathematical expressions. To develop fifth or sixth order equations on the basis of simple principles is not difficult in most cases. To obtain accurate solutions from such equations without tedious calculation, however, requires an appreciation of factors other than the simple arithmetic relationships of the various terms in a high order equation. In order to analyze a complex equilibrium situation it is desirable to employ "chemical intuition" before proceeding directly into the development of mathematical expressions. Keeping in mind also the experimental uncertainty present in K values as well as in measured pH values, it is reasonable to ignore calculation errors of less than 5%. This amounts to an error of 0.02 in log K or pH. Thus an appreciation of the experimental factors which limit the accuracy of equilibrium calculations brings about a more realistic and balanced approach to these calculations.

The use of graphical methods also helps bring the problem of significance of various terms in complicated expressions into proper focus. This book includes the first attempt to treat ionic equilibria by the graphical methods of Sillen. These are very powerful techniques because they simplify the solutions to otherwise extremely complex problems. Also, the pictorial representation permits the reader to see at a glance how the concentrations of various species involved in an equilibrium system change with conditions. For these reasons graphical representations are used throughout the book.

The series of chapters on acid-base equilibria are developed in considerable detail. This has been done for two reasons. First, such calculations are of widespread importance, and second, the same techniques are applicable to the equilibria discussed in successive chapters.

In keeping with the growing recognition of the importance of separation processes other than precipitation we have included a section describing equilibrium calculations employed in liquid-liquid extraction and ion-exchange.

We have tried to incorporate in this book many ideas from our collective teaching experience. Essentially all of the material presented in this text represents an outgrowth of the ideas of other authors, particularly T. B. Smith, Herbert A. Laitinen and Lars Gunnar Sillen. We gratefully acknowledge the help of Gary S. Kozak in proof reading.

Henry Freiser
Quintus Fernando

1963
University of Arizona

Table of Contents

I

Introduction

This chapter is intended as a review of elementary chemical calculations. Before equilibrium calculations are discussed it is essential to become thoroughly familiar with various means of describing quantities of substances that take part in chemical reactions.

I-1. THE MOLE CONCEPT

Measurement of quantities of reacting substances must always be related to the fundamental chemical description of matter, namely to the number of molecules contained in the sample of the substance of interest. Surely there can be no more important characteristic of a sample of matter, from a chemical viewpoint, than the number of chemical units or molecules it contains. Although properties of the sample such as mass, volume, color, odor, etc. are of interest, the number of molecules in a substance permits us to predict easily, from the chemical equation, what quantities of the substances in the same reaction will be required or produced. Let us consider for example the reaction of sodium hydroxide with sulfuric acid according to the equation.

$$2\ NaOH + H_2SO_4 \rightleftharpoons Na_2SO_4 + 2\ H_2O$$

If we have a sample of sodium hydroxide containing 100 molecules, simple inspection of the equation will tell us that 50 molecules of sulfuric acid will be required to react with the sample of sodium hydroxide and that 50 molecules of sodium sulfate and 100 molecules of water will be produced in the reaction. Working with numbers of molecules is a little awkward because there are so many of them in even the minutest portions of matter. This awkwardness can be removed by using Avogadro's number, 6.023×10^{23}, of molecules, as a chemist's "dozen", so to speak. Let us call the amount of matter consisting of this number of molecules a MOLE.

The amount of a substance containing one Avogadro's number of molecules will have a weight in grams equal to its gram molecular weight (G.M.W.). This fact enables us to calculate the number of moles of a substance from its weight in grams. Thus

$$\text{Number of moles} = \frac{\text{weight in grams}}{\text{G.M.W.}}$$

Returning to our example we can now substitute moles for molecules without altering the sense of the statements since the number of moles is a measure of the number of molecules. One hundred moles of sodium hydroxide will require 50 moles of sulfuric acid to produce 50 moles of sodium sulfate and 100 moles of water.

I-2. CONCENTRATION UNITS

Since so much of our work is concerned with solutions we will find it necessary to discuss concentrations, which are defined as amounts of substances dissolved in a unit volume of solution. From among the various methods of expressing concentration we will choose those involving moles as having the greatest chemical interest.

I-2-(a). MOLARITY AND FORMALITY

Molarity (represented by M) is a unit of concentration expressed in terms of moles of solute per liter of solution. Thus if 58.5 g. NaCl are dissolved in enough water to give a solution volume of 2.0 liters, the molarity of the solution is seen to be:

$$M = \frac{58.5\ g./58.5\ g./mole}{2\ liters} = 0.50\ M$$

or, more generally,

$$M = \frac{\text{number of moles}}{V\ \text{liters}}$$

This definition may be rearranged to give

$$\text{Number of moles} = M \times V$$

We now have two methods of calculating the number of moles, namely

$$\text{Number of moles} = \frac{W(g)}{G.M.W.\ (g./mole)} = M \times V \quad (1.)$$

which serve to cover cases when we are dealing with substances either in the pure state or in solution.

It is of utmost importance to avoid confusing moles and molarity—a distinction which many beginning students find difficult to make. Remember that moles represent an amount whereas molarity signifies an amount per unit volume. For example the molarity of either 1.0 ml or 10 liters of 0.10 M HCl is the same; but the amount of HCl in 1.0 ml of 0.10 M HCl is 0.0001 mole and in 10 liters of 0.10 M HCl is 1.0 mole. A 0.1 M solution of NaCl or CH_3COOH or any other substance, is a solution that is made by dissolving one tenth of a mole of the substance in water and making the volume of the solution up to a liter with water. This describes the analytical rather than the equilibrium concentration of the NaCl or CH_3COOH. Some texts describe the analytical concentration by the term, formality, (i.e. the number of gram formula weights of solute per liter of solution), and reserve the molarity for the equilibrium concentration of a given species.

In this text whenever reference is made to an x M solution it will mean that the solution we are describing is made by dissolving x moles (strictly speaking, formula weights) of the substance in water and making the resultant solution up to a liter with water. Such analytical concentrations will be represented by the symbol C with a suitable subscript. Square brackets will be used only to denote the concentrations of the molecular or ionic species in the solution.

The most convenient unit of concentration for use in equilibrium calculations is the molarity. Although liberal use of molarity will be made throughout the text, it must be mentioned that this unit of concentration has certain disadvantages. First, there is no convenient way to calculate the exact amount of solvent present in the solution. Second, since the density of the solution varies with temperature the molarity will also vary.

For example, what is the water content in a 0.1000 M NaCl solution? We know that one liter of solution contains 0.1000 mole (58.45 × 0.1000 g.) of NaCl, but unless we know the weight of a liter of solution we cannot find the weight of water in one liter of the solution. If we find the value of the density of this NaCl solution to be 1.0028 g./ml at 20°C, then we can find the amount of water in the solution. Thus,

$$1.0028 \text{ g/ml} \times 1000 \text{ ml/1} = 1002.8 \text{ g./1}$$
$$1002.8 \text{ g./1} - 5.845 \text{ g./1 NaCl} = 997.0 \text{ g. } H_2O$$

However at 25° C this solution occupies a volume of 1.001 liters which means that the molarity has dropped by 0.1%. (With solutions more concentrated than 0.1000 M this temperature effect may be more pronounced). For most purposes such changes do not introduce any appreciable error in equilibrium calculations.

I-2-(b). MOLALITY

Molality is a unit of concentration expressed as moles of solute per kilogram of solvent. It will be immediately obvious that this method of expressing concentration avoids the drawbacks seen in molarity. The strength of a solution expressed in molality is temperature independent and the amount of solvent easily calculable. This is not achieved however without sacrificing our knowledge of the exact volume of the solution. This is a distinct disadvantage in analytical operations involving volumetric ware. In such operations use of molarity is preferred. However, in all equilibrium calculations, the advantage of having a concentration unit which is related to a definite composition makes molality the preferred choice. In this text, since most aqueous solutions described are relatively dilute, the molarity is sufficiently close to the molality for it to be used in its place as a reasonable approximation.

I-2-(c). MOLE FRACTION

Mole fraction is a unit of concentration expressed as the ratio of the number of moles of solute to the total number of moles of all components in the solution. This unit of concentration is listed here mainly for the sake of completeness. Mole fraction would be a desirable means of expressing concentrations for equilibrium purposes and indeed is often used for simple solutions containing a few components.

I-2-(d). NORMALITY

Normality is the unit of concentration expressed as the number of chemical equivalents of solute per liter of solution. This method of expressing concentrations is especially helpful in complex stoichiometric calculations. Normality is never to be used in equilibrium expressions.

I-2-(e). LOGARITHMIC EXPRESSION OF CONCENTRATION

In many places we will be using logarithmic expressions of concentrations and equilibrium constants. Whereas the theoretical significance of this will be described later, it will suffice now to mention its convenience by illustration with the well known concept of pH. The use of pH, a logarithmic relation of the hydrogen ion concentration, permits us to handle conveniently changes of many orders of magnitude in the hydrogen ion concentration. For example if the hydrogen ion concentration of a solution changes from 2.30×10^{-3} M to 3.80×10^{-11}M, the change can be represented more conveniently by saying that the pH changes from 2.64 to 10.42.

With pH as with a number of other logarithmic concentrations two place logarithms are usually adequate in relation to the accuracy of experimental measurements or the validity of the calculation that is carried out. For this reason as well as for its benefit in rapid mental calculations it is recommended that the student learn and memorize the two place logarithm table. This is not as formidable a task as it might appear at first. If the logarithms of 2, 3 and 7 are known to three places, the logarithms of the rest of the numbers may be easily derived.

$\log 2.0 = 0.301$
$\log 3.0 = 0.477$
$\log 7.0 = 0.845$

Hence, $\log 5.0 = \log 10 - \log 2.0 = 0.70$, $\log 6.0 = \log 3.0 + \log 2.0 = 0.78$ and $\log 9.0 = 2 \log 3.0 = 0.95$ (See problem 22 at the end of this chapter.)

SUGGESTIONS FOR FURTHER READING

W. F. Kieffer, The Mole Concept in Chemistry, Reinhold Publishing Corporation, New York, 1962.

L. F. Hamilton and S. G. Simpson, Calculations of Analytical Chemistry, McGraw Hill Book Co., Inc., New York 1960, pp. 1-128.

H. H. Willard, N. H. Furman and C. E. Bricker, Elements of Quantitative Analyses, D. Van Nostrand Co. Inc., New York, 1956, pp. 91-108 and pp. 299-361.

I. M. Kolthoff and S. B. Sandell, Textbook of Quantitative Analysis, The Macmillan Co., New York (1952), pp. 1-31.

PROBLEMS

1. What volume of 70.0% $HClO_4$,* (density 1.67 g/ml), is required to neutralize 125.0 ml of a 1.05 M NaOH solution?
2. A 0.5000 g. sample of an alloy gave 0.550 g. of nickel dimethylglyoxime complex, $Ni(C_4H_7N_2O_2)_2$ and 0.0332 g of of PbO_2. Calculate the percentage of nickel and lead in the alloy.
3. A 25.00 ml aliquot of H_2SO_4 solution is precipitated with $BaCl_2$ and the weight of the precipitate is 1.333 g. Calculate the molarity of the H_2SO_4.
4. Fuming sulfuric acid contains 20.0% free SO_3. If 1.250 g of this acid is dissolved in water and an excess of $BaCl_2$ added what is the weight of the precipitate formed?
5. A mixture contains only two constituents, NaCl and KCl. When a 1.25 g. sample of this mixture is precipitated with $AgNO_3$, 2.50 g. of AgCl were obtained. Calculate the percentage of Na in the original mixture.
6. A 5.00 g. sample of magnesium metal contains 2.50% of an inert impurity. What is the minimum volume of hydrochloric

*All percentages, unless otherwise specified, are weight by weight.

acid of specific gravity 1.18 and containing 36.0% of HCl that is required to react with all the magnesium in the sample?

7. What is the minimum volume of 0.100 M $AgNO_3$ required to precipitate the chromate and chloride quantitatively from a mixture containing 0.100 g K_2CrO_4 and 0.100 g. KCl?
8. A sample containing NaCl and KCl only and weighing 0.2022 g was treated with platinic chloride and a precipitate of K_2PtCl_6 weighing 0.0335 g was obtained. Calculate the percentage of K and Na in the original sample.
9. To 100.0 ml of a solution which is 0.150 M in $AgNO_3$ and 0.050 M in $Ba(NO_3)_2$, 50.0 ml of 0.400 M KCl and 50.0 ml of 0.300 M K_2SO_4 are added.
   (a) What is the total weight of the precipitate that is obtained?
   (b) What are the concentrations of K^+ and NO_3^- remaining in the solution?
10. How many ml of HCl (density 1.19 g/ml and containing 38.0% of HCl) are required to make 20.0 liters of a 0.160 M solution?
11. The density of 14.7 M phosphoric acid is 1.60 g/ml. Calculate the number of grams H_3PO_4 that are present in 1000 g of the acid.
12. The density of a 99.5% acetic acid solution is 1.057 g/ml. What is its molarity?
13. The density of a 28.0% ammonia solution is 0.900 g/ml. Calculate the molarity and the molality of the solution.
14. A solution containing 12.50 g methyl alcohol, 15.61 g. ethyl alcohol, 15.00 g acetone and 125.0 g water. Calculate the mole fraction of each component.
15. What is the mole fraction of the solute in a 1.50 molal aqueous solution?
16. Derive a general relationship between molality and mole fraction for an aqueous solution. At what molality will a simple proportionality to mole fraction be in error by 5.0%.
17. What is the molarity of an H_2SO_4 solution made by dissolving 85 g. SO_3 in water and making the resultant solution up to the mark, with water, in a 100 ml volumetric flask?
18. In what ratio must 98% H_2SO_4 and fuming sulfuric acid containing 20% free SO_3, be mixed in order to obtain 100% H_2SO_4?
19. Three solutions of sulfuric acid contain 75%, 80% and 85%

H_2SO_4. How many grams of 75% acid and 80% acid should be mixed in order to obtain 100 g of 78% acid? How many grams each of 80% acid and 85% acid should be mixed in order to obtain 100 grams of 83.3% acid?

20. Calculate the molality of the resulting solution if 100 g of 76.5% H_2SO_4 is mixed with 100 g of 83.3% H_2SO_4.
21. How many grams of fuming sulfuric acid containing 20% free SO_3 must be added to 100 grams of 95% H_2SO_4 solution to give a 98% H_2SO_4 solution?
22. Complete the following table as a two-place logarithm table without recourse to a logarithm table or slide rule.

Number	Logarithm	Number	Logarithm
1.2		4.0	
1.4		5.0	
1.6		6.0	
2.0	0.301	7.0	0.845
2.5		8.0	
3.0	0.477	9.0	

This table may be used to give logarithms correct to ± 0.02 (equivalent to an accuracy of 5% in the number). Intermediate values may be obtained by linear interpolation.

II

Chemical Equilibrium

The study of chemical reactions cannot be based on stoichiometry alone because many reactions are reversible and do not go to completion. For example if we were to mix at ordinary temperatures and pressures, one mole of nitrogen gas and three moles of hydrogen gas we would get far less than the expected two moles of ammonia.

$$N_2 + 3\ H_2 \rightleftharpoons 2\ NH_3$$

Stoichiometric principles tell us that for every mole of nitrogen that reacts, three moles of hydrogen will also react and two moles of ammonia will be formed. The amount of reaction that occurs, that is, the extent of reaction, depends in a complex fashion upon a number of reaction conditions, e.g. concentrations of each of the components (i.e. composition) in the reaction mixture, temperature, pressure, time, etc. It is useful to single out time and composition from all other variables by noting that under a given set of conditions the composition reaches a constant value after a certain time has elapsed. This composition defines the position of equilibrium (unless of course the reaction is irreversible).

II-1. REVERSIBLE REACTIONS

A reversible reaction is one that can be made to go either in the forward or the reverse direction by appropriately adjusting reaction conditions. For example the reaction $Cu^{++} + Zn \rightleftharpoons Cu + Zn^{++}$ goes in the forward direction in the Daniell cell. If a greater and opposing e.m.f. is applied across the Daniell cell then the above reaction will proceed in the reverse direction. Of course if the opposing e.m.f. is exactly equal to that of the Daniell cell then no current will flow and no chemical reaction will take place. At this point of balance only a slight change in the driving force i.e. the opposing e.m.f., will cause a reversal of the direction of the chemical reaction. Notice that reversibility is achieved only at (or very close to) the point of balance or equilibrium position.

In summary then a reversible reaction is one in which a slight change in driving force, or to put it another way, in conditions (temperature, pressure composition, etc.) will cause a reversal of the direction of the reaction.

II-2. FREE ENERGY AND CHEMICAL EQUILIBRIUM

The driving force of a chemical reaction occurring at constant temperature and pressure may be described in terms of the displacement of the system from equilibrium using a quantity called the free energy, F. Expressions for the free energy functions of various substances assume the following forms:

For one mole of an ideal substance in the gas phase:

$$F = F^\circ + RT \ln p \tag{II-1}$$

For one mole of a pure liquid or pure solid:

$$F = F^\circ \tag{II-2}$$

For one mole of a substance in solution:

$$F = F^\circ + RT \ln a \tag{II-3}$$

Where the value of F° depends not only on the nature of the substance but also on the state, gas, liquid, solid or solution, and in the latter upon the nature and amount of all of the other solution components present including the solvent. R is the universal gas constant, T the absolute temperature. The activity of the solute, a, is a dimensionless quantity which is a measure of the relative concentration of the solute; p has a similar significance for gases except that partial pressures are involved. If the change in free energy ($F_{final} - F_{initial}$) or ΔF for a process or reaction is negative the process or reaction will tend to proceed spontaneously, (not necessarily rapidly). A positive ΔF, on the other hand, corresponds to a tendency for the process to reverse spontaneously. A system is in equilibrium when the free energy change of ΔF is zero.

In summary then for chemical reactions at constant temperature and pressure:

$\Delta F = -$ reaction tends to proceed spontaneously
$\Delta F = 0$ reaction is at equilibrium
$\Delta F = +$ reaction tends to reverse itself.

II-3. FREE ENERGY AND THE EQUILIBRIUM CONSTANT

By applying the free energy functions to the general reaction system in solution:

$$aA + bB + \ldots\ldots \rightleftharpoons pP + qQ + \ldots\ldots \quad \text{(II-4)}$$

an expression which will relate the concentration variables in the equilibrium system can be derived. This expression called the equilibrium constant expression provides the basis for all calculations and predictions concerning the effects of changes in composition upon systems in chemical equilibrium.

For the reaction (II-4),

$$\Delta F = (pF_P + qF_Q + \ldots\ldots) - (aF_A + bF_B + \ldots\ldots) \quad \text{(II-5)}$$

In equation (II-5) each of the free energy terms (which represent free energies per mole) has been multiplied by the appropriate number of moles in the balanced chemical equation. Substituting from equation (II-3), since all of the reaction components are in solution, equation (II-5) becomes:

$$\Delta F = (pF^\circ_P + pRT \ln a_P + qF^\circ_Q + qRT \ln a_Q + \ldots\ldots)$$
$$- (aF^\circ_A + aRT \ln a_A + bF^\circ_B + bRT \ln a_B + \ldots) \quad \text{(II-6)}$$

where the a values correspond to the activities of the components in the reaction mixture. Separating the concentration dependent terms from the others we have:

$$\Delta F = pF^\circ_P + qF^\circ_Q + \ldots\ldots - aF^\circ_A - bF^\circ_B \ldots\ldots$$

$$+ RT \ln \frac{a_P^p \cdot a_Q^q \ldots}{a_A^a \cdot a_B^b \ldots} \quad \text{(II-7)}$$

The portion of the righthand side of the equation (II-7) which is independent of concentration is called $\Delta F°$, i.e.

$$\Delta F° = pF_P^° + qF_Q^° + \ldots - aF_A^° - bF_B^°$$

Incorporating this in Equation II-7

$$\Delta F = \Delta F° + RT \ln \frac{a_P^p \cdot a_Q \cdot \ldots}{a_A^a \cdot a_B^b \cdot \ldots} \qquad \text{(II-8)}$$

At equilibrium $\Delta F = 0$. Hence

$$-\Delta F° = RT \ln \frac{a_P^p \cdot a_Q^q \text{---}}{a_A^a \cdot a_B^b \text{---}} \qquad \text{(II-9)}$$

Taking the antilogarithm of equation (II-9) we obtain:

$$e^{-\frac{\Delta F°}{RT}} = K = \frac{a_P^p \cdot a_Q^q \cdot \text{---}}{a_A^a \cdot a_B^b \cdot \text{---}} \qquad \text{(II-10)}$$

Inasmuch as the F° terms and hence $\Delta F°$ do not vary with concentration,

$$e^{-\frac{\Delta F°}{RT}}$$

is a constant at constant temperature and is called the equilibrium constant, **K**.

Another useful application of equation (II-8) is for the prediction of the direction of reaction in a mixture of any initial composition. The activity quotient (the argument of the logarithmic term) in equation (II-8) of the mixture is calculated on the basis of the initial composition and ΔF evaluated. The direction of the reaction is now readily obtained using the sign of ΔF as criterion. A simpler corollary of this involves comparing the value of **K** for the reaction with the activity quotient.

This leads to an equivalent set of criteria for prediction of the direction of reaction based on the sign of ΔF:

K > activity quotient	reaction tends to proceed spontaneously (composition changes so as to increase activity quotient)
K = activity quotient	reaction is at equilibrium (composition does not change)

K $<$ activity quotient	reaction tends to reverse itself (composition changes so as to decreaseactivity quotient)

An illustration of the use of these criteria is in the development of the rules for precipitation (see Chap. VIII-4).

Although a full discussion of the concept of activity is given in Chapter III, some remarks about the relation between activity and concentration will be relevant at this stage. In general the activity of a substance in solution may be related to its concentration. Thus,

$$a_A = [A] \cdot \gamma_A \tag{II-11}$$

where γ_A is a factor which varies with the total composition of the solution and is called the activity coefficient of A. In extremely dilute solutions of most solutes the value of γ_A approaches unity and the activity is numerically equal to the concentration, although their dimensions are not the same. Hence for solutes in dilute solutions we may write equation (II-10) as follows:

$$K = \frac{[P]^p \cdot [Q]^q \cdot \text{---}}{[A]^a \cdot [B]^b \cdot \text{---}} \tag{II-12}$$

II-4. EQUILIBRIUM EXPRESSIONS FOR VARIOUS TYPES OF REACTIONS

In developing the equilibrium expression (II-10) a reaction (II-4) was described in which all of the components were in solution. From the nature of the free energy functions given in equation (II-1) and (II-2) it follows that any gaseous component will be represented in the equilibrium expression by its partial pressure and any pure liquid or solid by unity since the logarithmic term is absent in equation (II-2). Whenever the solvent appears in the chemical equation its free energy is considered to be sufficiently close to that of the pure liquid, provided the solutions are reasonably dilute so that it too is represented in the equilibrium expression by unity.

The following examples will illustrate the types of equilibrium expressions usually encountered.

(a) The Dissociation of Water.

$$2\ H_2O \rightleftharpoons H_3O^+ + OH^-$$

$$K_w = [H_3O^+][OH^-]$$

(b) The Dissociation of Ammonia.

$$NH_3 + H_2O \rightleftharpoons NH_4^+ + OH^-$$

$$K_b = \frac{[NH_4^+][OH^-]}{[NH_3]}$$

(c) The Stepwise Acid Dissociation of Carbonic Acid.

$$H_2CO_3 + H_2O \rightleftharpoons H_3O^+ + HCO_3^-$$

$$HCO_3^- + H_2O \rightleftharpoons H_3O^+ + CO_3^=$$

$$K_1 = \frac{[H_3O^+][HCO_3^-]}{[H_2CO_3]}$$

$$K_2 = \frac{[H_3O^+][CO_3^=]}{[HCO_3^-]}$$

(d) The Stepwise Formation of the Diaminosilver (1) Complex.

$$Ag^+ + NH_3 \rightleftharpoons Ag(NH_3)^+$$

$$Ag(NH_3)^+ + NH_3 \rightleftharpoons Ag(NH_3)_2^+$$

$$K_1 = \frac{[Ag(NH_3)^+]}{[Ag^+][NH_3]}$$

$$K_2 = \frac{[Ag(NH_3)_2^+]}{[Ag(NH_3)^+][NH_3]}$$

Note that formation constants, rather than dissociation constants, which are the reciprocals of the formation constants, will be used for complex ion equilibria in accord with the practice of most workers in this field.

(e) The Solubility of Silver Chromate.

$$Ag_2CrO_4\ (\text{solid}) \rightleftharpoons 2\ Ag^+ + CrO_4^=$$

$$K_{s.\,p.} = [Ag^+]^2[CrO_4^=]$$

(f) The Oxidation-Reduction Reaction between Fe(III)-Fe(II) and Ce(IV)-Ce(III) Couples.

$$Fe^{2+} + Ce^{4+} \rightleftharpoons Fe^{3+} + Ce^{3+}$$

$$K = \frac{[Fe^{3+}][Ce^{3+}]}{[Fe^{2+}][Ce^{4+}]}$$

(g) The Thermal Decomposition of Calcium Carbonate.

$$CaCO_3 \rightleftharpoons CaO + CO_2 \qquad K = p_{CO_2}$$

In all calculations it is essential to employ the appropriate value of the concentration equilibrium constant which is dependent on the ionic strength of the solution. In all exercises in this book beyond Chapter III, appropriate values of concentration equilibrium constants will be furnished. The manner in which such values are obtained from the thermodynamic equilibrium constant, and the ionic strengths of the solutions is shown in Chapter III.

II-5. FACTORS THAT AFFECT THE EQUILIBRIUM CONSTANT

II-5-(a). EFFECT OF COMPOSITION

Since the assumption that the values of the activity coefficients are unity is valid only in the special case of the extremely dilute solution, values of equilibrium constants as defined by equation (II-12) are not truly constant. Values of these constants, referred to as concentration constants, K, will vary with the composition of the solution. Since in most instances the composition of the solution can be fully described, such constants may be usefully employed for accurate calculations. It is imperative to employ that value of K which is appropriate for the particular composition. For solutions of electrolytes the most important composition parameter which determines the value of K is the ionic strength, the relationships between K and the ionic strength are given in Chapter III.

II-5-(b). EFFECT OF TEMPERATURE

Suppose we are dealing with a reaction which is endothermic, i.e. one which takes place with the absorption of heat energy. A change in the conditions by adding heat energy through a temperature rise would result in an increase in the extent of this reaction inasmuch as this increase tends to minimize the effect of the temperature change. Among the examples of endothermic reactions of interest is the self-ionization of water. For example:

$$H_2O \rightleftharpoons H^+ + OH^-, \ \Delta H° = 13.8 \text{ kcal./mole}$$

where $\Delta H°$ is the standard heat (enthalpy) of reaction; as might be predicted the dissociation of water is higher at 100°C than at 20°C. Conversely when the reaction is exothermic, the extent of reaction will decrease as the temperature increases. A number of weak acids have small heats of dissociation, e.g. $\Delta H°$ for acetic acid is -0.1 kcal/mole, for formic acid -0.01 kcal/mole, for boric acid is $+3.4$ kcal/mole. In such instances, the extent of reaction varies only slightly with temperature.

A quantitative expression of the effect of temperature upon the equilibrium constant is:

$$\mathbf{K} = A e^{-\Delta H°/RT} \qquad \text{(II-13)}$$

where the change in **K**, the equilibrium constant, with T, the absolute temperature is seen to be a function of $\Delta H°$. The factor A is usually constant over a small temperature range, and relates $\Delta H°$ to $\Delta F°$ (see equation II-10).

II-5-(c). EFFECT OF SOLVENTS

In the majority of reactions that we will be concerned with, water is the solvent that is used. Water is distinctive in having a very high dielectric constant of 78.5 at 25°C compared with 24.3 for ethyl alcohol and 4.2 for diethyl ether. The higher the dielectric constant of a medium the easier it is for ions to be separated in the medium, and therefore dissociation can occur more easily in water than in common organic liquids. This effect is illustrated by the manner in which the dissociation constant of acetic acid changes in a series of solvents having different dielectric constants (Table II-1). In the same connection note the effect of adding increasing amounts of dioxane to water on the dissociation constant of water, Table II-2. These two examples are typical of the effect of dielectric constant upon reactions which give rise to electrical charge separation. Naturally, the reverse type of reaction, one in which charge neutralization occurs will be favored by a reduction in the dielectric constant.

All ions in solution are solvated, some ions being solvated to a greater extent than others. It is to be expected that if we change the solvent the environment of the ions will also change, and the specific properties of the ions as well as the solvent molecules will determine to a large extent the size and type of solvation shell that will be formed around the ion. For example if mixed solvents are used, such as a mixture of dioxane and water, the specific properties of the electrolyte ions may very well exert a "sorting" effect on the solvent molecules. This would mean that the electrolyte ions may be surrounded by more solvent molecules of water than dioxane. These considerations point to the fact that it may be easier for ions to exist in one type of solvent rather than in another, and therefore, the equilibrium constant for the dissociation of a substance will be dependent on solvation effects.

Table II-1

Effect of Solvent on the Dissociation Constant of Acetic Acid at 25°C

Solvent	Dielectric Constant	Dissociation Constant
Water	78.5	2×10^{-5}
Methyl Alcohol	32.6	5×10^{-10}
Ethyl Alcohol	24.3	5×10^{-11}

Table II-2

Ion Product of Water in Dioxane-Water Mixtures at 25°C

Weight Percentage of Dioxane	Dielectric Constant	Ion Product Constant
0	78.5	1.01×10^{-14}
20	60.8	2.40×10^{-15}
45	38.5	1.81×10^{-16}
70	17.7	1.40×10^{-18}

II-6. IMPORTANCE OF EQUILIBRIUM CALCULATIONS

Complex mixtures, in which a whole series of chemical reactions can occur, are frequently encountered. In such mixtures the final composition is quite different from that of the initially prepared mixture. A detailed knowledge of the composition of such mixtures is essential for the understanding of their chemistry. Short of actually measuring the amount of each and every species present, the only way to obtain this knowledge is from a consideration of the appropriate equilibria.

In planning analytical procedures, questions of the following sort can be answered by information obtained through equilibrium calculations: (a) Will a substance precipitate (or separate in another manner such as by solvent extraction or ion exchange) from solution, and if so, to what extent? Will other substances separate under the same conditions? If so, can we alter conditions by using complexing agents or controlling the acidity of the solution, to obtain a selective separation? (b) What are the characteristics of reactions that are of practical use in methods of determination? Do they proceed to a sufficient extent? What are the nature and extent of any side reactions? In titrimetry, what are the criteria for the selection of suitable indicators?

In trying to interpret the physical, chemical, or biological properties of solutions, it is equally essential to know their detailed composition since each species that is present makes a unique contribution to these properties.

The determination of equilibrium constants is naturally of utmost importance for the solution of problems of the type outlined above. In addition, the equilibrium constant (and ΔF°) provides a means of characterizing a chemical reaction and is useful in theoretical considerations. A study of equilibrium constant values provides a basis for evaluation of the influence on chemical reactions of the many factors related to chemical structure. In this connection, ΔH° as well as ΔF°, is of importance.

The orientation of this book is largely based on the assumption that all appropriate equilibrium constant values are available for use in the calculations described. Descriptions of the means of the accurate determination of equilibrium constants is beyond the scope of this book. However, the essential features of such determinations are implicit in the expressions derived in the text. That is, instead of having the concentration variables dependent on K, the situation can be readily reversed.

SUGGESTIONS FOR FURTHER READING

E. J. King, Qualitative Analysis and Electrolytic Solutions, Chapter 9, Harcourt, Brace and World, Inc., New York (1959).

T. S. Lee, Chapter 7, Part 1, Vol. 1, Treatise on Analytical Chemistry, Interscience, New York (1959), I. M. Kolthoff and P. J. Elving, Editors.

II-7. PROBLEMS

1. Write expressions for the equilibrium constants for each of the following systems. Unless otherwise designated all species are in aqueous solution. (s = solid, g = gas)
   (a) $CaCO_3(s) \rightleftharpoons Ca^{++} + CO_3^{=}$
   (b) $NH_3 + H_2O \rightleftharpoons NH_4^{+} + OH^{-}$
   (c) $CO_2(g) + H_2(g) \rightleftharpoons CO(g) + H_2O(g)$
   (d) $Bi_2S_3(s) + 6\,H^{+} \rightleftharpoons 3\,H_2S + 2\,Bi^{3+}$
   (e) $H_2S \rightleftharpoons 2\,H^{+} + S^{=}$
   (f) $BaSO_4(s) + CO_3^{=} \rightleftharpoons BaCO_3(s) + SO_4^{=}$
   (g) $Ag(CN)_2^{-} \rightleftharpoons AgCN(s) + CN^{-}$
   (h) $Ag_2O(s) + H_2O \rightleftharpoons 2\,Ag^{+} + 2\,OH^{-}$
   (i) $2\,H_2O_2 \rightleftharpoons 2\,H_2O + O_2(g)$
   (j) $Hg_2Cl_2(s) \rightleftharpoons Hg_2^{++} + 2\,Cl^{-}$
2. Write expressions for the equilibrium constants for the following systems which represent the reactions between hydrogen and iodine in the vapor phase.
   (a) $H_2 + I_2 \rightleftharpoons 2\,HI$
   (b) $\frac{1}{2}H_2 + \frac{1}{2}I_2 \rightleftharpoons HI$

   Should the equilibrium constants for the systems (a) and (b) be the same or different? Will a mixture initially containing 1 mole/1. of each of the components reach the same equilibrium composition in (a) (b)? Explain.
3. Consider the following system that is in equilibrium:

   $2\,SO_3 \rightleftharpoons 2\,SO_2 + O_2 - 42$ kcals.

   (a) What is the effect of adding more oxygen to the system?
   (b) What is the effect of adding more SO_3 to the system?
   (c) What is the effect of adding more SO_2 to the system?

(d) What is the effect of increasing the total pressure of the system?
(e) What is the effect of decreasing the temperature of the system?

4. Describe systems, i.e., write chemical equations, for which each of the following equations hold.

(a) $K = \frac{[HCO_3^-]}{[H^+][CO_3^=]}$ (c) $K = \frac{[H_2CO_3]}{[CO_2]}$

(b) $K = \frac{[H^+]^2[CO_3^=]}{[H_2CO_3]}$ (d) $K = p_{CO_2}$

5. If at 0°C, 100 g. of water dissolves 42 g. of $ZnSO_4$ and at 100°C, 100 g. of water dissolves 81 g of $ZnSO_4$, is heat evolved or absorbed when $ZnSO_4$ dissolves in water?
6. The solubility of $CdSO_4$ in water at 0°C is greater than its solubility at 100°C. Is the dissolution of $CdSO_4$ in water endothermic or exothermic?
7. The equilibrium constant of a reaction is 1.5×10^{-5} at 25° C and 7.5×10^{-5} at 45°C. Calculate $\Delta H°$ for the reaction (i.e. the standard heat of reaction).
8. The solubility of calcium sulfate is 2.09 g/l at 30° C and 1.62 g/l at 100° C. Calculate its heat of solution.
9. At 800°C sufficient CO_2 is introduced into a closed vessel containing equal quantities of solid $CaCO_3$ and CaO to give an initial pressure of 0.30 atmospheres. At this temperature, for the reaction $CaCO_3 \rightleftharpoons CaO + CO_2$, $\Delta F° = 3.24$ kilocalories per mole.
   (a) Calculate ΔF for the initial system
   (b) What is the partial pressure of CO_2 at equilibrium?
   (c) How will the answer to (b) change if the quantity of $CaCO_3$ is doubled?
10. A solution is prepared by mixing 0.01 moles solid $NaHSO_4$, and 0.01 moles of solid Na_2SO_4 with 1.0 liter of 1.0×10^{-3} M HCl.
    For the reaction $HSO_4^- \rightleftharpoons H^+ + SO_4^=$, $K = 1.0 \times 10^{-2}$
    At equilibrium will the hydrogen ion concentration be greater than or less than 1.0×10^{-3} M?
11. The ion product of water at 0°C is 1.1×10^{-15}, at 25°C is 1.0×10^{-14} and at 60°C is 9.6×10^{-14}. Calculate $\Delta H°$.

III

The Application of the Activity Concept

III-1. ACTIVITY AND ACTIVITY COEFFICIENT

In Chapter II the free energy of a substance A was defined in terms of the activity a_A as follows:

$$F_A = F_A^\circ + RT \ln a_A \tag{III-1}$$

where a_A is the activity is given by the equation:

$$a_A = \gamma_A \cdot C_A \tag{III-2}$$

The activity is a measure of the concentration of a substance related to its concentration at a universally adopted reference or standard state. Since we are dealing with real rather than ideal substances in which concentrations as such do not accurately represent the behavior of these substances, the activity is a corrected or "effective" concentration ratio. The activity coefficient, γ, incorporates both the concentration at the standard state and the correction arising from non-ideal behavior, and hence has units of reciprocal concentration.

III-2. ACTIVITY AND THE EQUILIBRIUM EXPRESSION

Using equation III-1 to express the free energy we may now derive, in the manner used for equation II-10, the equilibrium expression for the reaction:

$$aA + bB \rightleftharpoons cC + dD \tag{III-3}$$

$$\mathbf{K} = \frac{a_C^c \times a_D^d}{a_A^a \times a_B^b} \tag{III-4}$$

K is a true constant being independent of all variables except temperature and is called the thermodynamic equilibrium constant.

The exact expression for the thermodynamic equilibrium constant in terms of molar concentrations and activity coefficients can now be obtained by substituting for each a term in equation III-4.

$$\mathbf{K} = \frac{C_c^c \times C_D^d}{C_A^a \times C_B^b} \times \frac{\gamma_C^c \times \gamma_D^d}{\gamma_A^a \times \gamma_B^b} \tag{III-5}$$

In equation II-12 a constant K was defined in terms of concentrations of reactants and products. Let us call this constant K to distinguish it from **K** the true or thermodynamic constant. Hence it can be seen that equation III-5 may also be written as follows:

$$\mathbf{K} = K \left\{ \frac{\gamma_C^c \times \gamma_D^d}{\gamma_A^a \times \gamma_B^b} \right\} \tag{III-6}$$

Values of **K** not K are usually tabulated since K values will vary with the activity coefficient factor. In the appendix of this book will be found a tabulation of **K** values for a wide variety of useful reactions.

Implicit in equation III-6 is the basis for a very important approach to the incorporation of activity coefficients in all equilibrium calculations. As will be discussed later, it will be possible to arrive at a sufficiently good value of the activity coefficient quotient, $\gamma_C^c \times \gamma_D^d / \gamma_A^a \times \gamma_B^b$, in terms of a single experimental parameter, namely the ionic strength. Hence by using the ionic strength and the tabulated value of **K** it will be a simple matter to arrive at a value of K which applies to the problem at hand. The significance of this lies in the ability to tackle the remainder of the equilibrium calculation without the need of referring ever again to activity coefficients until the final answer is obtained.

In equation III-5 the activity coefficient quotient, $\gamma_C^c \times \gamma_D^d / \gamma_A^a \times \gamma_B^b$, should be unity if the ions in solution behaved in an ideal manner, i.e., if they had no effect on each other and were not restrained in their movement in solution in any way by neighboring ions of solute. Since ions carry a charge there is a coulombic force that exists between them, but in extremely dilute solutions these forces are minimized and the solution approaches ideal behavior. As a consequence of this behavior the activity coefficients all approach unity as the solution approaches infinite dilution.

Since it is recognized that interionic forces do exist, their effects have to be corrected for in some manner if the behavior of ions in solution is to be described accurately. Since interionic forces will depend on ionic charge and concentration in solution, let us introduce the ionic strength of a solution which measures this.

III-3. THE IONIC STRENGTH OF A SOLUTION

The ionic strength of a solution is defined as follows:

$$\mu = \frac{1}{2} \sum C_i \times z_i^2 \qquad \text{(III-7)}$$

where C_i is the concentration in moles/liter of an ion i, and z_i is the charge on the ion.

In an aqueous solution which is 0.1 M in KCl, the concentration of K^+ is 0.1 M and that of Cl^- is 0.1 M, and the K^+ and Cl^- ions carry unit charges. Therefore the ionic strength of the solution is given by:

$$\mu = \frac{1}{2} (0.1 \times 1^2 + 0.1 \times 1^2)$$

$$\mu = 0.1$$

It is readily seen that the ionic strength of any electrolyte is proportional to its concentration. The ionic strength of 0.1 M K_2SO_4 is equal to $\frac{1}{2}(0.1 \times 2 \times 1^2 + 0.1 \times 2^2) = 0.3$

Example 1.

Show that the ionic strength of a solution containing C moles/liter of a strong electrolyte $Fe_2(SO_4)_3$ is proportional to its concentration.

Since the substance is a strong electrolyte we can assume that it is completely dissociated into 2 Fe^{3+} ions and $3(SO_4)^{2-}$ ions. Since the concentration of the electrolyte is C molar,

The molar concentration of $Fe^{3+} = C \times 2$

The molar concentration of SO_4^{2-} = C × 3

Therefore the ionic strength = $1/2\,(2C \times (3)^2 + 3C \times (2)^2)$

$$\mu = \frac{(18 + 12)C}{2} = 15C$$

Therefore the ionic strength is proportional to C.

The student should work out for himself what this proportionality constant is for electrolytes of other charge types.

Example 2.

Calculate the ionic strength of a solution which is 0.100 M in NaCl, 0.030 M in KNO_3 and 0.050 M in K_2SO_4.

Method A: $Na^+ = Cl^- = 0.100M$
$K^+ = 0.030 + 0.050 \times 2 = 0.130$ M
$NO_3^- = 0.030$ M
$SO_4^{=} = 0.050$ M

Therefore $\mu = 1/2(0.100 + 0.100 + 0.130 + 0.030 + 0.050 \times 2^2) = 0.280.$

Method B: Since the ionic strength is given by: = k × M where k is a constant for each type of electrolyte, k for NaCl and KNO_3 is 1 and for K_2SO_4 is 3, therefore the ionic strength = 1 × 0.100 + 1 × 0.030 + 3 × 0.050
$\mu = 0.280.$

III-4. THE DEBYE-HÜCKEL THEORY OF STRONG ELECTROLYTES

It is possible to derive an expression for the activity coefficient of an ion by considering the balance between the effect of coulombic attraction on ions which serves to constrain their movement and the thermal agitation which counteracts this restraint. It was on this basis that Debye and Hückel developed a theory which gave the activity coefficient, γ_i, of an ion i, having a charge z_i in a solution of ionic strength μ.

$$-\log \gamma_i = \frac{Az_i^2\sqrt{\mu}}{1 + Ba\sqrt{\mu}} \qquad \text{(III-8)}$$

A and B are constants and equal to 0.51 and 3.3 × 10^{+7} respectively at 25° in aqueous solution; $\underline{a}$ is the ion size parameter which is a measure of the diameter of the hydrated ion. Values of $\underline{a}$ for a number of common ions are given in Table III-1.

Table III-1

Ion Size Parameters

Ions	Ion Size Parameters $a \times 10^8$ cm
Sn^{4+}, Ce^{4+}	11
H^+, Al^{3+}, Fe^{3+}, Cr^{3+}.	9
Mg^{2+}.	8
Li^+, Ca^{2+}, Cu^{2+}, Zn^{2+}, Sn^{2+}, Mn^{2+}, Fe^{2+}, Ni^{2+}, Co^{2+}.	6
Sr^{2+}, Ba^{2+}, Cd^{2+}, Hg^{2+}, $S^=$, CH_3COO^-, $(COO)_2^=$	5
Na^+, $H_2PO_4^-$, Pb^{2+}, $CO_3^=$, $SO_4^=$, $CrO_4^=$, $HPO_4^=$, PO_4^{3-}.	4
OH^-, F^-, CNS^-, SH^-, ClO_4^-, Cl^-, Br^-, I^-, NO_3^-, K^+, NH_4^+, Ag^+.	3

The combined effect of the charges on ions and their concentrations is expressed as the ionic strength of a solution. It must be noted that the ionic strength of a solution depends on the concentration of all the ions that are present in the solution and not on just the ionic species that are involved in a particular equilibrium. In equation III-8 the value of $\mu^{1/2}$ can be calculated for any solution, but the constants A, and B and the ion size parameter, a, have to be known before the activity coefficient of an ion can be calculated.

Example 3.

Calculate the activity coefficient of the Pb^{2+} ion in (a) a solution which is 0.005 M in $Pb(NO_3)_2$ and (b) a solution which is 0.005 M in $Pb(NO_3)_2$ and 0.040 M in KNO_3.

(a) The ionic strength of 0.005 M $Pb(NO_3)_2$ is $3 \times 0.005 = 0.015$.

Therefore,

$$-\log \gamma_{Pb^{2+}} = \frac{0.51 \times 4 \times (0.015)^{1/2}}{1 + 3.3 \times 10^7 \times 4 \times 10^{-8} \times (0.015)^{1/2}}$$

$$= 0.22$$

Hence, $\gamma_{Pb^{2+}} = 0.60$

(b) The ionic strength of a solution which is 0.005 M in $Pb(NO_3)_2$ and 0.040 M in KNO_3 is $3 \times 0.005 + 1 \times 0.040 = 0.055$.

Therefore,

$$-\log \gamma_{Pb^{2+}} = \frac{0.51 \times 4 \times (0.055)^{1/2}}{1 + 3.3 \times 10^7 \times 4 \times 10^{-8} \times (0.055)^{1/2}}$$

$$= 0.37$$

Hence $\gamma_{Pb^{2+}} = 0.43$

In many calculations the ion size parameter may be taken to be about 3 Å and therefore $a \times B = 3 \times 10^{-8} \times 0.33 \times 10^8$ which makes $a \times B$ approximately equal to unity. Equation III-8 then becomes:

$$-\log \gamma_i = \frac{A \times z_i^2 \times \mu^{1/2}}{1 + \mu^{1/2}} \qquad \text{(III-9)}$$

Example 4.

Calculate the activity coefficient of Pb^{2+} in the same solutions as in Example 3.

(a) $$-\log \gamma_i = \frac{0.51 \times 4 \times (0.015)^{1/2}}{1 + (0.015)^{1/2}}$$

$$= 0.22$$

$$\gamma_i = 0.60$$

(b) $$-\log \gamma_i = \frac{0.51 \times 4 \times (0.055)^{1/2}}{1 + (0.055)^{1/2}}$$

$$= 0.39$$

$$\gamma_i = 0.41$$

The theoretical predictions made by Debye and Hückel agreed with experiment in only very dilute solutions and as a consequence equations III-8 and III-9 apply to dilute solutions. However it may be assumed for purposes of calculation that equations III-8 and III-9 can be applied even when the ionic strength of a solution is as high as 0.1. In extremely dilute solutions the denominator in equation III-9 is unity, since $\mu^{1/2}$ is less than 0.01, and since in aqueous solutions at 25°C the constant A is equal to 0.51, equation III-9 is reduced to its simplest possible form called the Limiting Law:

$$-\log \gamma_i = 0.51 \times z_i^2 \ \mu^{1/2} \qquad \text{(III-10)}$$

Example 5.

Calculate the activity coefficients of Pb^{2+} in the same solutions as in Example 3 using the Debye-Hückel Limiting Law.

(a) $-\log \gamma_{Pb^{2+}} = 0.51 \times 4 \times (0.015)^{1/2}$

$= 0.25$

$\gamma_{Pb^{2+}} = 0.56$

(b) $-\log \gamma_{Pb^{2+}} = 0.51 \times 4 \times (0.055)^{1/2}$

$= 0.48$

$\gamma_{Pb^{2+}} = 0.33$

The Limiting Law gives an accurate representation (± 0.02 log units) of the activity coefficient of a singly charged ion at ionic strengths of 0.05 or under of a doubly charged ion at $\mu \leq 0.014$ and of a triply charged ion at $\mu \leq 0.005$.

The significant influence of the charge of an ion and of the ionic strength upon activity coefficient values is obvious from equations III-8, III-9 and III-10. However there are two other factors that affect the values of activity coefficients which have not yet been mentioned that merit our interest, namely temperature and nature of solvent. In order to understand these effects let us return to equation III-8 and mention the fact that the parameter A is inversely proportional to the product of the dielectric constant, D, and the absolute temperature T raised to the 3/2 power.

$$A \propto \frac{1}{D^{3/2} \times T^{3/2}} \qquad \text{(III-11)}$$

The parameter A is not too sensitive to temperature changes. A change in temperature of 5°C from 25°C will cause a change in A of less than 1%. Changing solvents would give rise to much more serious changes in activity coefficients since the resulting change in dielectric constant is relatively large. In a 50% v/v mixture of ethanol and water, the dielectric constant has a value of 49. Hence the parameter A in this mixture is about twice its value in water.

III-5. ACTIVITY COEFFICIENTS OF IONS

In the preceding section a theoretical approach to the evaluation of ionic activity coefficients was developed and was based on the Debye-Hückel Theory which was experimentally verified in dilute solutions of electrolytes. In solutions of ionic strengths greater than 0.1, empirical values of activity coefficients replace those calculated from the Debye-Hückel Theory, for calculations that demand the highest accuracy. For our purposes equation III-9 provides an accuracy that is adequate.

Activity coefficients of single ions cannot be measured directly. Instead on the basis of experiments in which the free energy of an electrolyte is determined by various methods, (such as by measuring freezing point depressions of solutions or by the measurement of the e.m.f's of cells), a quantity called the mean ionic activity coefficient is obtained. For an electrolyte A_mB_n the mean ionic activity coefficient is defined as follows:

$$(\gamma_{\pm})^{m+n} = (\gamma_A)^m \times (\gamma_B)^n$$

where γ_A and γ_B are the ionic activity coefficients. However Kielland* has developed a series of values of single ion activity coefficients from which values of $\gamma_{\pm}$ for various electrolytes may be calculated which are in reasonable agreement with experimental values up to an ionic strength of about 0.1

III-6. ACTIVITY COEFFICIENTS OF MOLECULAR SOLUTES

The activity coefficients of uncharged solutes are described fairly accurately up to unit ionic strength by the equation:

$$\log \gamma = k \times \mu$$

in which γ and μ have their usual meaning, and k is a proportionality constant called the salting coefficient which depends on the nature of both the solute and the ions in solution. Since values of k vary for the most part between 0.01 and 0.10, activity coefficients of molecular solutes can be considered to be effectively unity in solutions having ionic strengths at least as high as 0.2.

III-7. ACTIVITIES AND EQUILIBRIUM CALCULATIONS

In solving equilibrium problems, we see from equation III-5 that both concentrations and activity coefficients of the

---

*J. Kielland, J. Am. Chem. Soc., 59, 1675 (1937).

substances participating in the equilibrium must be evaluated. The Debye-Hückel Theory (Equations III-8, 9 and 10) provides us with a means of calculating activity coefficients of ions as functions of ionic strength. Instead of obtaining values for individual activity coefficients of components of a system in equilibrium, we may achieve a greater simplicity and convenience by combining the factors in the activity coefficient quotient in the equilibrium equation to give a single factor whose value depends upon the ionic strength. On this basis the value of K for any equilibrium reaction may be written as a function of ionic strength.

In solving problems then, our concern with activity will be directed towards the effort of obtaining the appropriate K value. Beyond this, the calculation will proceed in the "classical" way using concentrations.

The method for obtaining K is as follows:

Rewriting equation III-6 to give a solution for K we have:

$$K = \mathbf{K} \times \left\{ \frac{\gamma_A^a \times \gamma_B^b}{\gamma_C^c \times \gamma_D^d} \right\} \qquad \text{(III-12)}$$

Considering the logarithmic form of equation III-12 we may write,

$$pK = p\mathbf{K} + c \log \gamma_C + d \log \gamma_D - a \log \gamma_A - b \log \gamma_B$$

where $pK = -\log K$ and $p\mathbf{K} = -\log \mathbf{K}$

Substituting equation III-9 for values of γ

$$pK = p\mathbf{K} + 0.51 \left\{ az_A^2 + bz_B^2 - cz_C^2 - dz_D^2 \right\} \times \frac{\mu^{1/2}}{1 + \mu^{1/2}} \qquad \text{(III-13)}$$

Note that if A, B, C or D is uncharged, $z = 0$ and $\log \gamma = 0$

From equation III-13, the usefulness of the tabulation of values of the function

$\dfrac{\mu^{1/2}}{1 + \mu^{1/2}}$ versus μ, (Table III-2) in calculating K values is evident since the term $0.51(az_A^2 + bz_B^2 - cz_C^2 - dz_D^2)$ will be a constant multiplier having values depending on the type of equilibrium that is under consideration.

For example consider the solubility of a slightly soluble electrolyte $A_m B_n$.

$$A_m B_n \rightleftharpoons mA^{+n'} + nB^{-m'}$$

In this reaction, $c = m$, $d = n$, $z_C = n'$, $z_D = m'$ and $a = 0$, $b = 0$, $z_A = 0$ and $z_B = 0$.
Equation III-13 then becomes:

$$pK = \mathbf{pK} - 0.51(mn'^2 + nm'^2) \times \frac{\mu^{1/2}}{1 + \mu^{1/2}} \qquad \text{(III-14)}$$

For a 1 : 1 electrolyte such as AgCl,

$$pK = \mathbf{pK} - \frac{1.0 \times \mu^{1/2}}{1 + \mu^{1/2}} \qquad \text{(III-15)}$$

For a 1 : 2 electrolyte such as Ag_2CrO_4 or $PbCl_2$,

$$pK = \mathbf{pK} - \frac{3.0 \times \mu^{1/2}}{1 + \mu^{1/2}} \qquad \text{(III-16)}$$

For a 2 : 2 electrolyte such as $BaSO_4$,

$$pK = \mathbf{pK} - \frac{4.0 \times \mu^{1/2}}{1 + \mu^{1/2}} \qquad \text{(III-17)}$$

A similar series of expressions may be derived for a polyprotic acid such as H_3X.

$$pK_1 = \mathbf{pK}_1 - \frac{\mu}{1 + \mu^{1/2}}$$ (also applicable to a monoprotic acid.)

$$pK_2 = \mathbf{pK}_2 - \frac{2(\mu)^{1/2}}{1 + \mu^{1/2}}$$

$$pK_3 = \mathbf{pK}_3 - \frac{3(\mu)^{1/2}}{1 + \mu^{1/2}}$$

Note: In most accurate calculations involving the ion H^+, the use of the function

$$\frac{\mu^{1/2}}{1 + \mu^{1/2}}$$

is not entirely adequate since the ion size parameter for H^+ is approximately three times the size assumed in the equation. The error, most noticeable in monoprotic acids, amounts to slightly less than 0.04 log units in pK at an ionic strength of 0.1. While this error usually need not be considered, equation III-8 may be used if necessary.

For a cationic acid such as the ammonium ion, K is equal to **K** at all ionic strengths.

In general, $$pK = p\mathbf{K} - \frac{N \times \mu^{1/2}}{1 + \mu^{1/2}} \qquad \text{(III-18)}$$

where N is an integer whose value depends upon the nature of the equilibrium involved as illustrated above.

Values of pK are obtained from the tabulated value of p**K** and from the value using the Table III-2 where the function

$$\frac{\mu^{1/2}}{1 + \mu^{1/2}}$$

is tabulated as a function of ionic strength.

Example 6.

What is the pK_2 of carbonic acid in a solution of ionic strength 0.050 at 25° C?

$$pK_2 = p\mathbf{K}_2 - \frac{2\,\mu^{1/2}}{1 + \mu^{1/2}}$$

From Table III-2

$$pK_2 = p\mathbf{K}_2 - 2(0.183).$$

Since $p\mathbf{K}_2 = 10.33$, $pK_2 = 9.96$

It is of interest to note that in all of the equations relating pK and p**K**, the value of pK decreases with increasing ionic strength. This will be true in all equilibria in which there is a charge separation. In all equilibria of this type therefore, the extent of the reaction will always increase with increasing indifferent electrolyte concentration, i.e. with the concentration of an electrolyte that does not participate in the equilibrium. This effect although significant, is swamped whenever the electrolyte is directly involved in the equilibrium.

In subsequent chapters most calculations will be based on the concentration constant values, i.e. K values.

By using K rather than **K** values, attention will be focused on general techniques involving equilibrium problems without the added complexity of activity corrections. An attempt has been made to provide K values appropriate for the prevailing ionic strength. This will explain the variations in K values from problem to problem. The student will be able to verify his grasp of the material presented in this chapter by checking the K values himself, starting from the tabulated **K** values.

Table III-2

Table for the Calculation of Values of K
as a Function of Ionic Strength in Water at 25° C.

$$pK = \mathbf{pK} - \frac{N \times \sqrt{\mu}}{1 + \sqrt{\mu}}$$

(N is calculated as described in Sec. III-7)

μ	$\sqrt{\mu}/1 + \sqrt{\mu}$	μ	$\sqrt{\mu}/1 + \sqrt{\mu}$
0.0001	0.010	0.0240	0.134
0.0003	0.017	0.0260	0.139
0.0005	0.022	0.0280	0.143
0.0007	0.026	0.0300	0.148
0.0010	0.031	0.0320	0.152
0.0020	0.043	0.0340	0.156
0.0030	0.052	0.0360	0.160
0.0040	0.060	0.0380	0.163
0.0050	0.066	0.0400	0.167
0.0060	0.072	0.0500	0.183
0.0070	0.077	0.0600	0.197
0.0080	0.082	0.0700	0.209
0.0090	0.087	0.0800	0.220
0.0100	0.091	0.0900	0.231
0.0120	0.099	0.1000	0.240
0.0140	0.106	0.1500	0.279
0.0160	0.112	0.2000	0.309
0.0180	0.118	0.3000	0.355
0.0200	0.123	0.4000	0.387
0.0220	0.129	0.5000	0.414

SUGGESTIONS FOR FURTHER READING

T. S. Lee, Chapter 7, Part I, Vol. I, Treatise on Analytical Chemistry, Interscience, New York (1959), I. M. Kolthoff and P. J. Elving, Editors.

H. A. Laitinen, Chemical Analysis, McGraw Hill Book Co., Inc., (1960), Chapter 2.

III-8. PROBLEMS

1. Calculate the ionic strengths of each of the following solutions:
   (a) 0.05 M KNO_3.
   (b) a mixture of 0.05 M KNO_3 and 0.01 M $NaNO_3$.
   (c) a mixture of 0.02 M $MgCl_2$, 0.01 M NaCl and 0.03 M $ZnSO_4$.
   (d) 0.1 M H_2SO_4.
   (e) 0.01 M C_2H_5OH.
   (f) 0.10 M CH_3COOH.
2. Show that the ionic strength of any strong electrolyte is given by the equation, $\mu = k \times c$, where μ is the ionic strength of the solution, c the concentration of the strong electrolyte and k, a constant. (a) Is k the same for all strong electrolytes? Will this equation hold for (b) a mixture of different types of strong electrolytes? (c) a solution of a weak electrolyte?
3. Calculate the activity coefficients of the sodium ion in the following solutions:
   (a) 0.01 M NaCl
   (b) a mixture of 0.02 M NaCl and 0.03 M $NaNO_3$
   (c) a mixture of 0.05 M $NaClO_4$ and 0.03 M HCl
   (Refer to Table III-1 for ion size parameters.)
4. Calculate the activity of the hydrogen ion in a solution which is 0.01 M in HCl and 0.05 M in $KClO_4$. What is the percentage error in the activity of the hydrogen ion if the ion size parameter of H^+ is assumed to have the average value of 3×10^{-8}? What is the corresponding error in the logarithm of the activity of the hydrogen ion?
5. Calculate the activities as well as the activity coefficients of Cl^-, K^+ and ClO_4^- in the solution in question 4.
6. A solution contains ceric perchlorate, sodium perchlorate, lithium perchlorate and perchloric acid, the molarities of all of which are 0.01. Arrange the cations in order of decreasing activity.
7. Derive relationships between K, (concentration constants) and **K**, (thermodynamic constants) for the following systems:
   (a) $H_2O \rightleftharpoons H^+ + OH^-$
   (b) $AgCl$ (solid) $\rightleftharpoons Ag^+ + Cl^-$
   (c) Ag_2CrO_4 (solid) $\rightleftharpoons 2Ag^+ + CrO_4^=$
   (d) $BaSO_4$ (solid) $\rightleftharpoons Ba^{++} + SO_4^=$

Which of these K values will be affected most by a 10% increase in the ionic strength of the solution?

8. Derive relationships between K and **K** for all the dissociation steps involved in each of the following aqueous systems:
   (a) CH_3COOH
   (b) $H_2C_2O_4$
   (c) H_3PO_4
   (d) $C_5H_5NH^+$ (pyridinium ion)
   (e) $C_6H_5NH_3^+$ (anilinium ion)
   (f) sodium acetate
   (g) ammonium carbonate

   (In all cases assume an ion size parameter of 3×10^{-8}.)

9. Calculate the ion product of water, (K) at a given temperature in 0.10 M NaCl assuming that the K_w value for water at the same temperature is 1.0×10^{-14}. What is the hydrogen ion concentration in this solution? Is this solution acidic, neutral or basic?

10. What is the activity of the hydrogen ion in 0.01 M $HClO_4$ in 45 weight % dioxane-water at 25°C? (Use the data in Table II-2, the relation III-11, and the limiting law approximation.)

11. Derive a relationship between the solubility product of AgCl (K) and the ionic strength of a solution which is saturated with AgCl. Will this relationship be valid if the ionic strength of the solution is altered by adding, (a) NaCl, (b) $NaNO_3$?

12. Repeat problem 11 with $PbCl_2$, $BaSO_4$, $Ca_3(PO_4)_2$, and $AlPO_4$. Tabulate values of N for each charge type.

IV

Acid-Base Equilibria

IV-1. THEORIES OF ACIDS AND BASES

Of the many theories that have been proposed through the years to explain the properties of acids and bases, the Bronsted-Lowry or proton transfer theory and the Lewis theory, are most generally useful.

IV-1-(a). THE BRONSTED-LOWRY THEORY

In 1923 Bronsted and Lowry independently developed an acid-base theory based on the central role of the proton. They defined an acid as a proton donor and a base as a proton acceptor. Thus an acid-base reaction is one in which proton transfer occurs.

$$\text{i.e. Acid} \rightleftharpoons \text{Base} + H^+ \qquad \text{(IV-1)}$$

According to this definition, neutral molecules such as H_3PO_4 or H_2O, cations such as NH_4^+ and anions like $H_2PO_4^-$, all behave as acids.

$$\text{e.g. } NH_4^+ \rightleftharpoons NH_3 + H^+ \qquad \text{(IV-2)}$$

Similarly cations, anions and neutral molecules can all act as bases. Certain substances such as H_2O, and SH^- behave as acids as well as bases and are called ampholytes or amphoteric electrolytes.

$$\underset{\text{acid}}{SH^-} \rightleftharpoons H^+ + \underset{\text{base}}{S^=}$$

$$\underset{\text{base}}{SH^-} + H^+ \rightleftharpoons \underset{\text{acid}}{H_2S}$$

Equation IV-2 is a simplification of the proton transfer reaction that takes place if the reaction is carried out in a solvent such as water. Bare protons do not exist in any solvent. The characterization of a substance as an acid or a base may be made in terms relative to the solvent water. That is to say, an acid is a substance capable of donating a proton to water, and a base is a substance capable of accepting a proton from water. Reaction IV-2 is therefore more correctly represented as follows:

$$NH_4^+ + H_2O \rightleftharpoons NH_3 + H_3O^+ \qquad \text{(IV-3)}$$

The hydrated proton* is represented as H_3O^+. As was pointed out earlier it is, strictly speaking, incorrect to write H^+ to represent a hydrated proton, although this is generally accepted for the sake of convenience. Throughout this book, H^+ or H_3O^+ will be used almost interchangeably.

It is evident from equation IV-3 that there is a proton transfer from the acid, NH_4^+ to the water molecule H_2O. Therefore there are two substances that behave as acids and two substances that behave as bases in this reaction. The cation NH_4^+ and the hydrated proton H_3O^+ are both acids while the neutral molecule H_2O as well as the ammonia molecule NH_3, are bases.

$$NH_4^+ + H_2O \rightleftharpoons NH_3 + H_3O^+ \qquad \text{(IV-4)}$$

$$\text{acid}_1 + \text{base}_2 \rightleftharpoons \text{base}_1 + \text{acid}_2$$

The pairs of compounds $NH_4^+ - NH_3$ and $H_3O^+ - H_2O$ are called conjugate acid-base pairs. Further examples of acid-base reactions are:

Dissociation of weak acids:

$$CH_3COOH + H_2O \rightleftharpoons CH_3COO^- + H_3O^+ \qquad \text{(IV-5-a)}$$

$$NH_4^+ + H_2O \rightleftharpoons NH_3 + H_3O^+ \qquad \text{(IV-5-b)}$$

Dissociation of weak bases:

$$NH_3 + H_2O \rightleftharpoons NH_4^+ + OH^- \qquad \text{(IV-6-a)}$$

$$CH_3COO^- + H_2O \rightleftharpoons CH_3COOH + OH^- \qquad \text{(IV-6-b)}$$

*There is some evidence that the hydrated proton is $H_9O_4^+$ and not H_3O^+.

It will be seen later in this book that the methods of treating these reactions are very similar. This is apparent from the fact that it is not necessary to describe the equilibria in equations IV-5 and 6 by means of different types of equilibrium constants. One type of equilibrium constant, namely the acid dissociation constant, will suffice to describe exactly, all the systems represented by equations IV-5 and 6. The reactions of the type represented by equations IV-5-b and IV-6-b are commonly referred to as hydrolysis reactions. Since this term is still in common use, we will refer to it when convenient. It must be kept in mind that "hydrolysis" is merely an acid-base reaction.

The equilibrium constant for equation IV-5-a is given by

$$K_a = \frac{[CH_3COO^-][H_3O^+]}{[CH_3COOH]}$$

where K_a is the acid dissociation constant of the acid. The equilibrium constant for equation IV-6-a is given by:

$$K_b = \frac{[NH_4^+][OH^-]}{[NH_3]}$$

However there is really no necessity for introducing a new type of constant, the basic dissociation constant, K_b, since equation IV-6-a may be recognized as a result of subtracting the acid dissociation constant of NH_4^+, from the dissociation constant of water. Thus:

$$H_2O \overset{K_w}{\rightleftharpoons} H^+ + OH^-$$

$$-\ (NH_4^+ \overset{K_a}{\rightleftharpoons} H^+ + NH_3)$$

$$NH_3 + H_2O \rightleftharpoons NH_4^+ + OH^-$$

This subtraction of equations is equivalent to dividing the corresponding equilibrium expressions.

$$\frac{K_w}{K_a} = \frac{[H^+][OH^-]}{[NH_3][H^+]/[NH_4^+]}$$

$$= \frac{[NH_4^+][OH^-]}{NH_3}$$

Hence we see that K_b is identical to $\frac{K_w}{K_a}$.

Although this relation was derived from a specific consideration of the NH_4^+/NH_3 conjugate pair, it will be recognized as a generally valid reaction, viz.,

$$K_a \times K_b = K_w \quad (IV\text{-}7)$$

From this equation it follows readily that if we list a series of acids in increasing strength, we have automatically listed the conjugate bases in the order of decreasing strength.

IV-1-(b). THE LEWIS THEORY

The Lewis theory of acids and bases eliminated the special role that the proton has in the Bronsted-Lowry theory. This was accomplished by defining an acid as any electron-pair deficient species. A base from this viewpoint is a species capable of furnishing electron pairs. Thus, acid-base reactions are considered as coordination reactions. This theory is of great value in understanding metal coordination complex formation.

In the following reactions:

$$H^+ + NH_3 \rightleftharpoons NH_4^+$$

$$Ag^+ + 2NH_3 \rightleftharpoons Ag(NH_3)_2^+$$

the Ag^+ ion is seen to act in a manner that is similar to a proton, and the Ag^+ ion behaves as a dibasic acid. The subject of metal coordination complexes which involves reactions of this type, will be considered at length in chapter IX.
The Lewis Theory has also found extensive use in explaining reactions in non-aqueous media, involving non-protonic acids such as BF_3, $AlCl_3$, etc.

IV-2. STRENGTHS OF ACIDS AND BASES

The strengths of acids, that is their proton donating tendencies, naturally depend upon the proton accepting tendency of the base with which they react. An ordering of acids according to their strengths is obtained by the use of a reference base. In aqueous solutions the reference base is water. Thus the strengths of a series of acids, HSO_4^-, H_2CO_3 and HCN can be compared by measuring the extents of their dissociation in water.

$$HSO_4^- + H_2O \rightleftharpoons H_3O^+ + SO_4^=$$

$$H_2CO_3 + H_2O \rightleftharpoons H_3O^+ + HCO_3^-$$

$$HCN + H_2O \rightleftharpoons H_3O^+ + CN^-$$

$$(\text{acid}_1 + \text{base}_2 \rightleftharpoons \text{acid}_2 + \text{base}_1)$$

Similarly the comparison of the strengths of bases in aqueous solutions involves the use of water as the reference acid.

$$NH_3 + H_2O \rightleftharpoons NH_4^+ + OH^-$$

$$\underset{\text{aniline}}{C_6H_5NH_2} + H_2O \rightleftharpoons C_6H_5NH_3^+ + OH^-$$

$$\underset{\text{pyridine}}{C_5H_5N} + H_2O \rightleftharpoons C_5H_5NH^+ + OH^-$$

$$\underset{\text{trimethylamine}}{(CH_3)_3N} + H_2O \rightleftharpoons (CH_3)_3NH^+ + OH^-$$

In these reactions there is a competition for protons between the two bases or, looking at it from another point of view, there is a tendency for the two acids to lose protons. Therefore the relative amounts of the conjugate acid-base pairs that exist at equilibrium will be a measure of the strengths of the acids and bases. This is equivalent to saying that the dissociation constants measure strengths of acids and bases.

The acid dissociation constants for the acids HSO_4^-, H_2CO_3 and HCN are written as follows

$$K_a = \frac{[H_3O^+][SO_4^=]}{[HSO_4^-]} = 1.0 \times 10^{-2}$$

$$K_a = \frac{[H_3O^+][HCO_3^-]}{[H_2CO_3]} = 4.4 \times 10^{-7}$$

$$K_a = \frac{[H_3O^+][CN^-]}{[HCN]} = 4.0 \times 10^{-10}$$

It is obvious that the larger the numerical value of K_a, the stronger the acid, i.e. the tendency to lose a proton is greater. Therefore the acids can be arranged in order of decreasing strength as follows:

$$HSO_4^- > H_2CO_3 > HCN$$

It is of interest to note that values of pK_a (defined as $pK_a = -\log_{10} K_a$), increase with decreasing acid strength. This means that pK_a values increase with increasing strengths of the conjugate bases. For this reason, values of pK_a are often used as measures of basic strength or basicity.

Similarly, the base dissociation constants K_b, are written as follows:

$$K_b = \frac{[NH_4^+][OH^-]}{[NH_3]} = 1.8 \times 10^{-5}$$

$$K_b = \frac{[C_6H_5NH_3^+][OH^-]}{[C_6H_5NH_2]} = 3.9 \times 10^{-10}$$

$$K_b = \frac{[C_5H_5NH^+][OH^-]}{[C_5H_5N]} = 2.0 \times 10^{-9}$$

$$K_b = \frac{[(CH_3)_3NH^+][OH^-]}{[(CH_3)_3N]} = 6.3 \times 10^{-5}$$

The larger the value of K_b the stronger the base, and therefore the bases arranged in order of decreasing base strength are:

$$(CH_3)_3N > NH_3 > C_5H_5N > C_6H_5NH_2$$

Unfortunately it is not possible to establish the relative strengths of strong acids or strong bases in this manner. Strong acids such as HCl, HNO_3 and $HClO_4$, all appear equally strong when dissolved in water, because they react quantitatively with water to yield the ion H_3O^+ in each case. This is referred to as the leveling effect. The relative strengths of these acids can however be determined in solvents less basic than water in which incomplete reaction occurs.

IV-3. THE CONCEPT OF pH

In a great many examples of practical importance it is necessary to deal with small concentrations of hydrogen or hydroxyl ions, and the method of writing these concentrations is of necessity rather awkward. To overcome this, Sorenson in 1909 proposed a more convenient method of expressing small concentrations of hydrogen or hydroxyl ions. In this method the H^+ or OH^- concentrations were written as their negative logarithms. The reason the negative logarithm was chosen by Sorenson is that the most frequently encountered concentrations are lower than unity. For such concentrations, the negative logarithm gives a positive number. However, the student must be aware always of the fact that concentration changes and the corresponding changes in the negative logarithms of these concentrations have the opposite sense. That is, a decrease in the pH corresponds to an increase in H^+.

$$-\log [H^+] = pH$$

and

$$-\log [OH^-] = pOH$$

In a number of later instances this "p-notation" will be used to indicate the negative logarithm of the term that is preceded by p. For example

$$pK_a = -\log K_a$$

$$pM = -\log [M]$$

Another method of expressing the relationship between $[H^+]$ and pH is:

$$[H^+] = 10^{-pH}$$

Also,

$$[OH^-] = 10^{-pOH}$$

Therefore,

$$[H^+] \times [OH^-] = 10^{-(pH + pOH)} = K_w = 10^{-pK_w}$$

Hence,

$$pK_w = pH + pOH$$

Example 1.

Calculate the pH and pOH of a 2.5×10^{-2} M solution of $HClO_4$. In this problem assume that $pK_w = 13.86$.*

---

*This value of pK_W has been calculated according to methods described in chapter III section 7. In most subsequent problems this practice of presenting appropriate K values will be followed.

$$[H^+] = 2.5 \times 10^{-2}$$
$$= 10^{\log 2.5} \times 10^{-2}$$
$$= 10^{0.40} \times 10^{-2}$$
$$= 10^{-1.60}$$

Therefore pH = 1.60

$$pOH = 13.86 - 1.60$$
$$= 12.26$$

Example 2.

Calculate the pH of a 2.00×10^{-3} M solution of NaOH. Assume $pK_w = 13.96$.

$$[OH^-] = 2.00 \times 10^{-3}$$
$$= 10^{\log 2.00} \times 10^{-3}$$
$$= 10^{-2.70}$$

Therefore, pOH = 2.70

and $pH = 13.96 - 2.70$

$$= 11.26$$

An examination of the experimental quantity that Sorenson took to be the pH, in the light of modern theory, reveals that this quantity is closer to the negative logarithm of the activity, rather than the concentration of the hydrogen ion. The negative logarithm of the hydrogen ion activity defines the p_aH as seen in the following equation:

$$p_aH = -\log a_{H+} = -\log H^+ - \log \gamma_{H+} \qquad \text{(IV-8)}$$

Strictly speaking, it is impossible to determine single ion activity coefficients which makes exact experimental evaluation of p_aH impossible (See Section III-5). However, from a practical viewpoint, the p_aH values of solutions are defined in terms of standard buffers; the p_aH values of these standard buffers are calculated using the best available data for mean activity coefficients (Section III-5) that are applicable to the components of these solutions.

When the "p" notation is used in the following chapters in this book, it will be understood that we are dealing with the negative logarithm of concentrations unless the subscript, a, is added to indicate that we are dealing with the negative logarithm of the activity. It is useful to note again that pK will represent the negative logarithm of the concentration constant, whereas **pK** stands for the negative logarithm of the thermodynamic constant.

We are accustomed to seeing pH values that lie within the range 0 - 14. However these values do not represent limiting extremes. A few examples will suffice to illustrate this point.

Example 3.

Calculate the pH of a 2 M solution of $HClO_4$.

$[H^+] = 2$

Therefore $pH = -\log[H^+] = -\log 2 = -0.3$

Example 4.

Calculate the pH of a 4 M solution of NaOH. Assume that $pK_w = 13.9$

$[OH^-] = 4$

Therefore, $-\log[OH^-] = pOH = -0.60$
and since $pH + pOH = 13.9$

$$pH = 14.5$$

Ordinarily, the reliability of measured pH values, using equipment that is available in most laboratories, is not better than ± 0.02 pH. Therefore all calculations in this chapter on acid-base equilibria will be based on this assumption. That is, all pH values in problems will be calculated correct to ± 0.02.

If the difference in two pH values is 0.02, then the corresponding difference in the hydrogen ion concentration can be readily calculated as follows:

$pH - pH' = 0.02$

i.e. $\log[H^+]' - \log[H^+] = 0.02$

$$\log\frac{[H^+]'}{[H^+]} = 0.02$$

Therefore, $\dfrac{[H^+]'}{[H^+]} = 1.05$

This means that for a difference in pH of 0.02 there is a corresponding difference of 5% in the hydrogen ion concentrations. Therefore in problems on acid-base equilibria, terms that make less than a 5% difference in $[H^+]$ can be omitted when calculating the pH correct to ± 0.02.

It must be emphasized that the quantity 5% has no special theoretical significance and it is not difficult to give reasons for choosing some other quantity such as 1%, 2% or even 10% in making approximations. Also it must be realized that unless activity coefficients are taken into account in the calculation of proper values of K (or in some other suitable fashion), the answers obtained by these calculations may easily be as much as 100% or more in error. Nevertheless the only way to treat problems in equilibria in a rigorous manner is to adopt a definite and consistent method of making approximations, and throughout this text, 5% is the quantity that has been adopted as the upper limit for excluding a term from a given expression.

IV-4. THE CONCEPT OF A NEUTRAL SOLUTION

Although the dissociation equilibrium of water has been discussed previously, its central importance in acid-base equilibria in aqueous solutions merits further attention. As shown in Section IV-1, the dissociation of acids and bases in aqueous solutions involve water as a reactant. Before proceeding to calculations involving solutions of acids, bases or their mixtures, let us consider the case of pure water which is neutral. That is, in pure water,

$$H_2O \rightleftharpoons H^+ + OH^-$$

$$[H^+] = [OH^-] \qquad \text{(IV-9)}$$

This relationship defines a neutral solution whether we are dealing with pure water or with solutions of acids or bases. Just as neutrality is defined by equation IV-9, an acid solution is readily seen to be one in which

$$[H^+] > [OH^-]$$

and conversely a basic solution one in which $[OH^-] > [H^+]$. Since in pure water at 25°C, $K_w = 1.00 \times 10^{-14}$, the pH of a neutral solution is $\frac{1}{2}pK_w$ or 7.00. As a useful approximation, in aqueous solutions at 25°C, a pH value of 7 represents a dividing line with lower pH values indicating acidic and higher pH values, basic solutions. It must not be inferred from this however that a pH of 7.00 defines a neutral solution in all cases. The value of K_w, a concentration constant, will change with ionic strength, dielectric constant and temperature. Therefore the pH of a neutral solution will change accordingly.

Example 5.

What is the pH of pure water at 100°C?

$K_w = 5.0 \times 10^{-13}$ at 100°C

$\frac{1}{2}pK_w = pH = \frac{1}{2}(12.30) = 6.15.$

It is of interest to note that although this solution is neutral, it has a significantly higher hydrogen ion concentration than water at 25°C.

As an aid to visualizing the way in which $[H^+]$ and $[OH^-]$ vary in all aqueous solutions let us examine Figure IV-1 in which

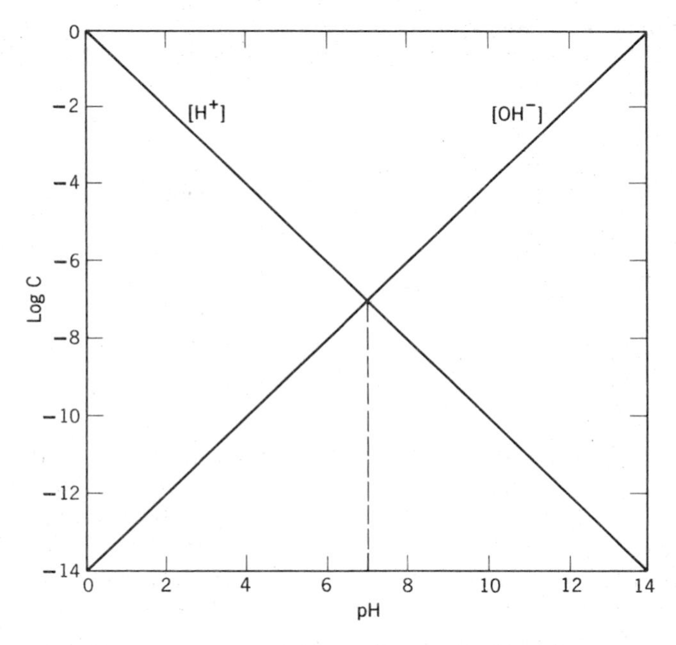

Fig. IV-1—Variation of log $[H^+]$ and log $[OH^-]$ in Aqueous Solutions ($pK_W = 14.0$)

the logarithms of the concentrations of these ions are plotted against the pH. It will be noticed that the line representing log $[H^+]$ is the graphical form of the equation: $\log[H^+] = -pH$

Similarly the line representing log $[OH^-]$ is obtained from the equation:

$$\log [OH^-] = -\text{pOH} = \text{pH} - \text{p}K_w$$

An additional value of this graph lies in its use in the solving of acid-base problems. For instance the pH of a neutral solution as defined by equation IV-9 is simply the value at which these two lines cross (where their values are equal). This graphical approach will be seen to be a powerful aid in the solution of rather complex calculations.

IV-5. SOLUTIONS OF STRONG ACIDS AND STRONG BASES

It is a fairly simple matter to calculate the hydrogen ion concentration of a dilute solution of a strong acid; but in extremely dilute solutions of strong acids, the hydrogen ions contributed by water must be taken into account. Failure to do so will lead to absurd answers as will be illustrated by the following example.

Example 6.

Calculate the hydrogen ion concentration and the pH of a 1.00×10^{-1} M solution of $HClO_4$ on progressive dilution with pure water.

Molarity of $HClO_4$	$[H^+]$	pH
1.00×10^{-1}	1.00×10^{-1}	1.00
1.00×10^{-3}	1.00×10^{-3}	3.00
1.00×10^{-5}	1.00×10^{-5}	5.00
1.00×10^{-9}	?	?

If the solution is diluted further until the molarity of the $HClO_4$ is 1.00×10^{-9} the pH of the solution is certainly not 9.00, which would mean that a basic solution has been obtained by diluting a solution of a strong acid! The pH of the solution can never be greater than 7.00, since the $[H^+]$ cannot be less than that present in pure water. Therefore on progressive dilution of the $HClO_4$ solution its $[H^+]$ finally reaches a limiting value equal to 1.00×10^{-7} and the pH of the solution levels off at a value of 7.00.

Let us now consider the method necessary for solving a problem in which the strong acid is sufficiently dilute to require that the $[H^+]$ from water be taken into account.

Example 7.

Calculate the hydrogen ion concentration of a 5.00×10^{-8} M solution of HCl. Since the HCl is so dilute, the $[H^+]$ from water will make a significant contribution to the total $[H^+]$.

The $[H^+]$ produced by the dissociation of water must be necessarily equal to the $[OH^-]$ present in solution as prescribed by the equation for the dissociation of water; the $[H^+]$ from the hydrochloric acid must of course be equal to the $[Cl^-]$ present in solution since HCl is completely dissociated. Therefore the total $[H^+]$ obtained at equilibrium is given by

$$[Cl^-] + [OH^-] = [H^+] \tag{IV-10}$$

and since $[OH^-] = K_w / [H^+]$

$$[H^+] = [Cl^-] + \frac{K_w}{[H^+]} \tag{IV-11}$$

In the above problem, $[Cl^-] = 5.00 \times 10^{-8}$
Assume that $K_w = 1.00 \times 10^{-14}$
If these values are substituted in equation IV-11, a quadratic equation is obtained:

$$[H^+]^2 - 5.00 \times 10^{-8} [H^+] - 1.00 \times 10^{-14} = 0$$

Therefore

$$[H^+] = \frac{5.00 \times 10^{-8} + (25.00 \times 10^{-16} + 4.00 \times 10^{-14})^{1/2}}{2}$$

$$= 1.28 \times 10^{-7}$$

corresponding to pH 6.89

Note that only one of the roots of the quadratic equation was taken since $[H^+]$ must be real and positive.

In this problem, 5.00×10^{-8} moles/liter of H^+ was introduced as HCl and 7.80×10^{-8} moles/liter of H^+ were formed from dissociation of water. (Why was this not 1.00×10^{-7} moles/liter H^+?)

In order to find the molarity of a strong acid at which the dissociation of water must begin to be taken into account let us examine Fig. IV-2. In this figure the situation that applies to a solution 10^{-4} M HCl is illustrated. In addition to the lines representing $[H^+]$ and $[OH^-]$ we notice the horizontal line which represents the concentration of Cl^- in the solution. Since HCl is completely dissociated, $[Cl^-]$ is a constant at any pH and hence this line has a slope of zero. The equation:

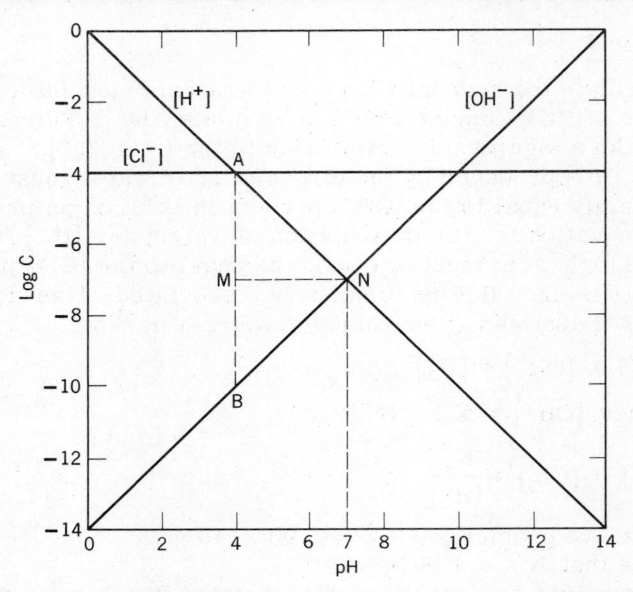

Fig. IV-2—Graphical representation of the components of a 10^{-4} M solution of HCl

$$[H^+] = [Cl^-] + [OH^-] \qquad \text{(IV-10)}$$

developed in example 7 applies here as it does to all solutions of HCl. As may be seen from an examination of Fig. IV-2 the point A at which the $[H^+]$ lines crosses the $[Cl^-]$ line represents the solution to the equation because the OH^- concentration at this pH, (which is represented by the value of log $[OH^-]$ at point B) is negligible in comparison to log $[Cl^-]$. As we deal with smaller and smaller concentrations of HCl, the horizontal line representing the Cl^- concentration would naturally drop lower, bringing the point of intersection A closer to point N and therefore making the distance AB smaller. When $[OH^-]$ (at point B) reaches 5% of $[Cl^-]$ the distance AB would represent log 0.05 and therefore would be equal to 1.3 log units in length. When this point is reached AM = 0.65 log units (since the $[H^+]$ and $[OH^-]$ lines have equal and opposite slopes and the triangle ANB is an isoceles triangle). From this analysis it is apparent that when the concentration of HCl falls to within 0.65 log units of neutrality, i.e. pH $= \frac{1}{2} pK_w$, in this case 7.00, it will be necessary to consider the dissociation of water as in problem 7. Similarly when the concentration of HCl is 0.65 units below the point N, then the pH of the solution is essentially that of pure water.

In summary then:

(a) When $C_a > 10^{-6.35}$ i.e. $> 4.5 \times 10^{-7}$ M the hydrogen ion concentration of the strong acid solution is effectively equal to its molarity.

(b) When $C_a < 10^{-7.65}$ i.e. $< 2.2 \times 10^{-8}$ M, the hydrogen ion concentration of the strong acid solution is effectively equal to that of neutral water.

(c) When C_a is between these two limits, the method employed in Problem 7 may be used.

The hydrogen or hydroxyl ion concentration of a solution of a strong base such as KOH or NaOH can be calculated in a similar manner. The analogous equations for a strong base such as NaOH are:

$$[Na^+] + [H^+] = [OH^-] \qquad \text{(IV-12)}$$

Therefore $C_b + [H^+] = \dfrac{K_w}{[H^+]}$, where C_b is the molar concentration of the strong base. i.e. $[H^+]^2 + C_b[H^+] - K_w = 0$ This equation is analogous to the quadratic equation that was obtained for the general case of a strong acid. This equation can be rewritten in terms of OH^-:

$$[OH^-]^2 - C_b[OH^-] - K_w = 0 \qquad \text{(IV-13)}$$

It can be readily seen that this equation applies to the same concentration limits as for strong acids.

A graphical solution of equation IV-10:

$$[H^+] = [Cl^-] + [OH^-]$$

could be carried out as an alternative route to solving problems such as Example 7. Figure IV-3 is constructed in the usual fashion. Since $[OH^-] > [Cl^-]$ we might take the intersection of the $[H^+]$ and $[OH^-]$ lines as a first approximation of the pH. At this point however $[Cl^-]$ is smaller than $[OH^-]$ by a factor of $10^{0.30}$ (this is the length of line NE). A corrected value of $[H^+]$ would then be:

$$[H^+] = [OH^-] + [OH^-] \times 10^{-0.30}$$

in which the second term represents the substitution for $[Cl^-]$. Therefore:

$$[H^+] = [OH^-]\,(1 + 10^{-0.30})$$

$$= [OH^-] \times 1.50$$

We now draw a new line parallel to the [OH⁻] line but 0.18 log units higher (log 1.50 = 0.18). Where this crosses the [H⁺] line, represents the pH of the solution. As may be seen from the Fig. IV-3, this happens at a pH of 6.91. The small difference between this graphically obtained value and the value of 6.89 from the exact equation in problem 7 arises from our addition of [Cl⁻] and [OH⁻] at a pH of 7.00 instead of at the pH of the solution, namely 6.91. Such errors which can be large (for example in 10^{-7} M HCl) require successive approximations for accurate results. This sets some limitations to the convenient use of the graphical method.

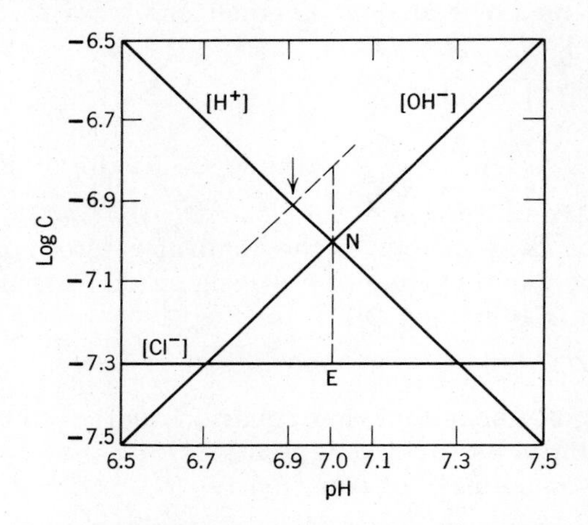

Fig. IV-3—Graphical Determination of the pH of a 5.00×10^{-8} M Solution of HCl

IV-6. CHARGE BALANCE

Equation IV-10 derived in Example 7 is an algebraic representation of the fact that any solution containing an electrolyte has no excess positive or negative charge. All solutions are electrically neutral and the sum of the positive charges in solution must be equal to the sum of the negative charges in solution. With this principle, a useful relation between the concentrations of positively charged and negatively charged ions can be developed. A few examples will show the universal applicability of this principle.

In the case that has been just considered in Example 7, the solution contains the following ions, H^+ and OH^- from H_2O and H^+ and Cl^- from HCl. The positive charges in solution arise from H^+ and the sum of the positive charges in solution is proportional to the total concentration of the hydrogen ions, i.e. $[H^+]$. Similarly the sum of the negative charges in solution is proportional to $[Cl^-] + [OH^-]$. Therefore using the principle of electroneutrality or the charge balance the following equation is obtained:

$$[H^+] = [Cl^-] + [OH^-]$$

Consider a solution of H_3PO_4 in pure water. This solution contains multivalent ions such as $HPO_4^=$ and PO_4^{3-}. The only positively charged ions present in this solution are hydrogen ions. The negative ions that are present are OH^-, from the dissociation of water, $H_2PO_4^-$, $HPO_4^=$ and PO_4^{3-}, from the dissociation of H_3PO_4. It is important to realize that each mole of $HPO_4^=$, a dinegatively charged ion, has the same amount of negative charge as 2 moles of an ion carrying a single negative charge. Also one mole of the phosphate ion, PO_4^{3-}, has the same amount of negative charge as 3 moles of an ion carrying a single negative charge. Therefore the charge balance equation can be written as follows:

$$[H^+] = [OH^-] + [H_2PO_4^-] + 2[HPO_4^=] + 3\,[PO_4^{3-}]$$

where all the terms in square brackets denote molar concentrations. It will be seen in later examples that the charge balance gives a simple and powerful tool for the solution of many problems in chemical equilibria.

IV-7. PROTON BALANCE

A useful modification of the electroneutrality principle is obtained by equating the sums of the concentration terms of species resulting from proton releasing and proton consuming reactions. For example in an HCl solution our starting point would be HCl and H_2O. In the dissociation of HCl, for every Cl^- formed one H^+ must be released. For every OH^- formed from H_2O one H^+ must be released. For every H^+ (really H_3O^+) formed from water one H^+ is consumed. Therefore:

$$[H^+] = [Cl^-] + [OH^-]$$

Similarly in a solution of H_3PO_4 we can see that for every $H_2PO_4^-$, $HPO_4^=$ and PO_4^{3-} formed, one, two and three protons respectively are released. Thus:

$$[H^+] = [OH^-] + [H_2PO_4^-] + 2[HPO_4^{=}] + 3[PO_4^{3-}]$$

As a further illustration let us consider a solution of Na_2HPO_4. If this substance is dissociated into $2\,Na^+$ and $HPO_4^{=}$ without further reaction, the proton balance could be written as $[H^+] = [OH^-]$. However $HPO_4^{=}$ may be transformed into other species. The formation of PO_4^{3-} from $HPO_4^{=}$ and of OH^- from H_2O are the only proton releasing reactions. Hence:

$$[H^+] + 2[H_3PO_4] + [H_2PO_4^-] = [PO_4^{3-}] + [OH^-]$$

Notice that since the Na^+ is not involved in any proton exchange reaction it does not affect the proton balance.

SUGGESTIONS FOR FURTHER READING

E. J. King, Qualitative Analysis and Electrolytic Solutions, Harcourt, Brace and World, Inc., New York(1959) Chapter 11.

R. P. Bell, The Proton in Chemistry, Cornell University Press, Ithaca, New York, (1959), Chapters 1, 2, 3 and 4.

L. G. Sillen, Part I, Vol. I, Chapter 8, Treatise on Analytical Chemistry, Interscience, (1959), I. M. Kolthoff and P. J. Elving, Editors.

IV-8. PROBLEMS

1. Classify the following ions or molecules as Bronsted acids, Bronsted bases or ampholytes: H_3O^+, NH_4^+, NH_3, CN^-, HSO_4^-, HCO_3^-, $S^{=}$, OH^-, $C_6H_5NH_2$, $(CH_3)_3NH^+$, and H_2CO_3.
2. (a) Write equilibrium constants for the protolytic reactions of the following weak acids: CH_3COOH, HCN, NH_4^+, HSO_4^- and HCO_3^-.

   (b) Write equilibrium constants for the protolytic reactions of the following weak bases which are conjugates of the weak acids in (a). CH_3COO^-, CN^-, NH_3, $SO_4^{=}$ and $CO_3^{=}$.

   (c) Derive the relationship between the two protolytic reactions of each conjugate acid-base pair. Hence show that the equilibrium constants for both types of protolytic reactions can be expressed in terms of the acid dissociation constants.

3. The following substances when dissolved in water form conjugate acid-base pairs: HCN, CH_3COOH, $C_6H_5NH_3^+$, NH_4^+, C_6H_5COOH.
   (a) Arrange the acids in order of increasing acid strength.
   (b) Arrange the conjugate bases in order of increasing base strength.
4. Calculate the pH of the following solutions:
   (a) 2.8×10^{-4} M $HClO_4$
   (b) 3.95×10^{-3} M HNO_3
   (c) 9.86×10^{-2} M HCl
   (d) 5 M $HClO_4$
5. The ionic product of water at 0°C is 1.1×10^{-15}, at 25°C is 1.0×10^{-14} and at 60°C is 9.6×10^{-14}. Calculate the pH of a neutral solution of pure water at 0°C, 25°C and 60°C.
6. Calculate the pH of the following solutions:
   (a) 2.00×10^{-3} M HCl
   (b) pure water at 60°C. (K_w for water at 60°C is 9.5×10^{-14})
   (c) 2.00×10^{-3} M NaOH (Assume that K_w for water is 1.09×10^{-14})
   (d) 5.00×10^{-8} M NaOH
7. Calculate the hydrogen ion concentrations in the following solutions of HCl:
   (a) 5.05×10^{-8} M
   (b) 2.00×10^{-9} M
   (c) 3.86×10^{-3} M
8. Calculate the hydroxyl ion concentrations in the following KOH solutions:
   (a) 5.32×10^{-3} M
   (b) 9.95×10^{-8} M
   (c) 1.00×10^{-10} M
9. Write a mathematical equation which relates the concentrations of the positively and negatively charged ionic species obtained when the following substances are dissolved in pure water: H_2SO_4, H_2S, Na_3PO_4, CH_3COOH and $NaHCO_3$.
10. Write the charge balance equations for the following solutions:
   (a) 0.1 M Na_2SO_4
   (b) a mixture which is 0.45 M in $NaClO_4$ and 0.3 M in $HClO_4$
   (c) 1.00×10^{-3} M H_3PO_4

(d) a mixture which is 0.5 M in NH_3 and 0.15 M in NH_4Cl.

11. Write the proton balance equations for the solutions in problem 9.
12. Solve problems 7(a) and (b), and 8(b) and (c) by the graphical method outlined in IV-5.

V

Monobasic Bronsted Acids and Monoacidic Bronsted Bases

V-1. THE DEGREE OF DISSOCIATION

The difference between a strong and weak electrolyte is that the former is completely dissociated into ions in solutions of moderate concentration (~ 0.1 M), whereas the latter is not. Therefore the solution of a weak acid such as acetic acid, CH_3COOH, contains acetate ions, hydrogen and hydroxyl ions and in addition undissociated acetic acid molecules. In general let us use the abbreviation HA for any neutral monoprotic acid, i.e. an acid that is capable of releasing one proton; H_2A will be used to designate a diprotic acid and H_nA a polyprotic acid.

The extent to which a weak acid is dissociated in water will depend on its acid dissociation constant K_a, and on its total concentration.

$$HA \rightleftharpoons H^+ + A^-$$

$$K_a = \frac{[H^+][A^-]}{[HA]}$$

The larger the numerical value of K_a, the more H^+ and A^- ions are formed at a given concentration, and the stronger the acid; whereas the smaller K_a is, the weaker the acid.

The extent of dissociation of a weak acid depends on the concentration of the acid as may be seen from the following example. In a solution of a weak acid HA whose initial concentration is C_a moles/liter, in which x moles/liter dissociates, the equilibrium concentrations of both H^+ and A^- ions are each x moles/liter (assuming that the H^+ from water may be neglected) and that of the undissociated HA is $(C_a - x)$ moles/liter

$$\underset{(C_a - x)}{HA} \rightleftharpoons \underset{x}{H^+} + \underset{x}{A^-}$$

Therefore

$$K_a = \frac{X^2}{C_a - X} \qquad \text{(V-1)}$$

As may be seen from this expression, X, the amount that dissociates varies inversely with the concentration, reaching a value equal to C_a as C_a itself approaches zero. This may be seen more clearly by repeating this example after introducing the term, α, the fraction dissociated. Using the fraction α, which is defined as the ratio of X to C_a, the equilibrium expression (V-1) becomes

$$K_a = \frac{C_a \alpha^2}{1 - \alpha} \qquad \text{(V-1a)}$$

If two weak acids of the same concentration C_a, are taken, then the acid with the smaller K_a will have a smaller value of α, i.e. it will be less dissociated than the acid with the larger K_a. For any acid, K_a is essentially constant. Therefore as C_a decreases α must increase. This means that the degree of dissociation of an acid will increase when the concentration of the acid decreases. Equations (V-1) and V-2) show that if the concentration of the acid, C_a, and its dissociation constant, K_a, are known it is possible to calculate the extent of dissociation of the acid.

Equations exactly analogous to (V-1) and (V-1a) can be derived for a weak base such as ammonia.

$$NH_3 + H_2O \rightleftharpoons NH_4^+ + OH^-$$

A solution of ammonia in water is represented as $NH_3 + H_2O$ rather than NH_4OH, since the concentration of NH_4OH molecules in solution is extremely small. Whether the dissociation of ammonia is represented as above or as:

$$NH_4OH \rightleftharpoons NH_4^+ + OH^-$$

will make no difference to the calculation of the degree of dissociation of ammonia in water. If K_b is the dissociation constant or ammonia, α, its degree of dissociation, and C_b, the concentration of ammonia in water, it can be shown in a manner analogous to that used for a weak acid, that,

$$K_b = \frac{[NH_4^+][OH^-]}{[NH_3]} \quad \frac{C_b \times \alpha^2}{(1-\alpha)}$$

Equations V-1 and V-1a may be simplified in cases where X is less than 5% of C_a or α is less than 0.05. For such cases, we may write

$$K_a = \frac{X^2}{C_a} \quad \text{or} \quad C_a\alpha^2 \qquad \text{(V-2)}$$

V-2. SOLUTIONS CONTAINING A WEAK ACID

Figure V-1 describes the manner in which the concentrations of a weak acid HA and its conjugate base A^- vary with pH, and can be constructed in the following manner.

First draw the two lines* corresponding to $[H^+]$ and $[OH^-]$ having slopes of -1 and $+1$ respectively and intersecting at $pH = \frac{1}{2}pK_w$.

Mark a point, S, whose ordinate is $\log C_{HA}$, the log of the analytical concentration of HA, and whose abscissa is at a

---

*The basis on which these lines were drawn arises from the following equations (derived by solving for [HA] and $[A^-]$ from (V-3) and (V-7).

$$[HA] = \frac{[H^+] \cdot C_{HA}}{K_a + [H^+]} \quad \text{and} \quad [A^-] = \frac{K_a \cdot C_{HA}}{K_a + [H^+]}$$

At $[H^+] \gg K_a$, $[HA] = C_{HA}$ (Log [HA] independent of pH) and
$\log [A^-] = pH - pK_a + \log C_{HA}$ (line of slope + 1)

At $[H^+] \ll K_a$, $[A^-] = C_{HA}$ (Log $[A^-]$ independent of pH) and
$\log [HA] = \log C_{HA} + pK_a - pH$ (line of slope -1)

At $[H^+] = K_a$ i.e. $pH = pK_a$
$[HA] = [A^-] = \frac{1}{2} C_{HA}$

Hence the two lines intersect at $\log [HA] = \log [A^-]$
$= \log C_{HA} - \log 2 = \log C_{HA} - 0.30$

This point P is 0.30 directly below S.

value of pH = pK_a . At pH values below this point, HA is the predominant species and is represented by the horizontal line at log C_{HA}, whereas at high pH values the continuation of this line represents $[A^-]$ which is the predominant species. Next, draw two lines of slope +1 and −1 parallel to the H^+ and OH^- lines to pass through the point S terminating at 1 pH unit before the point S. Mark another point P exactly 0.3 log unit directly below S. Connect the appropriate horizontal and sloping line fragments free hand using point P as a guide.

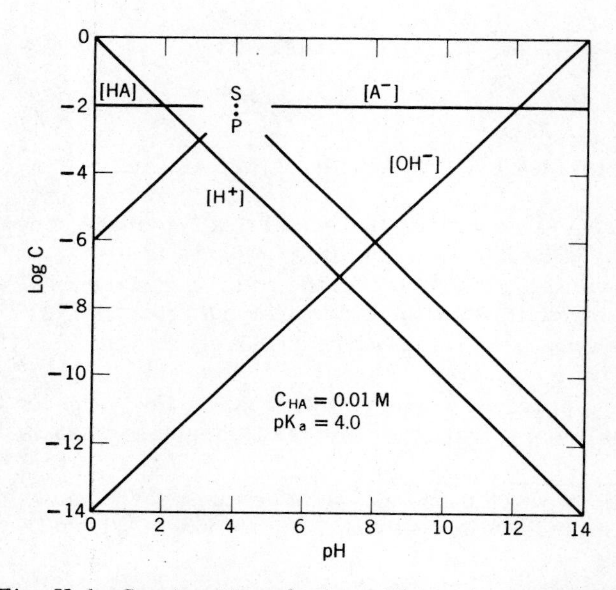

Fig. V-1—Construction of a logarithmic diagram for the conjugate acid-base pair, HA − A^-.

This figure (Fig. V-2) applies to cationic and anionic acids as well as neutral acids.

The equation for the charge balance or proton balance in a solution of HA in water is

$$[H^+] = [A^-] + [OH^-]$$

From Figure V-2 we can see that the $[H^+]$ line crosses the $[A^-]$ line at point Q long before it crosses the $[OH^-]$ line. That is to say that the $[OH^-]$ is negligible in comparison to the

$[A^-]$, and hence $[H^+] = [A^-]$. Hence the pH of the solution will be given by the value at point Q. This will be true so long as $[A^-] > 20\,[OH^-]$ or the line QR is 1.30 log units in length.

We may now describe the conditions in which the $[OH^-]$ becomes of sufficient importance to warrant its consideration. The first of these, increasingly dilute solutions of HA, we have already recognized in the case of strong acids (see section IV-5). That is, as the solution of HA gets increasingly dilute the $[A^-]$ line (as well as the [HA] line) shifts downwards so that the point Q approaches the $[OH^-]$ line. Similarly, the crossing point Q approaches the $[OH^-]$ line as the pK_a of the acid under consideration increases (shifts to the right). An analysis similar to that carried out in section IV-5 shows that when Q occurs at a pH no greater than 6.35, the $[OH^-]$ may be neglected without incurring an error greater than 5%.

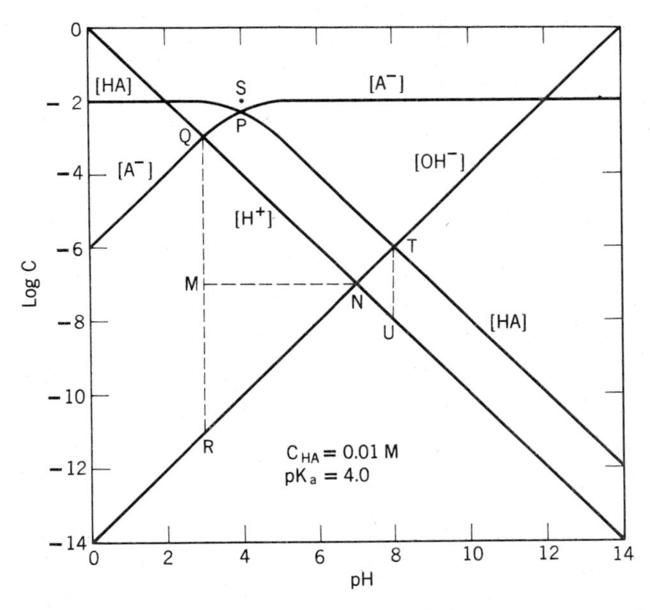

Fig. V-2—Logarithmic variation of the components of the conjugate acid-base pair, HA – A^-, with pH.

In cases where we may not neglect the $[OH^-]$ the calculation of the $[H^+]$ in a solution of a weak acid, HA, proceeds in the following manner.

Since the solution contains four species H^+, OH^-, HA, and A^- we shall need four equations in order to obtain an explicit solution for $[H^+]$. These are

(a) Acid dissociation, $K_a = \dfrac{[H^+][A^-]}{[HA]}$ (V-3)

(b) Water dissociation, $K_w = [H^+][OH^-]$ (V-4)

(c) Charge or proton balance, $[H^+] = [A^-] + [OH^-]$ (V-5)*

(d) Mass or Material balance, $C_a = [HA] + [A^-]$ (V-6)

The fourth equation is obtained by considering the mass or material balance of the acid. If the initial concentration of the acid is C_a moles/liter, then after equilibrium is attained the sum of all the species in solution originating from the acid must be equal to C_a

Hence using equations V-3, V-4, V-5, and V-6, the hydrogen ion concentration, $[H^+]$, can be solved for in terms of C_a, K_w and K_a. The simplest method of eliminating the quantities [HA], $[A^-]$ and $[OH^-]$ from these four equations is to substitute for [HA] and $[A^-]$ in equation (V-3). From equation (V-5)

$$[A^-] = [H^+] - [OH^-] \qquad \text{(V-7)}$$

From equations V-7 and V-6

$$[HA] = C_a - ([H^+] - [OH^-])$$

Substituting these values in equation V-3,

*Note that although this charge balance equation would be different if the acid were charged the proton balance equation would be analogous to (V-5). For example for the NH_4^+ ion in a solution of NH_4Cl the charge balance would be:

$$[NH_4^+] + [H^+] = [OH^-] + [Cl^-]$$

But the proton balance would be:

$$[H^+] = [OH^-] + [NH_3] \qquad \text{(V-8)}$$

Hence the final equation V-9 applies to charged as well as uncharged monoprotic acids.

$$K_a = \frac{[H^+] \times ([H^+] - [OH^-])}{C_a - ([H^+] - [OH^-])} \qquad (V-9)$$

Substituting $K_w/[H^+]$ for $[OH^-]$ in equation V-9

$$K_a = \frac{[H^+] \times ([H^+] - K_w/[H^+])}{C_a - ([H^+] - K_w/[H^+])}$$

i.e. $[H^+]^3 + K_a[H^+]^2 - [H^+](K_w - C_a \cdot K_a) - K_a K_w = 0$ (V-10)

Equation (V-10) is an exact equation which gives the hydrogen ion concentration for a monoprotic acid. The concentrations of all ions and molecules in a solution of the acid HA is governed by the two equilibrium constants K_a and K_w, and the equation that gives the hydrogen ion concentration of the solution is a cubic equation.*

As was mentioned earlier, it is not usually necessary to solve the cubic equation (V-10) in order to calculate the hydrogen ion concentration of a weak acid in aqueous solution. In almost all cases much simpler forms of equation (V-10) can be used. Alternatively, the graphical method described above might prove most convenient. The conditions under which simpler versions of equation V-10 can be used, are discussed next.

In most acid solutions that are encountered the concentration of OH^- ions is very small. Adopting 5% as the upper limit for excluding $[OH^-]$ when compared with $[H^+]$,

i.e. If $[OH^-] < 5\%$ of $[H^+]$ then $[H^+] - [OH^-] = [H^+]$

Equation (V-9) now becomes:

$$K_a = \frac{[H^+] \times [H^+]}{C_a - [H^+]} \qquad (V-11)$$

This is a quadratic equation in $[H^+]$:

$$[H^+]^2 + [H^+] \times K_a - K_a \times C_a = 0 \qquad (V-12)$$

---

*It will be seen later that if there are three equilibria in solution a fourth power equation in $[H^+]$ will be obtained. This is in fact a general rule. The hydrogen ion concentration of a solution is given by an equation in which the $[H^+]$ is raised to a power equal to $(n + 1)$, where n = the number of equilibria that govern the concentrations of ions and molecules in solution.

Furthermore, if $[H^+] < 5\%$ of C_a, equation (V-11) is further simplified and becomes:

$$K_a = \frac{[H^+]^2}{C_a}$$

$$\text{i.e. } [H^+] = (K_a \cdot C_a)^{\frac{1}{2}} \qquad \text{(V-13)}$$

Equation (V-13) is the simplest equation for the calculation of the hydrogen ion concentration of a weak acid. This equation is readily converted to the logarithmic form:

$$pH = \tfrac{1}{2}(pK_a - \log C_a) \qquad \text{(V-13a)}$$

Therefore in calculating the hydrogen ion concentration of a weak acid equations (V-10), (V-12) or (V-13) can be used. It is however important to use the appropriate equation in any given case, and the following example shows the manner in which a typical calculation is carried out.

Example 1.

Calculate the hydrogen ion concentration of a 5.00×10^{-3} M solution of acetic acid. The acid dissociation constant of acetic acid = 1.90×10^{-5}.

$$[H^+] = (K_a \cdot C_a)^{\frac{1}{2}} = (9.50 \times 10^{-8})^{\frac{1}{2}} = 3.08 \times 10^{-4}\ M$$

Therefore $[OH^-]$ which is equal to $K_w/[H^+]$ is obviously less than 5% of $[H\ ]$ and may be neglected. However this approximate value of $[H\ ]$ is greater than 5% of C_a.

$$\text{i.e. } [H^+]_{\text{approx}} > 2.5 \times 10^{-4}$$

Therefore the quadratic equation (V-11) must be used in this case.

$$[H^+]^2 + 1.90 \times 10^{-5}\,[H^+] - 9.50 \times 10^{-8} = 0$$

$$[H^+] = \frac{-1.90 \times 10^{-5} + (3.62 \times 10^{-10} + 38.0 \times 10^{-8})^{\frac{1}{2}}}{2}$$

$$= 2.99 \times 10^{-4}\ M$$

Note that the above quadratic equation can have only one real and positive root. (Why?)

The calculation of the hydrogen ion concentration of a monoprotic acid must therefore be carried out in a stepwise manner as follows:

1. Calculate an approximate value of $[H^+]$ from the equation:

$$[H^+]_{approx} = (K_a \cdot C_a)^{\frac{1}{2}}$$

2. Calculate the value of $[OH^-]$ from:

$$[OH^-] = K_w/[H^+]_{approx}$$

If this value of $[OH^-] < 5\%$ of $[H^+]_{approx}$ and also if $[H^+]_{approx} < 5\%$ of C_a.

Then $[H^+]_{approx}$ is sufficiently accurate.

3. If $[OH^-] < 5\%$ of $[H^+]_{approx}$ but $[H^+]_{approx} > 5\%$ of C_a, then $[H^+]$ is obtained by solving the quadratic equation:

$$K_a = \frac{[H^+]^2}{C_a - [H^+]}$$

4. If the solution of the weak acid is approximately neutral, then the difference between $[H^+]$ and $[OH^-]$ is very small. Therefore if $([H^+]_{approx} - [OH^-]) < 5\%$ of C_a, equation (V-9) becomes

$$K_a = \frac{[H^+] \times ([H^+] - [OH^-])}{C_a}$$

$$\text{i.e. } K_a = \frac{[H^+]^2 - K_w}{C_a}$$

$$\text{Therefore, } [H^+] = (K_a \cdot C_a + K_w)^{\frac{1}{2}} \qquad \text{(V-14)}$$

5. If $[OH^-] > 5\%$ of $[H^+]_{approx}$ and $[H^+]_{approx} > 5\%$ of C_a. It is necessary to solve the cubic equation (V-10) to obtain an answer for $[H^+]$. Equation (V-10) can be solved for $[H^+]$ by a method that involves trial and error. The coefficients of every term in equation (V-10) are known; the value of $[H^+]_{approx}$ is substituted in this equation and adjusted appropriately until the left hand side of the equation becomes equal to zero.

Example 2.

Calculate the hydrogen ion concentration of a 1.00×10^{-4} M solution of HCN. K_a for HCN is 4.00×10^{-10}.

$$[H^+]_{approx} = (K_a \cdot C_a)^{\frac{1}{2}} = (4.00 \times 10^{-14})^{\frac{1}{2}}$$

$$= 2.00 \times 10^{-7}$$

$$\text{Hence } [OH^-] = \frac{1.00 \times 10^{-14}}{2.00 \times 10^{-7}} = 5.00 \times 10^{-8}$$

Since this solution is approximately neutral and $([H^+] - [OH^-])$ $< 5\% \; C_a$, then equation V-14 applies.

$$[H^+] = (K_a C_a + K_w)^{\frac{1}{2}} = (4.0 \times 10^{-14} + 1 \times 10^{-14})^{\frac{1}{2}}$$
$$= 2.24 \times 10^{-7}$$

Example 3.

Calculate the hydrogen ion concentration of a 2.6×10^{-2} M solution of NH_4Cl. Assume that K_b for ammonia is 2.5×10^{-5} and K_w for water is 1.4×10^{-14}.

Equation (V-9) can be used to solve this problem as will be seen below. If C_a moles/liter is the concentration of the cationic acid that is initially added to water, when equilibrium is reached.

$$C_a = [NH_4^+] + [NH_3] = [Cl^-]$$

Also, from charge balance $[NH_4^+] + [H^+] = [OH^-] + [Cl^-]$ or from proton balance $[H^+] = [OH^-] + [NH_3]$ and

$$K_b = \frac{[OH^-] \times [NH_4^+]}{[NH_3]} = \frac{K_w}{K_a}$$

where

$$K_a = \frac{[NH_3] \times [H^+]}{[NH_4^+]}$$

Substituting for $[NH_3]$ and $[NH_4^+]$

$$K_b = \frac{K_w}{K_a} = \frac{[OH^-] \times (C_a + [OH^-] - [H^+])}{([H^+] - [OH^-])}$$

i.e.

$$\frac{[H^+]}{K_a} = \frac{(C_a + [OH^-] - [H^+])}{([H^+] - [OH^-])}$$

Hence,

$$K_a = \frac{[H^+] \times ([H^+] - [OH^-])}{C_a - ([H^+] - [OH^-])} = \frac{K_w}{K_b}$$

which is the same as equation (V-9).

$$[H^+]_{approx} = (K_a \times C_a)^{\frac{1}{2}}$$
$$= \left(\frac{2.6 \times 10^{-2} \times 1.4 \times 10^{-14}}{2.5 \times 10^{-5}} \right)^{\frac{1}{2}} = 3.8 \times 10^{-6}\,M$$

and $[OH^-] = 3.7 \times 10^{-9}$ which is less than 5% of 3.8×10^{-6}. Also $[H^+]_{approx}$ is less than 5% of 2.6×10^{-2}; hence $[H^+]_{approx}$ is sufficiently accurate.

In the previous section dealing with the dissociation of weak acids, expressions that relate pH, concentration terms and pK were developed. In practice, the unknown of interest is more often than not the pK value rather than the pH, which after all, may be measured conveniently. The questions that have been discussed up to now must therefore be restated as follows: if $[H^+]$, $[OH^-]$, C_a and K_w are known or can be measured, what method should be adopted for calculating K_a? A detailed discussion of this problem is outside the scope of this book. It is readily seen however that the equations that have already been derived can be rearranged in a simple manner to give K_a. For a monoprotic acid, K_a could be obtained by substituting the measured values of $[H^+]$, $[OH^-]$ and C_a in equation (V-9) for a series of solutions of HA at varying C_a. For every value of the concentration of the acid that is taken there will be corresponding values of $[H^+]$ and $[OH^-]$ that can be measured and calculated. When these values are substituted in

$$K_a = \frac{[H^+] \times ([H^+] - [OH^-])}{C_a - ([H^+] - [OH^-])}$$

a series of values for K_a will be obtained and they can be averaged to give the best value for K_a. A variation of this method is to determine K_a graphically. For example the following equation can be readily derived from equation (V-9):

$$pK_a = pH + \text{Log}\, \frac{(C_a - [H^+] + [OH^-])}{[H^+] - [OH^-]} \qquad \text{(V-15)}$$

If a series of solutions of varying concentration C_a are taken and their pH measured, and if these values are plotted against the log term in the above equation, a straight line of unit negative slope should be obtained and the intercept of this straight line on the y-axis should give the value of pK_a.

Another obvious variation of the method that has been described for determining pK_a values is to titrate a solution of a weak acid with NaOH. It can be readily shown that for any point on the titration curve

$$K_a = \frac{[H^+] \times ([H^+] + [Na^+] - [OH^-])}{C_a - ([H^+] + [Na^+] - [OH^-])} \qquad \text{(V-16)}$$

The initial concentration of the acid is C_a and values of $[H^+]$, $[Na^+]$ and $[OH^-]$ can be obtained experimentally and substituted in equation V-16 to give K_a. Note that in this case only one value of C_a is necessary, but as is true with all other terms in equation V-16 the concentration term C_a will change from one point to another in the titration since the total volume of the solution changes on the addition of NaOH solution to the acid solution. A reasonable value of K_a can be obtained from a single point on the titration curve since at 50% neutralization of the acid, $pH = pK_a$. (See chapter XIII)

The same considerations apply to solutions of weak bases and the student could work out for himself the manner in which K_b values for weak bases are obtained.

V-3. SOLUTIONS CONTAINING A WEAK BASE

Equations analogous to those derived for weak acids can be derived for weak bases. It should be remembered that anionic bases are sometimes considered in a separate category called salts of strong bases and weak acids. The application of the Bronsted Theory eliminates the need for such additional complication. However the student should be prepared to interpret references to hydrolysis of salts in terms of weak base (or weak acid, see Equation V-5) calculations which now follow.

If B represents a neutral monoacidic weak base, C_b, its molar concentration and K_b its dissociation constant,

$$B + H_2O \rightleftharpoons BH^+ + OH^-$$

$$K_b = \frac{[BH^+][OH^-]}{[B]} \tag{V-17}$$

$$K_w = [H^+][OH^-]$$

Mass Balance: $C_b = [B] + [BH^+]$ (V-18)

Charge or Proton Balance: $[H^+] + [BH^+] = [OH^-]$ (V-19)*

---

*If the weak base were charged, e.g. the acetate ion in a solution of sodium acetate, it would be useful to use the proton balance equation:

$$[H^+] + [HOAc] = [OH^-] \tag{V-19a}$$

As was pointed out earlier this equation also leads to equation (V-20).

Substituting for BH^+ and B from equations (V-18) and (V-19) in (V-17),

$$K_b = \frac{[OH^-] \times ([OH^-] - [H^+])}{C_b - ([OH^-] - [H^+])} \qquad \text{(V-20)}$$

This equation is analogous to equation (V-8) for a weak monoprotic acid and is a cubic equation in $[OH^-]$. Simplified versions of equation (V-20) can be used under the following conditions:

If $[H^+] < 5\%$ of $[OH^-]$, $K_b = \dfrac{[OH^-]^2}{C_b - [OH^-]}$ (V-21)

If $[H^+] < 5\%$ of OH^- and also $[OH^-] < 5\%$ of C_b

$$K_b = \frac{[OH^-]^2}{C_b} \quad \text{i.e. } [OH^-] = (K_b \cdot C_b)^{\frac{1}{2}} \qquad \text{(V-22)}$$

Finally if the solution is approximately neutral, i.e. $([OH^-] - [H^+]) < 5\%$ of C_b

$$[OH^-] = (K_b \cdot C_b + K_w)^{\frac{1}{2}} \qquad \text{(V-23)}$$

Example 4.

Calculate the hydrogen ion concentration in a 0.05 M solution of sodium acetate. Assume that K_a for acetic acid is 2.8×10^{-5} and K_w for water is 1.5×10^{-14}.

$$OAc^- + H_2O \rightleftharpoons HOAc + OH^-$$

If C_b is the concentration of the sodium acetate added,

$$C_b = [OAc^-] + [HOAc] = [Na^+]$$

$$[H^+] + [Na^+] = [OAc^-] + [OH^-]$$

or

$$[H^+] + [HOAc] = [OH^-]$$

Therefore

$$\frac{K_w}{K_a} = \frac{[HOAc] \times [OH^-]}{[OAc^-]} = \frac{[OH^-] \times ([OH^-] - [H^+])}{C_b - ([OH^-] - [H^+])}$$

which, as pointed out earlier, is the same as equation (V-20).

$$[OH^-]_{approx} = \left(\frac{1.5 \times 10^{-14} \times 5 \times 10^{-2}}{2.8 \times 10^{-5}}\right)^{\frac{1}{2}}$$

$$= \left(\frac{C_b \cdot K_w}{K_a}\right)^{\frac{1}{2}}$$

$$= 5.2 \times 10^{-6}$$

Hence

$$[H^+] = \frac{1.5 \times 10^{-14}}{5.2 \times 10^{-6}} = 2.9 \times 10^{-9}\ M$$

Since $[H^+] < 5\%$ of $[OH^-]$ and also $[OH^-] < 5\%$ of C_b, the value of $[OH^-]_{approx}$ is sufficiently accurate. Therefore the hydrogen ion concentration of the solution is 2.9×10^{-9} M.

Figure V-2, developed earlier for weak acids, can be recognized as describing equally well the variation of the components of a solution of a weak base with pH. Let us consider a solution of NaA, where A^- is the conjugate base of HA. The proton balance for this solution is

$$[H^+] + [HA] = [OH^-]$$

In most cases $[H^+]$ is sufficiently smaller than [HA] so that it can be neglected. Hence the pH of the solution is that of the point of intersection T for the [HA] and $[OH^-]$ lines. The assumption of the negligible $[H^+]$ is valid when the line TU is at least 1.30 log units in length. (Why?) We have seen in previous examples how to deal with cases when $[H^+]$ may not be neglected.

V-4. SOLUTIONS CONTAINING A CONJUGATE ACID-BASE PAIR

It might seem strange at first to find a section entitled, "solutions containing a conjugate acid-base pair" since in the preceding sections V-1 and V-2 we have seen that any solution of a weak acid (or a weak base), contained some of the conjugate base (or acid). That is, acetic acid solutions contained acetate ions, ammonia solutions contained ammonium ions and so on. However in all these cases there was only one source of the conjugate acid-base pair. As a result of this restriction the concentrations of the conjugate acid-base pair could not vary independently of each other. If we wished we could prepare a solution containing any desired ratio of acid to conjugate base by mixing appropriate quantities of the acid, e.g. HOAc or NH_4Cl and the conjugate base NaOAc or NH_3. The great practical significance of such a procedure is to provide a means of obtaining solutions of known pH containing significant amounts of both acid- and base-neutralizing components. Small amounts of acids or bases can be added to such solutions without materially affecting the pH. A detailed consideration of the behavior and characteristics of such solutions which are said to have buffer action will be given in the next section.

Let us now consider a solution which contains C_a moles/liter of a weak monoprotic acid (either charged or uncharged) and C_b moles/liter of its conjugate base. The concentration of the acid HA (or BH^+) is given by the equation:

$$[HA] \text{ or } [BH^+] = C_a - [H^+] + [OH^-]$$

and the concentration of the conjugate base $[A^-]$ or $[B]$ is given by the equation:

$$[A^-] \text{ or } [B] = C_b + [H^+] - [OH^-]$$

These equations have been previously derived. (See Sections V-1 and V-2.) The equilibrium expression is given by:

$$K_a = \frac{[H^+][A^-]}{[HA]} \text{ or } \frac{[H^+][B]}{[BH^+]}$$

Substituting in this equilibrium expression for the concentrations of the conjugate acid and conjugate base,

$$K_a = \frac{[H^+]\{C_b + ([H^+] - [OH^-])\}}{\{C_a - ([H^+] - [OH^-])\}} \qquad \text{(V-24)}$$

Equation (V-24) has been derived without making any simplifying assumptions. Note that if $C_b = 0$, the equation must reduce to that which gives the hydrogen ion concentration of a monoprotic acid,

$$\text{i.e.} \quad K_a = \frac{[H^+] \times ([H^+] - [OH^-])}{C_a - ([H^+] - [OH^-])}$$

which is the same as equation V-9.

If $C_a = 0$, the equation must reduce to that which gives the hydrogen ion concentration of the conjugate base of the monoprotic acid,

$$\text{i.e.} \quad K_a = \frac{[H^+] \times \{C_b + ([H^+] - [OH^-])\}}{([OH^-] - [H^+])}$$

$$= \frac{K_w\{C_b + ([H^+] - [OH^-])\}}{[OH^-] \times ([OH^-] - [H^+])}$$

Therefore,

$$\frac{K_w}{K_a} = \frac{[OH^-] \times ([OH^-] - [H^+])}{C_b - ([OH^-] - [H^+])} = K_b$$

which is the same as equation (V-20).

Equation (V-24) is cubic in $[H^+]$ and in many instances can be simplified. If $[OH^-] < 5\%$ of $[H^+]$, and if $[H^+] < 5\%$ of C_a and $< 5\%$ of C_b, we get the simplest form of the equation,

$$K_a = \frac{[H^+] \times C_b}{C_a} \qquad \text{(V-25)}$$

Equation (V-24) is therefore a general equation that gives the concentration of the hydrogen ion in a solution made by dissolving a monoprotic acid (HOAc or NH_4Cl) in water or by dissolving a monoacidic base (NH_3 or NaOAc) in water or by dissolving a monoprotic acid as well as its conjugate base (HOAc and NaOAc or NH_4Cl and NH_3) in water. All the equations that have been derived in Sections V-2 and V-3 can be obtained from equation V-24. It is important to keep in mind that in all instances C_a and C_b refer to the initial, or "analytical" concentration of the conjugate acid and the conjugate base.

Example 5.

Calculate the pH of a solution which is 0.1 M in formic acid and 0.05 M in potassium formate. K_a for formic acid is 2.6×10^{-4} -

This solution consists of the weak acid HCOOH and its conjugate base $HCOO^-$.

$C_a = 0.1$ M and $C_b = 0.05$ M

Therefore the hydrogen ion concentration of such a solution is given by the simplest equation:

$$K_a = [H^+] \times \frac{C_b}{C_a}$$

$$\text{i.e. } 2.6 \times 10^{-4} = [H^+] \times \frac{0.05}{0.1}$$

Therefore $[H^+] = 5.2 \times 10^{-4}$

It is obvious that both $[OH^-]$ and $[H^+]$ are less than 5% of C_a as well as C_b. Therefore the simplest form of the equation (V-24) can be used. The pH of the solution is therefore 3.28.

Example 6.

What is the concentration of sodium sulfamate that should be added to a 0.1 M solution of sulfamic acid to give a pH of 1.55? The pK_a of sulfamic acid, HNH_2SO_3, is 0.65.

In this problem the simplified equation (V-25) may not be used because the hydrogen ion concentration in solution is very much greater than 5% of C_a. However, since the solution is highly acidic,

i.e. $[H^+] \gg [OH^-]$, and equation V-24 can be modified to give:

$$K_a = \frac{[H^+] \cdot (C_b + [H^+])}{(C_a - [H^+])}$$

$$10^{-0.65} = \frac{10^{-1.55} \times (C_b + 10^{-1.55})}{(10^{-1.0} - 10^{-1.55})}$$

$C_b = 0.55$ M.

Had equation V-25 been used in this problem, we would have obtained $C_b = 0.80$ M--which is significantly in error.)

V-5. BUFFER SOLUTIONS

Since the vast majority of reactions taking place in aqueous solutions are affected by changes in pH, it is of vital importance in the study or control of these reactions to have them occur in solutions in which pH changes are kept to a minimum. Such solutions, to which small quantities of either acids or bases may be added without materially altering their pH, are called buffer solutions.

In the extremes of the pH range, solutions of strong acids or bases serve as buffers. This may be seen from the following example.

Example 7.

What is the change in pH of 100 ml of 0.5 M HCl, upon the addition of 1 ml of 0.1 M NaOH?

The initial pH of the solution is given by:

$$pH = -\log 0.5 = 0.30$$

The final concentration of HCl and therefore of H^+ is

$$\frac{(100 \times 0.5 - 1 \times 0.1)}{101} = 0.499$$

Only 0.2% of the acid was required to neutralize the NaOH added. The final pH is $-\log 0.499 = 0.30$. Therefore the change in pH is seen to be less than 0.01 units.

The next example shows what occurs if the acid were more dilute.

Example 8.

What is the change in pH of 100 ml of 0.005 M HCl upon the addition of 1 ml of 0.1 M NaOH?

The initial pH $= -\log 0.005 = 2.30$.

The final concentration of $H^+ = \dfrac{(100 \times 0.005 - 1 \times 0.1)}{101}$

$$= 0.004$$

The final pH $= -\log 0.004 = 2.40$

Notice that 20% of the acid was required to neutralize the NaOH added. Here the change in pH is 0.10 units. With a solution of 0.001 M HCl, in this problem, the pH would have risen to about 7.0 since 100% of the acid would have been neutralized!

The significance of these results is that as the HCl solutions become increasingly dilute their neutralization capacity becomes increasingly small. In order that solutions of intermediate pH have sufficient neutralization capacity so that their pH values are reasonably stable to the effects of addition of acids and bases, they must contain reasonable quantities of a weak acid and its conjugate base. These function in the following way. Consider a solution containing an acid HA and its conjugate base A^-. If NaOH is added,

$$HA + OH^- \rightarrow A^- + H_2O$$

or if HCl is added, $A^- + H^+ \rightarrow HA$.

Hence the acid or base added is neutralized, the sole effect being a small change in the ratio of [HA] to [A^-].

As an illustrative example let us consider a solution of

0.085 M HF and 0.10 M NaF (assume pK_a of HF = 2.93), which, like the 0.001 M HCl, has a pH of 3.00. Upon the addition of 1 ml of 0.1 M NaOH to 100 ml of this solution, the pH changes only slightly since only a small percentage of the HF present is required to neutralize the small amount of base added. Viz:

Amount of HF initially = 100×0.085 = 8.50 mmoles
Amount of NaOH added = 1×0.1 = 0.10 mmoles
Amount of HF finally = $8.50 - 0.1$ = 8.40 mmoles
Amount of F^- finally = $100 \times 0.1 + 1 \times 0.1$ = 10.1 mmoles

From equation (V-25),

$$[H^+] = K_a \frac{C_a}{C_b}$$

$$= 10^{-2.93} \times \frac{8.40}{10.10} = 10^{-3.01}$$

$$pH = 3.01$$

This corresponds to a pH change of only 0.01 units.

An important property of buffer solutions is that their pH remains practically constant on dilution. If enough water were added to the 100 ml of HF-NaF mixture just considered to make exactly one liter of solution, the pH of the 10-fold more dilute solution would be 3.00*, since in a mixture of a conjugate acid-base pair the ratio of concentrations rather than the individual concentrations, determines the pH.

By way of comparison note that ten-fold dilution of a strong acid or strong base solution results in a pH change of 1.0 units. Thus the 0.5 M HCl whose pH is 0.30, upon tenfold dilution becomes 0.05 M HCl with a pH of 1.30. This limits the usefulness of strong acids and bases as buffers.

Whenever a buffer solution is used, these questions should be answered: (a) What is the pH of the buffer? (b) What is its buffer capacity? That is, when a given amount of acid or base is added to or liberated in the buffer medium, what will be the pH change?

*The value 3.00 was calculated on the assumption that the pK_a of HF remained constant as the ionic strength decreased from 0.1 to 0.01. Strictly speaking this is inaccurate since the appropriate pK_a value in the more dilute solution is 3.08. The correct value of pH is 3.15.

With regard to the first of these questions, a mixture of a weak acid and its conjugate base generally will have a pH in the vicinity of pK_a for the acid. For this reason, in selecting buffer components, acid-conjugate base pairs are chosen whose pK_a values are within one unit of the pH desired. Within this range, the ratios of acid-conjugate base concentration are within 10 to 0.1, and permit reasonable neutralization capacity in the buffer. Outside this range, such concentration ratios become so extreme that the solutions have little value as buffers.

As an illustration of this statement the following tabulation lets us compare the effects of the addition of 1 ml of 1.0 M NaOH to each of the following solutions containing the same concentration of each acid.

	System HX/NaX	System HY/NaY
	pK_{HX} = 3.00	pK_{HY} = 6.00
Initial Concentrations		
	[HX] = 0.10	[HY] = 0.10
	[X] = 0.10	[Y] = 0.0001
	pH = 3.00	pH = 3.00
Final Concentrations:		
	[HX] = 0.09	[HY] = 0.09
	[X] = 0.11	[Y] = 0.0101
	pH = 3.05	pH = 5.05

The ideal buffer system is one in which the pK_a of the acid is equal to the selected pH. In this ideal buffer, the efficiency is maximum because the concentrations of acid and conjugate base are equal. Although this situation is not encountered frequently in practice, one would always select buffer components having a pK_a as close as possible to the control pH. A further consideration of which one must always be aware is to be sure that the buffer components are compatible with the other materials present in the solution. For instance one would not use HF-NaF buffers in a solution containing any of the trivalent lanthanide ions since they form insoluble fluorides.

Let us now turn our attention to the question of buffer capacity. We have already seen how to calculate the pH change that occurs when a given amount of acid or base is added to a specific buffer. Now let us consider how to calculate the minimum concentrations of buffer components that are required to keep within a prescribed limit, the pH of a solution to which acid or base is added. This is illustrated in the following example:

Example 9.

It is desired to prepare a buffer whose pH remains within 0.20 units of 5.00 upon the release of 3 millimoles of H^+ in 100 ml of the solution. The total ionic strength of the solution is 0.50.

Step 1. Selection of Components:

Since a pH of 5.00 is desired, any weak acid with a pK_a between 4 and 6 may be used. From Table I in the Appendix, the following systems can be used:

	pK	pK(μ = 0.50)
Acetic acid	4.76	4.35
Propionic acid	4.87	4.46
Anilinium ion	4.59	4.59
Pyridinium ion	5.30	5.30

On the basis of favorable pK values, almost any one of these would be suitable. Since acetic acid is commonly available, let us proceed by selecting it.

Step 2. Selection of the Concentration Ratio of the Conjugate Acid-Base Pair:

This step involves solution of the equation that describes the pH of a conjugate acid-base pair.

$$pH = pK_a + \log \frac{[OAc^-]}{[HOAc]}$$

$$5.00 = 4.35 + \log \frac{[OAc^-]}{[HOAc]}$$

$$\frac{[OAc^-]}{[HOAc]} = 10^{+0.65} = 4.5$$

Note that the concentration ratio rather than the individual concentrations is fixed. In certain simple buffer calculations where the buffer capacity is not or need not be specified, the calculation would now be concluded. The concentration of either component may then be specified to suit one's own convenience, and that of the other is obtained from the fixed ratio. Thus if we select either 0.01 M HOAc or 0.02 M HOAc the corresponding NaOAc concentrations are 0.045 and 0.090 respectively.

Step 3: Calculation of the Concentrations of the Conjugate Acid-Base Pair.

Let the initial [HOAc] = x
From Step 2 then, the initial $[OAc^-]$ = 4.5 x

Since the buffer will be subjected to the addition of 3 millimoles of H^+ per 100 ml, this will decrease the $[OAc^-]$ and accordingly increase the [HOAc] by

$$\frac{3}{100} = 0.03.$$

Hence,

the final [HOAc] = x + 0.03
and $[OAc^-] = 4.5x - 0.03$

The final pH as given in the statement of the problem is 4.80. The solution of the equation,

$$4.80 = 4.35 + \log\frac{(4.5x - 0.03)}{x + 0.03}$$

for x in the following manner, gives the description of the desired buffer.

$$\frac{4.5x - 0.03}{x + 0.03} = 10^{0.45} = 2.8$$

$$4.5x - 0.03 = 2.8x + 0.084$$

Therefore x = 0.065.

Hence the desired buffer is a solution 0.065 M in HOAc and 4.5×0.065 or 0.29 M in NaOAc. This buffer mixture may be conveniently prepared by neutralizing 0.355 M HOAc, (the sum of 0.065 and 0.29) with enough NaOH to give a reading of pH 5.00 on the pH meter or to react with all but 0.065 moles/liter of the acid. Alternatively, acetic acid and sodium acetate may be mixed to give the desired concentrations.

In this section attention was focused on the use of monoprotic acid—conjugate base mixtures as buffer components. As will be shown in Sec. VI-7, exactly analogous considerations apply to conjugate acid-base mixtures of polyprotic acids. Therefore in problems dealing with the design of buffers, distinction is not made between mono- and polyprotic acid-base systems in the selection of buffer components.

SUGGESTIONS FOR FURTHER READING

T. B. Smith, Analytical Processes, Edward Arnold Publishers Ltd., London (1940), Chapter 10.

S. Bruckenstein and I. M. Kolthoff, Chapter 12, Part I, Vol. I, Treatise on Analytical Chemistry, Interscience, New York (1959). I. M. Kolthoff and P. J. Elving, Editors.

E. J. King, Qualitative Analysis and Electrolytic Solutions, Harcourt, Brace and World, Inc., (1959), Chapters 11 and 12.

L. G. Sillen, Part I, Vol. I, Chapter 8, Treatise on Analytical Chemistry, Interscience, (1959), I. M. Kolthoff and P. J. Elving, Editors.

V-6. PROBLEMS

1. (a) The pK_a of cyanic acid is 3.68. Calculate the degree of dissociation of a 1.00×10^{-3} M solution of the acid.
   (b) Calculate the hydrogen ion concentration of a 0.001 M solution of cyanic acid.
   (c) What is the effect of a tenfold increase in concentration on the percentage of dissociation of this acid? [Assume pK_a is the same as in (a).]
2. Calculate the hydrogen ion concentration and pH of the following solutions in pure water.
   (a) 0.10 M acetic acid (pK_a = 4.73)
   (b) 3.85×10^{-3} M acetic acid (pK_a = 4.74)
   (c) 2.0×10^{-3} M boric acid (pK_a = 9.24)
   (d) 0.001 M phenol (pK_a = 9.89)
   (e) 1.0×10^{-5} M hydrocyanic acid (pK_a = 9.40)
3. The pH of a 1.0×10^{-3} M solution of an acid is 3.50. Calculate the dissociation constant of the acid.
4. Calculate the pH of the following solutions:
   (a) 0.1 M NH_4Cl (pK_a = 9.24, pK_w = 13.76)
   (b) 0.2 M CH_3COONa (pK_a = 4.45, pK_w = 13.69)
   (c) 1.0×10^{-3} M C_6H_5COONa (pK_a = 4.17, pK_w = 13.97)
   (d) 0.02 M NH_3 (pK_a = 9.24, pK_w = 13.97)
   (e) 3.4×10^{-3} M KOH (pK_w = 13.95)
5. Calculate the concentration of the hydroxide ion in:
   (a) 0.05 M CH_3NH_2 (pK_a = 10.64, pK_w = 13.94)

5. (b) 0.05 M $C_2H_5NH_2$ (pK_a = 10.67, pK_w = 13.94)
   (c) 1.0×10^{-3} M NH_3 (pK_a = 9.24, pK_w = 13.99)
   (d) 2.5×10^{-3} M HN_3 (pK_a = 4.70, pK_w = 13.98)
6. An acid, HA, is 6% dissociated in a 0.03 M solution. What is its percentage dissociation in a 0.30 M solution? (Assume pK_a remains constant.)
7. Calculate the concentrations of all ionic and molecular species present in the following solutions:
   (a) 0.01 M KCN (pK_a = 9.28, pK_w = 13.88)
   (b) 0.03 M NH_3 (pK_a = 9.24, pK_w = 13.97)
   (c) 1.0×10^{-3} M NH_4Cl (pK_a = 9.24 pK_w = 13.97)
   (d) 0.2 M KCl (pK_w = 13.69)
8. Calculate the pH of the following solutions:
   (a) 0.01 M anilinium chloride (pK_a = 4.59, pK_w = 13.91)
   (b) 0.02 M pyridine (pK_a = 5.30, pK_w = 14.00)
   (c) 1.0×10^{-3} M monochloroacetic acid
   (pK = 2.83, pK_w = 13.97)
   (d) 2.00×10^{-8} M sodium hydroxide (pK_w = 14.00)
   (e) 1.00×10^{-7} M hydrochloric acid (pK_w = 14.00)
9. How much NH_3 will dissolve in 100 ml of water to raise the pH to (a) 7.50? (pK_a = 9.24, pK_w = 14.00)
   (b) 8.00?
   (c) Comment on the applicability of pH measurements on the problem of determining small quantities of ammonia.
10. A solution containing a mixture of sodium formate and formic acid has a pH of 3.0. Determine the ratio of the concentrations of formate ions and formic acid. pK_a = 3.52.
11. What weight of anhydrous sodium acetate should be added to 500 ml of a solution which is 0.05 M in acetic acid to give a buffer solution having a pH of 5.00? (pK_a = 4.63)
12. What weight of sodium hydroxide should be added to 1.0 liter of 0.1 M hydrofluoric acid to give a buffer solution having a pH of 2.5? (pK_a = 3.01)
13. How many ml of 0.10 M HCl should be added to 250 ml of a 0.05 M NH_3 solution to give a buffer solution having a pH of 9.0? (pK_a = 9.24)
14. Calculate the pH of the following solutions:
   (a) 0.20 M CH_3COOH + 0.10 M CH_3COONa
   (pK_a = 4.52)
   (b) 0.10 M HF + 0.20 M KF (pK_a = 2.86)
   (c) 0.1 M NH_3 + 0.05 M NH_4Cl (pK_a = 9.24)
15. (a) Calculate the pH of each of the solutions in question 4 if each solution is diluted ten times.

15. (b) Calculate the pH of each of the solutions in question (2) if each solution is diluted ten times. (Assume pK_a values remain constant.)
16. Determine the pH range in which a solution which is 0.1 M in NH_3 and 0.05 M in NH_4Cl can act effectively as a buffer. (pK_a = 9.24)
17. Using Table I in the appendix for pK_a values, select the appropriate buffer components and determine their concentrations for making buffer solutions having the following pH values:
(a) 2.5 (b) 4.7 (c) 6.4 (e) 13.5
18. Calculate the change in pH when 1.0 ml of 0.05 M NaOH is added to 50 ml of: (a) each of the solutions in question 14. (b) each of the solutions in question 2. (Assume pK_a values remain constant.)
19. Using NH_3 and NH_4Cl as the buffer components design a buffer having a pH initially of 8.8 whose pH will not change by more than 0.10 on the addition of
(a) 0.05 ml of 0.1 M NaOH to 100 ml of the buffer solution, or
(b) 0.05 ml of 0.1 M HCl to 100 ml of the buffer solution.
(c) Explain why answers to (a) and (b) might be different. pK_a = 9.24 for NH_4^+.
20. Calculate the concentrations of CH_3COOH and CH_3COONa that should be present in 200 ml of a solution whose pH is 4.90 and which will prevent a pH change in the solution of more than 0.15, on the addition of 5 ml. of 0.001 M NaOH or HCl. (A sufficient amount of indifferent electrolyte is present to adjust the ionic strength to a value of 0.5 so that pK_a = 4.35.)
21. Calculate the concentrations of all ionic and molecular species in the following solutions:
(a) 1.00×10^{-3} M KCN (pK_a = 9.37, pK_w = 13.97)
(b) 2.50×10^{-4} M KCN + 3.00×10^{-3} M HCN (pK_a = 9.38, pK_w = 13.98)
(c) 0.01 M KNO_2 and 2.10×10^{-2} M HNO_2 (pK_a = 3.21, pK_w = 13.91)
(d) 0.01 M boric acid (pK_a = 9.24, pK_w = 14.00)

(See also Problem Section VI-8 for additional buffer problems.)

VI

Polybasic Bronsted Acids and Polyacidic Bronsted Bases

VI-1. INTRODUCTION

In Chapter V we considered equilibria involving monoprotic acids, i.e. acids that can donate only one proton per molecule of acid. It is possible to have acids that are capable of donating more than one proton per molecule of acid. These acids are called polyprotic acids; H_3PO_4 is capable of donating three protons, H_2SO_4, $H_2C_2O_4$ and H_2S are examples of acids that can donate two protons per molecule of acid.

Similarly, many hydrated metal cations can be considered as polybasic acids in reactions that are sometimes referred to as hydrolysis of metal ions. For example:

$$Fe(H_2O)_6^{3+} \rightleftharpoons Fe(H_2O)_5OH^{2+} + H^+$$

$$Fe(H_2O)_5OH^{2+} \rightleftharpoons Fe(H_2O)_4(OH)_2^+ + H^+$$

The calculation of the hydrogen ion concentration in a solution of a polyprotic acid is complicated by the need to consider the various stepwise dissociations. H_3PO_4 dissociates in a stepwise manner as follows:

$$H_3PO_4 \rightleftharpoons H^+ + H_2PO_4^-$$

$$H_2PO_4^- \rightleftharpoons H^+ + HPO_4^=$$

$$HPO_4^= \rightleftharpoons H^+ + PO_4^{3-}$$

One simplifying feature is that the successive dissociation steps are suppressed by the hydrogen ions formed from the first dissociation step, as might have been predicted from Le Chatelier's Principle. Another simplifying feature arises from the fact that the successive dissociation constants decrease in value. Usually the decrease is so large that only the first dissociation need be considered in the calculation. In the above case of H_3PO_4 and also in many other cases of oxyacids the difference between any two successive dissociation constants is about 10^5. There are many examples of polyprotic acids in which the ratio of K_1 and K_2 (the first and second dissociation constants) lies between 10^4 and 10^5, (see Table I in Appendix.) This effect can be readily explained on the basis of at least two factors: (a) electrostatic and (b) statistical.

(a) We note that the successive protons must be removed from ions that are increasingly negatively charged i.e., the removal of successive protons from an acid H_nY becomes increasingly difficult.

(b) In the first dissociation step of H_nY there are n chances for a proton to leave if the hydrogen atoms are all equivalent but there is only one chance for it to recombine. In the second dissociation step the number of chances for protons to leave decreases to (n – 1) and the number of possible recombinations increases to 2. Therefore for a diprotic acid the ratio of K_1 to K_2 on the statistical effect alone would be

$$\frac{2 : 1}{1 : 2} = 4 : 1.$$

Hence it is obvious that the electrostatic effect predominates in most cases since the successive dissociation constants for diprotic acids differ by a much larger factor than that implied by the statistical effect alone. However, it must also be mentioned that there are many diprotic acids whose dissociation constants are not very far apart. For example, succinic acid has $K_1 = 6.2 \times 10^{-5}$ and $K_2 = 2.3 \times 10^{-6}$.

VI-2. SOLUTIONS CONTAINING A DIPROTIC ACID

Solutions of diprotic acids may be graphically described by diagrams such as Fig. VI-1 in which the variation of the concentration of the several species is shown as a function of pH. The construction of Fig. VI-1 is similar to that

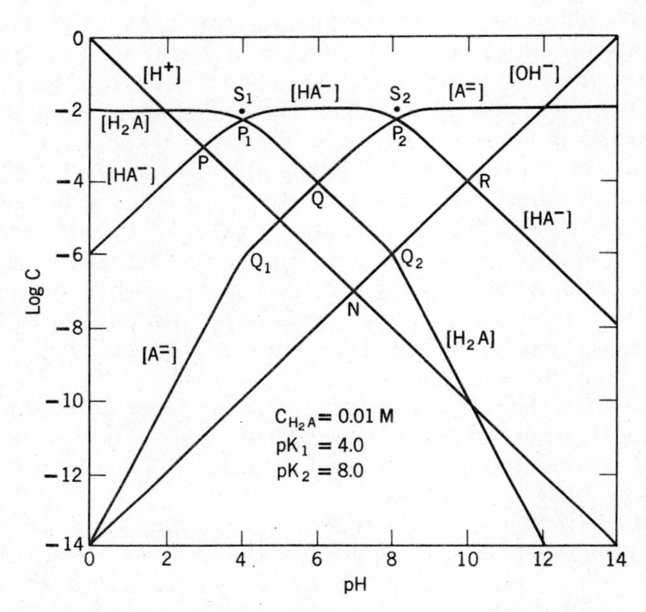

Fig. VI-1—Logarithmic Variation of the Components of a Diprotic Acid, H_2A, with pH.

described for the monoprotic acid in Chapter V. The steps may be summarized as follows.

(a) Draw the two lines corresponding to $[H^+]$ and $[OH^-]$ having slopes of -1 and $+1$ respectively and intersecting at $pH = \frac{1}{2} pK_w$.

(b) Mark points S_1 and S_2 having ordinate values equal log C_{H_2A} and abscissae at pH = pK_{a_1} and pK_{a_2}, respectively.

(c) Draw horizontal line segments at a value equal to log C_{H_2A} leaving blank those portions within one pH unit of S_1 and S_2.

(d) Draw two sets of two lines, each of slopes $+1$ and -1, (parallel to the $[H^+]$ and $[OH^-]$ lines), to pass through the points S_1 and S_2 and terminating at approximately one pH unit before the appropriate point (S_1 or S_2).

(e) Where the lines H_2A and $A^=$ reach pH values corresponding to points S_1 and S_2 (i.e. at points Q_2 and Q_1) continue these lines with slopes of -2 and $+2$.

(f) Mark points P_1 and P_2 at 0.30 log units directly below

S_1 and S_2 and connect the appropriate line fragments using the points P_1 and P_2 as guides.

In addition to aiding our visualization of the composition of solutions of the diprotic acid at various pH values, this diagram coupled with appropriate proton balance expressions provides the means for graphical solution to pH calculations.

For instance, in a solution of diprotic acid, H_2A, the proton balance is

$$[H^+] = [HA^-] + 2[A^=] + [OH^-]$$

Since the $[H^+]$ line usually crosses the $[HA^-]$ line long before it crosses the other two lines, the pH of this solution is the point of intersection $[H^+] = [HA^-]$.

For a solution of an ampholyte, NaHA

$$[H^+] + [H_2A] = [A^=] + [OH^-]$$

Since $[H_2A]$ and $[A^=]$ lines intersect first (at the highest log c) the other terms may be neglected. Therefore pH of this solution is represented usually by intersection point Q.

Finally for a diacidic base Na_2A

$$[H^+] + 2[H_2A] + [HA^-] = [OH^-]$$

Here the pH is given by the intersection of $[HA^-]$ and $[OH^-]$ lines at point R. It should be realized that the cases just considered represent the least complicated ones. Complications that arise when either the solutions become dilute or when the pK values become either too large or too small have already been discussed in the previous chapter. A further complication can arise with di- and poly-protic acids if the successive pK values are too close together. When this happens the assumptions involved in simplifying the proton balance equations do not apply. In such cases an inspection of the diagram will clearly show which terms in the proton balance equation need be considered. Graphical evaluation of the solution to a proton balance equation involving more than one term on each side is illustrated in Example 7, Chapter IV.

We will now turn to analytical means of handling such problems. The student is urged to prepare and refer to a graphical representation of the system being studied in order to develop a better appreciation of the relative magnitudes of the various terms in the equation describing the system.

The following equations describe the dissociation of a diprotic acid, H_2A, which is a neutral molecule. (It is of course possible to have a diprotic acid which is charged.)

$$H_2A \rightleftharpoons H^+ + HA^-$$

$$HA^- \rightleftharpoons H^+ + A^=$$

Therefore the dissociation constants K_1 and K_2 are given by the following equations:

$$K_1 = \frac{[H^+][HA^-]}{[H_2A]} \quad \text{(VI-1)}$$

$$K_2 = \frac{[H^+][A^=]}{[HA^-]} \quad \text{(VI-2)}$$

We may write the relations expressing mass and charge balances as follows:

$$C_a = [H_2A] + [HA^-] + [A^=] \quad \text{(VI-3)}$$

where C_a is the initial or "analytical" concentration of the acid,

and $[H^+] = [OH^-] + [HA^-] + 2[A^=]$ (VI-4)

Substituting for $[H_2A]$, and $[A^=]$ in terms of $[HA^-]$, K_1 and K_2 in equation (VI-3) we obtain:

$$C_a = \frac{[H^+] \times [HA^-]}{K_1} + [HA^-] + \frac{K_2 \times [HA^-]}{[H^+]}$$

$$\text{i.e. } C_a = [HA^-] \times \left[\frac{[H^+]}{K_1} + 1 + \frac{K_2}{[H^+]}\right] \quad \text{(VI-5)}$$

Equation VI-4 becomes

$$[H^+] - \frac{K_w}{[H^+]} = [HA^-]\left(1 + \frac{2K_2}{[H^+]}\right) \quad \text{(VI-6)}$$

Divide Equation (VI-5) by (VI-6) to eliminate $[HA^-]$

$$\frac{C_a \times [H^+]}{[H^+]^2 - K_w} = \frac{[H^+]^2 + K_1[H^+] + K_1K_2}{K_1([H^+] + 2K_2)}$$

Therefore,

$$[H^+]^4 + K_1[H^+]^3 + (K_1K_2 - K_w - K_1C_a)[H^+]^2 - (K_1K_w + 2K_1K_2C_a)[H^+] - K_1K_2K_w = 0 \quad \text{(VI-7)}$$

This is a quartic equation in $[H^+]$ and is quite formidable to solve, although if one is sufficiently interested, a trial and error method may be employed to solve it. A number of simplifying assumptions can usually be made.

(1) In almost all cases we can neglect the contribution of $[H^+]$ and $[OH^-]$ from water. This is tantamount to dropping all terms involving K_w from equation (VI-7). We should therefore obtain a cubic equation in $[H^+]$ since there are only two equilibrium constants, K_1 and K_2 that, control the concentrations of molecules and ions in solution. Therefore,

$$[H^+]^3 + K_1 \times [H^+]^2 + (K_1K_2 - K_1C_a) \times [H^+] - 2K_1K_2C_a = 0 \qquad (VI\text{-}8)$$

In equation (VI-8) the values of K_1, K_2 and C_a are all known and hence the coefficients of all the terms in $[H^+]$ can be calculated. The equation can be solved by a trial and error method in which an appropriate value of $[H^+]$ is substituted in equation (VI-8), and decreased or increased until the left-hand side of the equation becomes equal to zero. In many cases the calculation is further simplified.

(2) If the values of K_1 and K_2 are sufficiently separated, i.e. $K_2 \ll K_1$, the terms involving K_2 in equation (VI-8) can be dropped and we obtain a quadratic equation:

$$[H^+]^2 + K_1 \times [H^+] - K_1C_a = 0 \qquad (VI\text{-}9)$$

Note that this equation is identical with equation (V-10) that was obtained for a monoprotic acid. This equation can of course be further simplified if $[H^+]$ is less than 5% of C_a, to give:

$$[H^+] = (K_1 \times C_a)^{\frac{1}{2}}$$

Almost all diprotic acids of analytical importance have dissociation constants that differ by a factor which is greater than 10. Even in cases where the difference in K_1 and K_2 is about 10, the following method of calculation can be used to determine the hydrogen ion concentration of a solution containing the acid.

As a first approximation let us assume that the second dissociation step is negligible, i.e.

$$[H^+] = [HA^-]$$

If the second dissociation step does occur to an appreciable extent, then the concentration of $A^=$ ions formed is equal to K_2 since $[H^+] = [HA^-]$ in the equation:

$$K_2 = \frac{[H^+][A^=]}{[HA^-]}$$

For every $A^=$ ion that is formed in the second step an H^+ must also be formed. Therefore the total hydrogen ion concentration in the solution is given by:

$$[H^+]_{Total} = [H^+]_1 + [A^=]$$

i.e.

$$[H^+]_{Total} = [H^+]_1 + K_2 \qquad \text{(VI-10)}$$

where $[H^+]_1$ is the hydrogen ion concentration released in the first dissociation step. In most instances (VI-10) will give a sufficiently accurate answer for the hydrogen ion concentration of a diprotic acid. Therefore the method of calculation, as will be shown in the following examples, consists of the determination of the hydrogen ion concentration of the solution on the assumption that the acid is a monoprotic acid; the sum of this value of the hydrogen ion concentration and the numerical value of K_2 gives the total hydrogen ion concentration in solution. If K_2 is less than 5% of the hydrogen ion concentration obtained from the first step it is not necessary to add K_2 to $[H^+]$ from the first dissociation step.*

Example 1.

Calculate the hydrogen ion concentration of a 0.05M solution of H_2S in pure water. The two acid dissociation constants are $K_1 = 1.00 \times 10^{-7}$ and $K_2 = 1.3 \times 10^{-13}$.

$$H_2S \rightleftharpoons H^+ + SH^- \qquad K_1 = \frac{[H^+] \times [SH^-]}{[H_2S]}$$

$$SH^- \rightleftharpoons H^+ + S^= \qquad K_2 = \frac{[H^+] \times [S^=]}{[SH^-]}$$

*One of the more important uses to which equations (VI-7) and (VI-8) can be put is the calculation of K_1 and K_2 for a diprotic acid. In theory at least it is necessary to measure only the hydrogen ion concentrations of two solutions of the acid whose concentrations are C_{a_1} and C_{a_2}. Substituting these values in equation (VI-8), we will obtain two simultaneous equations which can be solved for K_1 and K_2. In practice of course we cannot rely on just two measurements for accurate values of K_1 and K_2, so somewhat different methods are used for such determinations.

Assumption (1). Neglect the hydrogen ion concentration obtained from the second dissociation step, and assume also that the $[OH^-]$ is less than 5% of the total $[H^+]$, i.e. $[H^+] = [SH^-]$.

This simplifies the problem and the following calculation is exactly the same as that for a monoprotic acid.

$$K_1 = 1.00 \times 10^{-7} = \frac{[H^+]^2}{0.05 - [H^+]}$$

Assumption (2). If $[H^+] < 5\%$ of 0.05, i.e. $[H^+] < 25.0 \times 10^{-4}$

$$\text{then, } [H^+] = (1.00 \times 10^{-7} \times 0.05)^{\frac{1}{2}}$$

$$= 7.1 \times 10^{-5}\,M$$

Let us now check our assumptions that were made in calculating this value of $[H^+]$. It is obvious that assumption (2) is valid, and also that the $[OH^-]$ which is equal to 1.4×10^{-10}, is less than 5% of $[H^+]$. The only assumption that remains to be checked is that we can neglect the hydrogen ion concentration formed from the second dissociation step.

Since $[H^+] = [SH^-]$, the concentration of the sulfide ion is given by

$$[S^=] = \frac{K_2 \times [SH^-]}{[H^+]} = K_2 = 1.3 \times 10^{-13}$$

For every sulfide ion produced in the second dissociation step we have a hydrogen ion that is also formed. Therefore the total concentration of H^+ is:

$$[H^+] = 7.1 \times 10^{-5} + 1.3 \times 10^{-13}$$

Since the quantity 1.3×10^{-13} is negligible, the concentration of H^+ in the solution is $7.1 \times 10^{-5}\,M$. Hence assumption (1) is valid.

We can also employ the mass balance for the acid and the charge balance for the solution to show conclusively that the approximations that have been made in this case are valid.

Mass balance:
$$0.05 = [H_2S] + [SH^-] + [S^=]$$
$$= [H_2S] + 7.1 \times 10^{-5} + 1.3 \times 10^{-13}$$
$$= [H_2S]$$

Charge Balance:
$$[H^+] = [OH^-] + [SH^-] + 2[S^=]$$
$$= 1.4 \times 10^{-10} + 7.1 \times 10^{-5} + 2.6 \times 10^{-13}$$
$$= 7.1 \times 10^{-5}$$

It is often useful to combine stepwise equilibrium expressions to obtain one that describes the overall reaction. Thus in the above example that we have considered, multiplying the expressions for K_1 and K_2 we obtain:

$$K_1 \times K_2 = \frac{[H^+] \times [SH^-]}{[H_2S]} \times \frac{[H^+] \times [S^=]}{[SH^-]}$$

$$= \frac{[H^+]^2 \times [S^=]}{[H_2S]} \qquad \text{(VI-11)}$$

In general, for a polyprotic acid H_nY whose successive dissociation constants are K_1, K_2, K_3, ----- K_n,

$$K_1 \times K_2 \times K_3 \times \text{-----} K_n = \frac{[H^+]^n \times [Y^{n-}]}{[H_nY]}$$

Equation (VI-11) is equivalent to adding the corresponding stepwise chemical equations, thus:

$$H_2S \rightleftharpoons H^+ + SH^-$$
$$SH^- \rightleftharpoons H^+ + S^=$$
$$H_2S + SH^- \rightleftharpoons 2H^+ + SH^- + S^=$$

or

$$H_2S \rightleftharpoons 2H^+ + S^= \qquad \text{(VI-12)}$$

It is of vital importance to avoid the mistake of assuming that the equation (VI-12) describes all the species that are present in solution. In this particular case, despite its absence in either equation (VI-11) or in (VI-12), SH^- is obviously present in the system. If this is forgotten it leads to the absurd conclusion from (VI-12) that $[H^+] = 2[S^=]$; we need only to remind ourselves of the relative magnitudes of $[H^+]$ and $[S^=]$ that were obtained in Example 1, to see the absurdity of this statement. Nevertheless equation (VI-11) is very useful. If any two of the three variables $[H^+]$, $[S^=]$ or $[H_2S]$ are specified, then this equation enables us to evaluate the third.

For example, in a 0.05M solution of H_2S whose hydrogen ion concentration is 7.1×10^{-5}M, the concentration of sulfide ions is given by:

$$[S^=] = K_1 \times K_2 \times \frac{[H_2S]}{[H^+]^2}$$

$$= \frac{1.00 \times 10^{-7} \times 1.3 \times 10^{-13} \times 0.05}{(7.1 \times 10^{-5})^2}$$

$$= 1.3 \times 10^{-13} M$$

Again, at what $[H^+]$ will the $[S^=]$ in a 0.05M solution of H_2S be equal to 1.3×10^{-13}?

$$[H^+] = \left(\frac{K_1 \times K_2 \times [H_2S]}{[S^=]}\right)^{\frac{1}{2}} = 7.1 \times 10^{-5} M$$

Example 2.

Calculate the sulfide ion concentration in a solution which is 0.05 M in H_2S and 0.30 M in HCl. Assume that $K_1 = 2.3 \times 10^{-7}$ and $K_2 = 6.8 \times 10^{-13}$ for H_2S.

Let us assume that the hydrogen ions produced by the dissociation of H_2S is negligible since the dissociation of H_2S is suppressed by the 0.30 M HCl.

Therefore,

$$\frac{[H^+]^2 \times [S^=]}{[H_2S]} = 2.3 \times 10^{-7} \times 6.8 \times 10^{-13}$$

$$[S^=] = \frac{15.7 \times 10^{-20} \times 0.05}{(0.30)^2}$$

$$= 8.7 \times 10^{-20} M$$

We can verify the assumption that the hydrogen ion concentration produced by the dissociation of H_2S is negligible by calculating the $[SH^-]$.

$$K_1 = 2.3 \times 10^{-7} = \frac{[H^+] \times [SH^-]}{[H_2S]} = \frac{0.30 \times [SH^-]}{0.05}$$

Therefore,

$$[SH^-] = 3.8 \times 10^{-8} M$$

The hydrogen ion concentration produced by the first dissociation step is only 3.8×10^{-8} M and that produced by the second dissociation step is equal to the sulfide ion concentration, and is 8.7×10^{-20} M. Hence the assumption that the concentration of hydrogen ions in solution is solely due to the hydrogen ions from the 0.30 M HCl, is justified.

From the foregoing examples it should be evident that in the case of a diprotic acid, in which the dissociation steps are well separated by a factor of 10^4 to 10^5, we can treat it as we do a monoprotic acid. Let us now take some examples in which the dissociation steps are much closer together.

Example 3.

Calculate the hydrogen ion concentration in a 1.0×10^{-3} M solution of succinic acid. The two acid dissociation constants for succinic acid are:

$$K_1 = 6.2 \times 10^{-5} \quad \text{and} \quad K_2 = 2.3 \times 10^{-6}.$$

Succinic acid, which is $H_2C_4H_4O_4$, is a diprotic acid and can be represented by H_2A.

$$K_1 = \frac{[H^+] \times [HA^-]}{[H_2A]} \quad \text{and} \quad K_2 = \frac{[H^+] \times [A^=]}{[HA^-]}$$

Assume that the formation of $A^=$ is negligible and that $[OH^-] < 5\%$ of $[H^+]$

i.e. $[H^+] = [HA^-]$

$$\text{Hence, } 6.2 \times 10^{-5} = \frac{[H^+] \times [H^+]}{C_a - [H^+]}$$

$$= \frac{[H^+]^2}{1.0 \times 10^{-3} - [H^+]}$$

It is obvious that we cannot make any further approximations and that we will have to solve the quadratic equation:

$$[H^+]^2 + 6.2 \times 10^{-5}[H^+] - 6.2 \times 10^{-8} = 0$$

$$[H^+] = \frac{-6.2 \times 10^{-5} + (38.4 \times 10^{-10} + 24.8 \times 10^{-8})^{\frac{1}{2}}}{2}$$

$$= 2.2 \times 10^{-4} \text{ M}$$

From this value of $[H^+]$ it is obvious that $[OH^-] < 5\%$ of $[H^+]$. However if the second dissociation step does take place to any appreciable extent, then as a first approximation, we can say that:

$$[H^+]_{Total} = [H^+]_1 + K_2$$
$$= 2.2 \times 10^{-4} + 2.3 \times 10^{-6}$$

Note that 2.3×10^{-6} is less than 5% of 2.2×10^{-4}, and therefore even in this case we are justified in neglecting the contribution of H^+ from the second dissociation step.

In Section VI-1 it was stated that a simplifying feature in the calculation of the hydrogen ion concentration in a solution of a diprotic acid is that the hydrogen ions formed in the first dissociation step suppress the second dissociation step. Let us apply this concept to the calculation of the hydrogen ion concentration of a diprotic acid, H_2A.

$$H_2A \rightleftharpoons H^+ + HA^-$$

$$HA^- \rightleftharpoons H^+ + A^=$$

Let $[H^+]_1$ be the concentration of hydrogen ions obtained from the first dissociation step, assuming that the second step does not occur, i.e. $[H^+]_1 = [HA^-]$. If $[H^+]_2$ is the concentration of the hydrogen ions obtained from the second dissociation step, then $[H^+]_2 = [A^=]$, since for every hydrogen ion obtained in the second step, one $A^=$ ion must be formed. Therefore the total hydrogen ion concentration in solution is

$$[H^+]_1 + [H^+]_2$$

the concentration of $[HA^-]$ in solution is

$$[H^+]_1 - [H^+]_2$$

and the concentration of $A^=$ is equal to $[H^+]_2$. When these values are substituted in the equilibrium expression for K_2, the following equation is obtained:

$$K_2 = \frac{[H^+]_1 + [H^+]_2}{[H^+]_1 - [H^+]_2} \times [H^+]_2 \qquad \text{(VI-13)*}$$

Since,

$$[H^+]_2 < 5\% \text{ of } [H^+]_1,$$
$$K_2 = [H^+]_2$$

*The contribution of $[H^+]$ from H_2O has been neglected.

However it is important to remember that this approximation is not valid in all cases, especially in solutions in which the concentration of the diprotic acid is of the same order of magnitude as the value of K_2. The following example illustrates this point.

Example 4.

Calculate the hydrogen ion concentration of a 1.0×10^{-2}M solution of H_2SO_4. Assume that K_2 for H_2SO_4 is 2.0×10^{-2}.

If we treat this in the usual manner as a diprotic acid, then the hydrogen ion concentration is given by equation (VI-10), i.e.

$$[H^+]_{Total} = [H^+]_1 + K_2$$

Also, since the first dissociation step takes place to completion,

$$[H^+]_1 = 1.0 \times 10^{-2}$$

Therefore

$$[H^+]_{Total} = 1.0 \times 10^{-2} + 2.0 \times 10^{-2}$$
$$= 3.0 \times 10^{-2} \text{ !!}$$

We should immediately recognize this as an absurd answer since we know that the maximum value of $[H^+]$ which results from the complete dissociation of H_2SO_4 is 2.0×10^{-2}. In fact since the HSO_4^- ions are not completely dissociated into $SO_4^=$ ions the total concentration of hydrogen ions in solution should be less than 2.0×10^{-2}M. It is therefore obvious that equation (VI-10), which is an approximate form of equation (VI-13) does not apply.

The problem can be solved by using equation (VI-13). Since $[H^+]_1 = 1.0 \times 10^{-2}$

$$K_2 = 2.0 \times 10^{-2} = \frac{(1.0 \times 10^{-2} + [H^+]_2)}{(1.0 \times 10^{-2} - [H^+]_2)} \times [H^+]_2$$

i.e. $[H^+]_2^2 + 3.0 \times 10^{-2}[H^+]_2 - 2.0 \times 10^{-4} = 0$

$$[H^+]_2 = \frac{-3.0 \times 10^{-2} + (9.0 \times 10^{-4} + 8.0 \times 10^{-4})^{\frac{1}{2}}}{2}$$

$$= .57 \times 10^{-2}$$

Therefore the total hydrogen ion concentration in solution

$$= 1.0 \times 10^{-2} + 0.57 \times 10^{-2}$$

$$= 1.6 \times 10^{-2} \text{ M}$$

This problem underlines the importance of always being aware of the assumptions involved in the expressions that are used in calculations.

VI-3. SOLUTIONS CONTAINING A POLYPROTIC ACID

The considerations outlined above in Section VI-2 apply to solutions of polyprotic acids as well as to diprotic acids since successive acid dissociation constants of a polyprotic acid are of continually decreasing importance. We may assert that only the first two dissociation constants in all acids of practical significance are of importance in calculating the hydrogen ion concentration in a solution of the acid. The following examples will demonstrate this fact.

Example 5.

Calculate the hydrogen ion concentration in a 3.0×10^{-3} M solution of phosphoric acid. The successive dissociation constants of phosphoric acid are:

$$K_1 = 7.4 \times 10^{-3};\ K_2 = 6.9 \times 10^{-8};\ K_3 = 5.1 \times 10^{-13}.$$

If it is assumed that only the first dissociation step is of importance the system can be treated as a monoprotic acid.

$$[H^+]_{Approx.} = (K_1 \cdot C_a)^{\frac{1}{2}} = (7.4 \times 10^{-3} \times 3.0 \times 10^{-3})^{\frac{1}{2}}$$
$$= 4.7 \times 10^{-3}\ M$$

The concentration of hydrogen ions contributed by the second dissociation step is given approximately by the numerical value of K_2 which is much less than 5% of $[H^+]_{Approx.}$ Therefore we can neglect H^+ ions that come from the second dissociation step and of course from the third dissociation step also.

The value of $[H^+]_{Approx.}$ would be sufficiently accurate if the following assumptions are correct: $[OH^-] < 5\%$ of $[H^+]$ and $[H^+] < 5\%$ of C_a. The former assumption is correct but the latter is not valid since 5% of C_a is 1.5×10^{-4}, and $[H^+]_{Approx.}$ is certainly greater than this. Therefore equation V-10 has to be employed in solving for $[H^+]$.

$$7.4 \times 10^{-3} = \frac{[H^+]^2}{3.0 \times 10^{-3} - [H^+]}$$

$$[H^+] = \frac{-7.4 \times 10^{-3} + (54.6 \times 10^{-6} + 88.8 \times 10^{-6})^{\frac{1}{2}}}{2}$$

$$= 2.3 \times 10^{-3}\ M$$

Example 6.

Calculate the concentration of hydrogen ions in a solution of 3.0×10^{-3} M citric acid. The successive acid dissociation constants for citric acid are: $K_1 = 8.0 \times 10^{-4}$; $K_2 = 2.0 \times 10^{-5}$ and $K_3 = 4.9 \times 10^{-7}$.

In this case the first two acid dissociation constants are close together and the third dissociation step will contribute a negligible amount of hydrogen ions.

Using equation V-10 to calculate the value of H^+ for a monoprotic acid,

$$8.0 \times 10^{-4} = \frac{[H^+]^2}{3 \times 10^{-3} - [H^+]}$$

Solving this quadratic equation we obtain:

$$[H^+] = 1.2 \times 10^{-3}\ M$$

The concentration of hydrogen ions produced by the second dissociation step is only 2.0×10^{-5} M and even in this case it is seen that the second dissociation step contributes a negligible quantity of hydrogen ions.

Therefore the hydrogen ion concentration in a solution which is 3.0×10^{-3} M in citric acid, is 1.2×10^{-3} M.

VI-4. THE COMPOSITION OF A SOLUTION CONTAINING A POLYPROTIC ACID AS A FUNCTION OF pH

It is often instructive to be able to describe the composition of solutions of polyprotic acids as a function of the hydrogen ion concentration. For example during the titration of a polyprotic acid with a base, the relative proportions of the various species present in solution will change with pH. The manner in which these changes occur is of importance in interpreting titration curves and in calculating titration errors.

The simplest method of determining the concentrations of the various species that are present in a solution of a polyprotic acid is to begin by defining a set of α values that represent the fractions of the total concentration present as each species. These α values are similar to those that were introduced for monoprotic acids (Section V-1). Thus in a solution of a polyprotic acid, H_nA whose initial or analytical concentration is C_a moles/liter, the concentration fraction of H_nA is α_0 and is defined as:

$$\alpha_0 = \frac{[H_nA]}{C_a}$$

Similarly the fractions corresponding to the species $H_{n-1}A^-$, $H_{n-2}A^=$, ----- A^{n-} are defined by:

$$\alpha_1 = \frac{[H_{n-1}A^-]}{C_a}$$

$$\alpha_2 = \frac{[H_{n-2}A^=]}{C_a}$$

$$\alpha_n = \frac{[A^{n-}]}{C_a}$$

Turning now to the successive dissociation equilibria:

$$K_1 = \frac{[H^+][H_{n-1}A^-]}{[H_nA]}, \quad K_2 = \frac{[H^+][H_{n-2}A^=]}{[H_{n\ 1}A^-]} \quad \text{etc.}$$

and introducing the α values, e.g.

$$K_1 = \frac{[H^+]C_a\alpha_1}{C_a\alpha_0} = \frac{[H^+]\alpha_1}{\alpha_0} \quad \text{etc.}$$

we arrive at a set of equations describing all the values in terms of $[H^+]$ and K values.

$$K_1 = \frac{[H^+]\alpha_1}{\alpha_0} \quad \text{or} \quad \alpha_1 = \frac{K_1}{[H^+]} \cdot \alpha_0$$

$$K_2 = \frac{[H^+]\alpha_2}{\alpha_1}$$

$$= \frac{[H^+]^2\alpha_2}{K_1\alpha_0} \quad \text{or} \quad \alpha_2 = \frac{K_1K_2 \cdot \alpha_0}{[H^+]^2}$$

$$K_3 = \frac{[H^+]\,\alpha_3}{\alpha_2}$$

$$= \frac{[H^+]^3 \cdot \alpha_3}{K_1 K_2 \alpha_0} \quad \text{or} \quad \alpha_3 = \frac{K_1 K_2 K_3 \cdot \alpha_0}{[H^+]^3} .$$

and in general,

$$K_n = \frac{[H^+]\,\alpha_n}{\alpha_{n-1}}$$

$$= \frac{[H^+]^n \alpha_n}{K_1 K_2 \cdot K_3 \ldots . K_{n-1}\,\alpha_0}$$

or

$$\alpha_n = \frac{K_1 \cdot K_2 \cdot K_3 \ldots . K_n \cdot \alpha_0}{[H^+]^n} \qquad \text{(VI-14)}$$

The mass balance for the acid is given by:

$$C_a = [H_nA] + [H_{n-1}A^-] + \ldots\ldots + [A^{n-}]$$

or,

$$1 = \alpha_0 + \alpha_1 + \alpha_2 + \ldots\ldots\ \alpha_n \qquad \text{(VI-15)}$$

Substituting from equations of the type (VI-14) we obtain an equation for α_0:

$$\frac{1}{\alpha_0} = 1 + \frac{K_1}{[H^+]} + \frac{K_1 K_2}{[H^+]^2} + \text{-----}\frac{K_1 K_2 \text{---} K_n}{[H^+]^n} \qquad \text{(VI-16)}$$

From this equation and equation (VI-14) it is possible to calculate all α values.

Example 7.

Calculate the concentration of acetate ions and undissociated acetic acid molecules in a 5.0×10^{-3} M solution of acetic acid in which the hydrogen ion concentration is 3.0×10^{-4} M. K_a for acetic acid is 1.9×10^{-5}.

In this example the concentrations of HOAc and OAc^- are given by:

$[OAc^-] = [H^+] = 3.0 \times 10^{-4}$ M (since $[OH^-]$ is less than 5% of $[H^+]$) and

$$[HOAc] = 5.0 \times 10^{-3} - 3.0 \times 10^{-4} = 4.7 \times 10^{-3}\ M$$

Therefore for monoprotic acids it is hardly necessary to define a set of α values in order to determine the concentrations of the various species present in solution. However let us apply equations VI-14, 15 and 16 to this simple case in order to familiarize ourselves with the method of calculation.

The concentration fractions of HOAc and OAc^- present in solution are defined by α_0 and α_1 respectively.

$$\alpha_0 = \frac{[HOAc]}{C_a} \quad \text{and} \quad \alpha_1 = \frac{[OAc^-]}{C_a}$$

From equation VI-16, $\dfrac{1}{\alpha_0} = 1 + \dfrac{K_1}{[H^+]} = 1 + \dfrac{1.9 \times 10^{-5}}{3.0 \times 10^{-4}} = 1.06$

Therefore $\alpha_0 = 0.94$ and $\alpha_1 = 1 - \alpha_0 = 0.06$

Hence $[HOAc] = C_a \times \alpha_0 = 0.94 \times 5.0 \times 10^{-3} = 4.7 \times 10^{-3}$ M

and $[OAc^-] = C_a \times \alpha_1 = 0.06 \times 5.0 \times 10^{-3} = 3.0 \times 10^{-4}$ M

Note that in the above example, if the pH of a 5.0×10^{-3} M solution of HOAc is increased by the addition of a base, the concentrations of HOAc and OAc^- can be readily calculated. Thus at a pH of 5.0 (assuming that K_a is constant and equal to 1.9×10^{-5}),

$$\frac{1}{\alpha_0} = 1 + \frac{1.9 \times 10^{-5}}{1.0 \times 10^{-5}} = 2.9$$

Therefore $\alpha_0 = 0.35$ and $\alpha_1 = 0.65$.

Hence $[HOAc] = 5.0 \times 10^{-3} \times 0.35 = 1.75 \times 10^{-3}$ M

and $[OAc^-] = 5.0 \times 10^{-3} \times 0.65 = 3.25 \times 10^{-3}$ M.

Example 8.

Calculate the concentration of undissociated succinic acid, $H_2C_4H_4O_4$ and the concentrations of the species $HC_4H_4O_4^-$ and $C_4H_4O_4^=$ in a 1.0×10^{-3} M solution of succinic acid at a pH of 6.0. Assume that $K_1 = 6.2 \times 10^{-5}$ and $K_2 = 2.3 \times 10^{-6}$ for succinic acid.

Let us represent succinic acid as H_2A.

$$\alpha_0 = \frac{[H_2A]}{C_a} \quad \alpha_1 = \frac{[HA^-]}{C_a} \quad \text{and} \quad \alpha_2 = \frac{[A^=]}{C_a}$$

Also

$$\frac{1}{\alpha_0} = 1 + \frac{K_1}{[H^+]} + \frac{K_1 K_2}{[H^+]^2}$$

$$= 1 + \frac{6.2 \times 10^{-5}}{1.0 \times 10^{-6}} + \frac{6.2 \times 2.3 \times 10^{-11}}{1.0 \times 10^{-12}}$$

$$= 206$$

Therefore $\alpha_0 = 4.85 \times 10^{-3}$

$$\alpha_1 = \frac{K_1 \alpha_0}{[H^+]} = \frac{6.2 \times 10^{-5} \times 4.85 \times 10^{-3}}{1.0 \times 10^{-6}} = 30.1 \times 10^{-2}$$

$$\alpha_2 = 69.4 \times 10^{-2}$$

Using these α values we get:

$[H_2A] = 1.0 \times 10^{-3} \times 4.85 \times 10^{-3} = 4.85 \times 10^{-6}$ M $= 4.9 \times 10^{-6}$ M

$[HA^-] = 1.0 \times 10^{-3} \times 30.1 \times 10^{-2} = 3.01 \times 10^{-4}$ M $= 3.0 \times 10^{-4}$M

and

$[A^=] = 1.0 \times 10^{-3} \times 69.5 \times 10^{-2} = 6.94 \times 10^{-3}$ M $= 6.9 \times 10^{-3}$M

Note that $[H_2A] + [HA^-] + [A^=] = 1.0 \times 10^{-3}$

Calculation of α values serve as convenient starting points for the construction of graphs representing more complex polyprotic acid systems such as EDTA which has two pK values that are very close together. In Fig. VI-2 is shown a log α vs pH plot for EDTA, which has been constructed in a manner very much like that described for the diprotic acid except for the pH region between 2 and 4. In this region selected values of α_0, α_1 and α_2 were calculated from the following expressions:

$$\alpha_0 = \frac{[H^+]^4}{[H^+]^4 + K_1[H^+]^3 + K_1K_2[H^+]^2 + K_1K_2K_3[H^+] + K_1K_2K_3K_4}$$

$$= \frac{[H^+]^4}{D}$$

where D is a convenient designation of the denominator

$$\alpha_1 = \frac{K_1[H^+]^3}{D}$$

$$\alpha_2 = \frac{K_1K_2[H^+]^2}{D}$$

Values of α_3 and α_4 are sufficiently well described by the construction techniques mentioned on page 81 for Fig. VI-1. The slope of the α_4 line in the region between pK_2 and pK_3 will

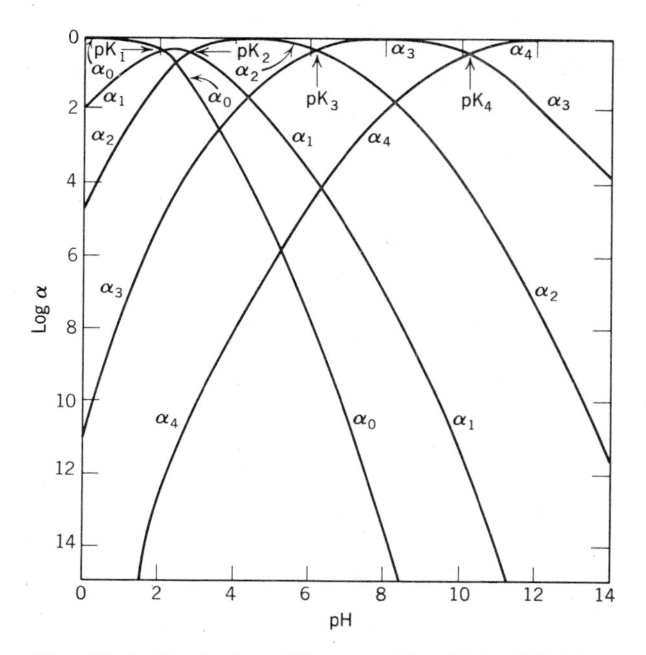

Fig. VI-2—Variation of log α with pH for EDTA.

be 2, in the region between pK_1 and pK_2 will be 3 and finally at pH values below pK_1 will become 4. The slopes of the other α lines will also change in different pK regions in the expected manner.

VI-5. SOLUTIONS CONTAINING A POLYACIDIC BASE

Since in our treatment of polyprotic acids it was recognized that only two dissociation equilibria need be considered in calculating the $[H^+]$ in a solution of any polyprotic acid, we may safely conclude that a similar argument applies to any solution of any polyacidic base. Accordingly we will limit this discussion to the development of equations for diacidic bases such as sulfide, carbonate ions, and ethylenediamine, $(H_2N.CH_2.CH_2.NH_2)$.

In a solution of C_b molar diacidic base we may write the following equations:

$$B + H_2O \rightleftharpoons BH^+ + OH^- \qquad K_{B_1} = \frac{[BH^+][OH^-]}{[B]} \qquad (VI\text{-}17)$$

$$BH^+ + H_2O \rightleftharpoons BH_2^{++} + OH^- \qquad K_{B_2} = \frac{[BH_2^{++}][OH^-]}{[BH^+]} \qquad (VI\text{-}18)$$

Since these equations are exactly analogous to equations VI-1 and 2 that were developed for a diprotic acid we can write the equation that is analogous to equation VI-7 as follows:

$$[OH^-]^4 + K_{B_1}[OH^-]^3 + (K_{B_1}K_{B_2} - K_w - K_{B_1}C_b) \times [OH^-]^2 - (K_{B_1}K_w + 2K_{B_1}K_{B_2}C_b) \times [OH^-] - K_{B_1}K_{B_2}K_w = 0 \qquad (VI\text{-}19)$$

Note that in this equation $[OH^-]$ has been substituted for $[H^+]$, K_{B_1} and K_{B_2} for K_1 and K_2 and C_b for C_a.

Equation VI-19 could be simplified in the same manner in which equation VI-7 was simplified. If the contribution of $[H^+]$ and $[OH^-]$ from water is neglected all terms involving K_w can be dropped and a cubic equation in $[OH^-]$ is obtained:

$$[OH^-]^3 + K_{B_1}[OH^-]^2 + [OH^-](K_{B_1}K_{B_2} - K_{B_1} \cdot C_b) - 2K_{B_1}K_{B_2}C_b = 0 \qquad (VI\text{-}20)$$

Furthermore if $K_{B_2} < K_{B_1}$, the terms involving K_{B_2} in equation VI-20 can be dropped to give a quadratic equation:

$$[OH^-]^2 + K_{B_1}[OH^-] - K_{B_1}C_b = 0 \qquad (VI\text{-}21)$$

If $[OH^-] < 5\%$ of C_b, this equation is further simplifed to:

$$[OH^-] = (K_{B_1}C_b)^{1/2} \qquad (VI\text{-}22)$$

Equations (VI-21 and 22) are the same as equations V-20 and 21 which were obtained for a monoacidic base. In instances where K_{B_1} and K_{B_2} are close together, the method described in Sec. VI-2 can be employed. As shown below, most cases of polyacidic bases of practical importance can be treated as simple monoacidic bases.

Example 9.

Calculate the hydroxyl ion concentration in a solution of 0.1 M Na_2S. Assume that $K_w = 1.6 \times 10^{-14}$; $K_1 = 1.6 \times 10^{-7}$ and $K_2 = 4.0 \times 10^{-13}$ where K_1 and K_2 are the acid dissociation constants of H_2S.

The hydroxyl ion concentration in a solution of Na_2S is governed by the following equilibria:

$$S^= + H_2O \rightleftharpoons SH^- + OH^-$$

$$SH^- + H_2O \rightleftharpoons H_2S + OH^-$$

$$K_{B_1} = \frac{[SH^-] \times [OH^-]}{[S^=]} = \frac{K_w}{K_2} = \frac{1.6 \times 10^{-14}}{4.0 \times 10^{-13}} = 0.04$$

$$K_{B_2} = \frac{[H_2S] \times [OH^-]}{[SH^-]} = \frac{K_w}{K_1} = \frac{1.6 \times 10^{-14}}{1.6 \times 10^{-7}} = 1.0 \times 10^{-7}$$

Since K_{B_2} is very much smaller than K_{B_1} there is only one equilibrium constant other than K_w, that governs the concentration of hydrogen ions in solution, namely, K_{B_1}. It is important to note that the second dissociation constant of the acid, H_2S, controls the numerical value of K_{B_1} the first dissociation constant of the polyacidic base, $S^=$.

Using the quadratic equation VI-21 to solve for OH , we obtain:

$$[OH^-]^2 + 0.04 \times [OH^-] - 0.04 \times 0.10 = 0$$

Therefore

$$[OH^-] = \frac{-0.04 + (16 \times 10^{-4} + 16 \times 10^{-3})^{1/2}}{2}$$

$$= 0.04 \text{ M}$$

The use of the quadratic equation rather than the cubic equation (VI-20) is justified since the concentration of $[H^+]$ is much less than 5% of $[OH^-]$.

Example 10.

Calculate the hydrogen ion concentration in a 1.0×10^{-4} M solution of Na_2CO_3, $K_w = 1.0 \times 10^{-14}$, $K_1 = 4.2 \times 10^{-7}$ and $K_2 = 4.8 \times 10^{-11}$ for H_2CO_3.

Let us assume that the only reaction of importance is:

$$CO_3^= + H_2O \rightleftharpoons HCO_3^- + OH^-$$

Therefore

$$K_{B_1} = \frac{[HCO_3^-] \times [OH^-]}{[CO_3^=]} = \frac{K_w}{K_2} = \frac{[OH^-] \times ([OH^-] - [H^+])}{C_b - ([OH^-] - [H^+])}$$

Furthermore if $[H^+]$ is less than 5% of $[OH^-]$ this equation reduces to the quadratic:

$$\frac{K_w}{K_2} = \frac{[OH^-]^2}{C_b - [OH^-]} = \frac{1.0 \times 10^{-14}}{4.8 \times 10^{-11}}$$

Therefore

$$[OH^-]^2 + 2.08 \times 10^{-4} \times [OH^-] - 2.08 \times 10^{-8} = 0$$

$$[OH^=] = \frac{-2.08 \times 10^{-4} + (4.35 \times 10^{-8} + 8.32 \times 10^{-8})^{\frac{1}{2}}}{2}$$

$$[OH^-] = 7.4 \times 10^{-5} \text{ M}$$

Therefore $[H^+]$ is less than 5% of $[OH^-]$ and the use of the quadratic equation is justified.

The first assumption that was made was that only K_{B_1} was of importance and it was thereby implied that the second dissociation of the polyacidic base, $CO_3^=$, was negligible. In other words the concentration of H_2CO_3 formed is assumed to be very small. In the following mass balance the concentration of H_2CO_3 is therefore omitted.

$$C_b = 1.0 \times 10^{-4} = [CO_3^=] + [HCO_3^-] \qquad \text{(VI-23)}$$

The first dissociation constant of H_2CO_3 is given by:

$$K_1 = \frac{[H^+] \times [HCO_3^-]}{[H_2CO_3]}$$

Since we have found that $[HCO_3^-] = [OH^-]$, it is now possible to calculate the concentration of H_2CO_3 in solution.

$$[H_2CO_3] = \frac{K_w}{K_1} = \frac{1.0 \times 10^{-14}}{4.2 \times 10^{-7}} = 2.4 \times 10^{-8} \text{ M}$$

Therefore the inclusion of the term $[H_2CO_3]$ in equation (VI-23) would have resulted in less than a 5% difference in the value of $[HCO_3^-]$ and of $[CO_3^=]$. This of course means that the amount of H_2CO_3 formed is very small and that the second dissociation step: $HCO_3^- + H_2O \rightleftharpoons H_2CO_3 + OH^-$ takes place to a negligible extent.

VI-6. SOLUTIONS OF AMPHOLYTES

An ampholyte may be defined as a substance which can gain or lose protons. As this definition implies, an ampholyte is an incompletely neutralized polyprotic acid or polyacid base. As examples we may cite $NaHCO_3$, NaH_2PO_4, Na_2HPO_4, the disodium salt of EDTA, ethylenediamine monohydrochloride, ($H_2N.CH_2.CH_2.NH_3^+ Cl^-$) and glycine.

In an aqueous solution the behavior of an ampholyte such as $NaHCO_3$ must be described in terms of reactions involving both the gain and loss of protons. Thus:

$$HCO_3^- + H_2O \rightleftharpoons H_2CO_3 + OH^-$$

and

$$HCO_3^- \rightleftharpoons H^+ + CO_3^=$$

The equilibrium constants for the above reactions are:

$$\frac{K_w}{K_1} = \frac{[H_2CO_3] \cdot [OH^-]}{[HCO_3^-]} \quad \text{and } K_2 = \frac{[H^+] \cdot [CO_3^=]}{[HCO_3^-]}$$

Mass Balance: $$C_s = [H_2CO_3] + [HCO_3^-] + [CO_3^=] \qquad \text{(VI-24)}$$

where C_s is the molarity of $NaHCO_3$.
Also

$$C_s = [Na^+] \qquad \text{(VI-25)}$$

Charge Balance: $$[H^+] + [Na^+] = [OH^-] + [HCO_3^-] + 2[CO_3^=] \qquad \text{(VI-26)}$$

Combining these equations:

$$[H^+] + [HCO_3^-] + [CO_3^=] + [H_2CO_3] = [OH^-] + [HCO_3^-] + 2[CO_3^=]$$

Therefore,

$$[H^+] + [H_2CO_3] = [OH^-] + [CO_3^=] \qquad \text{(VI-27)}$$

Since

$$[H_2CO_3] = [HCO_3^-] \times \frac{[H^+]}{K_1} \quad \text{and}$$

$$[CO_3^=] = [HCO_3^-] \times \frac{K_2}{[H^+]} \quad \text{and}$$

$$[OH^-] = \frac{K_w}{[H^+]}$$

Equation (VI-27) becomes:

$$[H^+] + [HCO_3^-] \times \frac{[H^+]}{K_1} = \frac{K_w}{[H^+]} + [HCO_3^-] \times \frac{K_2}{[H^+]}$$

Multiplying throughout by $K_1[H^+]$, we obtain:

$$[H^+]^2 (K_1 + [HCO_3^-]) = K_1 K_w + K_1 K_2 [HCO_3^-]$$

i.e.

$$[H^+]^2 = \frac{K_1 K_2 [HCO_3^-] + K_1 K_w}{K_1 + [HCO_3^-]} \qquad \text{(VI-28)}$$

Equation (VI-28) is exact but cannot be used in its present form. A reasonable approximation applicable to most cases is that the extent of the reactions undergone by the HCO_3^- is sufficiently small to warrant using C_s for the concentration of HCO_3^- at equilibrium. i.e.

$$[H_2CO_3] + [CO_3^=] < 5\% \text{ of } [HCO_3^-]$$

Equation (VI-28) now becomes:

$$[H^+]^2 = \frac{K_1 K_2 C_s + K_1 K_w}{K_1 + C_s} \qquad \text{(VI-29)}$$

Further simplification of this equation results when the solution is sufficiently concentrated, i.e. when $C_s \gg K_1$ and at the same time $C_s \gg K_w/K_2$ then the denominator $(K_1 + C_s) \approx C_s$ and in the numerator the term $K_1 K_w$ drops out to give:

$$[H^+]^2 = K_1 K_2 \qquad \text{(VI-30)}$$

It is interesting to note that according to equation (VI-30), the hydrogen ion concentration of an ampholyte is independent of its concentration. As a matter of fact under conditions where equation (VI-30) applies, the requirement that $[HCO_3^-] \approx C_s$ mentioned above is unnecessary since this quantity cancels out.

Although equation (VI-28) and its simplified versions were derived for $NaHCO_3$ solutions, they apply to other ampholytes as well, viz:

For a solution of Na_2HPO_4,

$$[H^+]^2 = \frac{K_2 K_3 [HPO_4^=] + K_2 K_w}{K_2 + [HPO_4^=]}$$

and for a solution of NaH_2PO_4,

$$[H^+]^2 = \frac{K_1 K_2 [H_2PO_4^-] + K_1 K_w}{K_1 + [H_2PO_4^-]}$$

where K_1, K_2 and K_3 are the successive dissociation constants of phosphoric acid.

Example 11.

Calculate the hydrogen ion concentration of 0.1 M potassium hydrogen succinate. Assume that $K_1 = 1.0 \times 10^{-4}$ and $K_2 = 1.0 \times 10^{-5}$ for succinic acid.

From equation (VI-29),

$$[H^+]^2 = \frac{K_1K_2C_s + K_1K_w}{K_1 + C_s}$$

where C_s is the concentration of the hydrogen succinate ion. Since $C_s = 0.1$, K_1K_w is very much less than $K_1K_2C_s$ and K_1 is very much less than C_s,

Hence

$$[H^+]^2 = K_1K_2$$

$$= 1.0 \times 10^{-9}$$

$$\therefore\ [H^+] = 3.2 \times 10^{-5}\ M$$

Note in this example that the ampholyte concentration is very large and it is unnecessary to make the assumption that the concentration of the hydrogen succinate ion is equal to 0.1 M. Furthermore K_1 and K_2 are quite close together and if a dilute solution of the ampholyte is taken the method of successive approximations will have to be used to determine the hydrogen ion concentration.

Example 12.

Calculate the hydrogen ion concentration of a 1.0×10^{-3} M solution of $NaHCO_3$.

$K_1 = 4.2 \times 10^{-7}$ and $K_2 = 4.8 \times 10^{-11}$ for H_2CO_3. and $K_w = 1.0 \times 10^{-14}$.

$$[H^+]^2 = \frac{K_1K_2[HCO_3^-] + K_1K_w}{K_1 + [HCO_3^-]} \qquad \text{(VI-28)}$$

Let us assume that $[HCO_3^-] = 1.0 \times 10^{-3}$

Then,

$$[H^+]^2 = \frac{4.2 \times 10^{-7} \times 4.8 \times 10^{-11} \times 1.0 \times 10^{-3} + 4.2 \times 10^{-7} \times 1.0 \times 10^{-14}}{4.2 \times 10^{-7} + 1.0 \times 10^{-3}}$$

$$= \frac{20.2 \times 10^{-21} + 4.2 \times 10^{-21}}{1.0 \times 10^{-3}}$$

$$= 24.4 \times 10^{-18}$$

Therefore $[H^+] = 4.9 \times 10^{-9}$ M

The original assumption that was made was that in the following equation which represents the mass balance for the bicarbonate ion,

$$0.1 = [HCO_3^-] + [H_2CO_3] + [CO_3^=] \qquad \text{(VI-31)}$$

the concentrations of H_2CO_3 and $CO_3^=$ are less than 5% of $[HCO_3^-]$. Since we have calculated $[H^+]$, it is possible to obtain $[H_2CO_3]$ and $[CO_3^=]$ from the equilibrium expressions for the dissociation of H_2CO_3.

$$K_1 = \frac{[H^+][HCO_3^-]}{[H_2CO_3]} \quad \text{and } K_2 = \frac{[H^+][CO_3^=]}{[HCO_3^-]} \qquad \text{(VI-32) and (VI-33)}$$

$$4.2 \times 10^{-7} = \frac{4.9 \times 10^{-9} \times 1.0 \times 10^{-3}}{[H_2CO_3]}$$

Therefore $[H_2CO_3] = 1.2 \times 10^{-5}$ M and

$$4.8 \times 10^{-11} = \frac{4.9 \times 10^{-9} \times [CO_3^=]}{1.0 \times 10^{-3}}$$

Therefore, $[CO_3^=] = 9.8 \times 10^{-6}$ M

Hence $[H_2CO_3] + [CO_3^=] < 5\%$ of $[HCO_3^-]$ and the assumption that

$[HCO_3^-] = 1.0 \times 10^{-3}$ M is justified.

In this example if the sum of $[H_2CO_3]$ and $[CO_3^=]$ happened to be greater than 5% of $[HCO_3^-]$, the method of successive approximations could be used to obtain $[H^+]$.

Briefly, this method consists of the following steps: the values of $[H_2CO_3]$ and $[CO_3^=]$ obtained are substituted in equation (VI-31) to give a better value of $[HCO_3^-]$ than was originally assumed. This value of $[HCO_3^-]$ is now used in equation (VI-28) to obtain a more correct value of $[H^+]$, which in turn is used to obtain more accurate values of $[H_2CO_3]$ and $[CO_3^=]$ from equations (VI-32 and 33). The latter values when substituted in equation (VI-31) will give a still better value of $[HCO_3^-]$ and the cycle of calculations can be repeated until two successive values of $[H^+]$ become almost equal. Usually it is necessary to carry out this cycle of calculations only once in order to obtain the correct value of $[H^+]$.

VI-7. SOLUTIONS CONTAINING A CONJUGATE ACID-BASE PAIR

Solutions containing a conjugate acid-base pair involving either a polyprotic acid or a polyacid base may be treated in a manner exactly analogous to cases involving monoprotic acids or monoacid bases. The reason for this is that in most solutions containing a mixture of a conjugate acid-base pair only one dissociation step need be considered. The main factor to keep in mind is the selection of the correct dissociation step and therefore the correct pK value.

Example 13.

What is the pH of a solution containing 0.1 M $NaHCO_3$ and 0.05 M Na_2CO_3?

$pK_1 = 6.21$ $pK_2 = 10.05$ $pK_w = 13.86$

The equations for the stepwise dissociation of carbonic acid are:

$$H_2CO_3 \overset{K_1}{\rightleftharpoons} H^+ + HCO_3^-$$

$$HCO_3^- \overset{K_2}{\rightleftharpoons} H^+ + CO_3^=$$

Since the major constituents of the mixture are HCO_3^- and $CO_3^=$, the dominant equilibrium is the second dissociation step. Thus

$$K_2 = \frac{[H^+][CO_3^=]}{[HCO_3^-]} = \frac{[H^+]C_b}{C_a} \qquad \text{(V-24)}$$

or

$$pH = pK_2 + \log\frac{0.05}{0.10}$$

$$= 10.05 - 0.30 = 9.75$$

From this answer it is obvious that both $[OH^-]$ and $[H^+]$ are less than 5% of either C_a or C_b. This justifies the use of the simplest equation (V-24).

Example 14.

What concentrations of Na_3PO_4 should be used in admixture with 0.01 M Na_2HPO_4 such that the solution has a pH of 12.0? ($pK_1 = 1.93$ $pK_2 = 6.76$ $pK_3 = 11.74$ $pK_w = 13.78$)

As in the previous example we select the third dissociation step as the dominant equilibrium since the major constituents of the mixture are $HPO_4^{=}$ and $PO_4^{\equiv}$.

In this problem the simplified equation V-24 may not be used since the $[OH^-]$ in solution is more than 5% of C_b. However, since the solution is highly alkaline $[OH^-] > [H^+]$ permitting equation V-24 to be modified as follows:

$$K_a = \frac{[H^+](C_b - [OH^-])}{(C_a + [OH^-])} = \frac{[H^+](C_b - K_w/[H^+])}{C_a + \frac{K_w}{[H^+]}}$$

Substituting then we have

$$10^{-11.74} = \frac{10^{-12.0}[C_b - 10^{-13.78 + 12.0}]}{10^{-2.00} + 10^{-1.78}}$$

$$C_b = 0.067 \text{ M } Na_3PO_4$$

Note that use of V-24 would have led to the answer C_b = 0.018 M which is grossly in error.

These two examples should be compared with examples 5 and 6 in Chapter V.

SUGGESTIONS FOR FURTHER READING

T. B. Smith, Analytical Processes, Edward Arnold Publishers Ltd., London (1940), Chapter 10.

S. Bruckenstein and I. M. Kolthoff, Chapter 12, Part I, Vol. I, Treatise on Analytical Chemistry, Interscience, New York (1959). I. M. Kolthoff and P. J. Elving, Editors.

E. J. King, Qualitative Analysis and Electrolytic Solutions, Harcourt, Brace and World, Inc., (1959), Chapters 11, 12, and 13.

H. A. Laitinen, Chemical Analysis, McGraw Hill (1960) New York, Chapter 3.

L. G. Sillen, Part I, Vol. I, Chapter 8, Treatise on Analytical Chemistry, Interscience, (1959), I. M. Kolthoff and P. J. Elving, Editors.

VI-8. PROBLEMS

1. Calculate the pH of 0.02 M solutions of the following:
   (a) H_2S ($pK_1 = 7.0$, $pK_2 = 12.9$)
   (b) $H_2C_2O_4$ ($pK_1 = 1.2$, $pK_2 = 4.14$)
   (c) H_3PO_4 ($pK_1 = 2.06$, $pK_2 = 7.03$, $pK_3 = 12.1$)
   (d) H_2CO_3 ($pK_1 = 6.35$, $pK_2 = 10.33$)
2. Calculate the hydrogen ion concentration of
   (a) 1.00×10^{-3} M solution of H_2SO_4 ($pK_2 = 1.87$)
   (b) 1.00×10^{-4} M solution of $H_2C_2O_4$
   ($pK_1 = 1.3$, $pK_2 = 4.25$)
3. What is the pH of a 5.0×10^{-3} M solution of glycine hydrochloride (or glycinium chloride)? ($pK_1 = 2.35$, $pK_2 = 9.71$)
4. Calculate the hydrogen ion concentrations of the following solutions:
   (a) 0.05 M Na_2S ($pK_1 = 6.8$, $pK_2 = 12.4$, $pK_w = 13.76$)
   (b) 1.00×10^{-3} M Na_3PO_4 ($pK_1 = 2.09$, $pK_2 = 7.08$, $pK_3 = 12.2$, $pK_w = 13.94$)
   (c) 0.15 M Na_2CO_3 ($pK_1 = 5.95$, $pK_2 = 9.53$, $pK_w = 13.60$)
   (d) 3.5×10^{-4} M Na_2CO_3 ($pK_1 = 6.32$, $pK_2 = 10.27$, $pK_w = 13.97$)
   (e) 1.0×10^{-3} M ethylenediamine
   ($pK_1 = 6.79$, $pK_2 = 9.90$, $pK_w = 14.00$)
5. Water exposed to the atmosphere will absorb atmospheric CO_2. Calculate the concentration of H_2CO_3 in such a solution whose pH values are 5.00, 5.50, and 6.00. Comment on the possible application of pH measurements to the problem of determining the amount of dissolved CO_2 in H_2O.
6. Calculate the pH of a 2.00×10^{-4} M solution of succinic acid. ($pK_1 = 4.20$, $pK_2 = 5.62$)
7. Determine the hydrogen ion concentration of a 0.01 M solution of EDTA. (ethylenediamine tetraacetic acid) ($pK_1 = 2.2$, $pK_2 = 2.97$, $pK_3 = 6.61$, $pK_4 = 10.86$)
8. The pH of a 0.01 M solution of $La(H_2O)_6 \cdot Cl_3$ is 5.40. What is the first acid dissociation constant of $La(H_2O)_6^{3+}$?
9. Solve problems 1, 2 and 4 graphically.
10. For each of the following substances in 0.01 M solution calculate the concentration of the most highly charged anion at pH 3.0, 6.0 and 9.0, if the ionic strength is 0.1 throughout.
    (a) H_2S ($pK_1 = 6.8$, $pK_2 = 12.4$)
    (b) H_3PO_4 ($pK_1 = 1.91$, $pK_2 = 6.72$, $pK_3 = 11.70$)

10. (c) EDTA (H_4Y) ($pK_1 = 2.0$, $pK_2 = 2.67$, $pK_3 = 6.16$, $pK_4 = 10.26$)
11. Calculate the concentrations of all molecular and ionic species in a 1.00×10^{-3} M solution of the following compounds:
    (a) Na_2CO_3 ($pK_1 = 6.30$, $pK_2 = 10.23$, $pK_w = 13.95$)
    (b) Na_2HPO_4 ($pK_1 = 2.10$, $pK_2 = 7.10$, $pK_3 = 12.2$, $pK_w = 13.95$)
    (c) $H_2C_2O_4$ ($pK_1 = 1.3$, $pK_2 = 4.21$, $pK_w = 13.97$)
    (d) H_2SO_4 ($pK_2 = 1.90$, $pK_w = 13.96$)
12. Calculate the fraction of the most highly charged anion of the following substances as a function of $[H^+]$. Plot the results logarithmically (i.e. Log α vs pH). Assume that the ionic strength is 0.1 throughout. What effect if any does the analytical concentration of the acid have on the α value.
    (a) 0.01 M H_2S
    (b) 0.01 M H_3PO_4
    (c) 0.01 M EDTA (H_4Y).
    See Problem 10 for pK values.
13. Calculate the hydrogen ion concentration in each of the following solutions:
    (a) 0.10 M glycine (assume ionic strength = 0.1) ($pK_1 = 2.35$, $pK_2 = 9.54$, $pK_w = 13.76$)
    (b) 0.01 M ethylenediamine monohydrochloride ($pK_1 = 6.88$, $pK_2 = 9.90$, $pK_w = 13.91$)
    (c) 1.0×10^{-3} M $NaHCO_3$ ($pK_1 = 6.32$, $pK_2 = 10.27$, $pK_w = 13.97$)
    (d) 2.5×10^{-2} M NaH_2PO_4 ($pK_1 = 2.01$, $pK_2 = 6.93$, $pK_3 = 12.0$, $pK_w = 13.86$)
    (e) 2.5×10^{-2} M Na_2HPO_4 $pK_1 = 1.94$, $pK_2 = 6.79$, $pK_3 = 11.8$, $pK_w = 13.79$)
14. Calculate the pH of a mixture containing 0.1 M H_2S and 0.05 M NaHS. ($pK_1 = 6.8$, $pK_2 = 12.5$, $pK_w = 13.82$)
15. What concentration of $NaHCO_3$ should be present in a 0.02 M solution of Na_2CO_3 to give a solution of pH 10.0? (Assume that the ionic strength is 0.10) ($pK_1 = 6.09$, $pK_2 = 9.85$, $pK_w = 13.76$)
16. Design a buffer using the phosphoric acid system whose initial pH is 7.0, such that the release of 1 millimole of H^+ per 100 ml would not lower the pH by more than 0.2 log units. (The ionic strength is 0.30). ($pK_1 = 1.79$, $pK_2 = 6.49$, $pK_3 = 11.3$, $pK_w = 13.64$)

VII

Acid-Base Mixtures

In analytical practice, mixtures of acids and bases are quite frequently encountered. Solutions in various industrial processes may contain a wide variety of acids and bases of all types. It will be seen in the following sections that the methods used previously can be readily adapted to the problem of calculating the hydrogen ion concentrations in such solutions.

VII-1. MIXTURES OF ACIDS

Let us dispose of some simple cases at the outset. Obviously in mixtures of strong acids the total hydrogen ion concentration is simply the sum of the concentrations of the acids in the mixture; a mixture of strong bases may be treated similarly. In problems involving mixtures of acids where one acid contributes at least twenty times more hydrogen ion than the rest, then one can safely ignore all but this one acid. This principle will guide us in dealing with mixtures of strong and weak acids as well as of weak acids alone. Of course if we have a mixture of a strong and a weak acid of comparable concentration then the contribution of the weak acid to the hydrogen ion concentration is of no importance.

Example 1.

Calculate the hydrogen ion concentration of a mixture of acids which is 0.02 M in HCl and 0.01 M in HCOOH. pK_a of HCOOH is 3.61.

In this solution, formic acid, if alone would contribute somewhat less than 10% of the $[H^+]$ that comes from HCl alone. The common ion effect however reduces the contribution from HCOOH to considerably less than 5% of the total H^+ as seen below.

Let $X = [H^+]$ from HCOOH in the mixture.

Total $[H^+] = 0.02 + X$

$$\text{Therefore } K_a = 10^{-3.61} = \frac{(0.02 + X) \cdot X}{(0.01 - X)}$$

$$\text{If } X < 5\% \text{ of } 0.01, \text{ then } 10^{-3.61} = \frac{0.02\,X}{.01}$$

Therefore $X = 10^{-3.91} = 1.3 \times 10^{-4}$M. This number is small enough to validate all the assumptions that have been made in this problem. Hence the hydrogen ion concentration of this solution may be taken to be that of the HCl alone, namely 0.02 M.

This example illustrates equally well the approach used when the contribution from the weak acid may not be neglected.

When we deal with a mixture of weak acids, the same principle, that if one of the acids is significantly stronger than the rest it will probably contribute at least twenty times more H^+ than the others, will apply.

Example 2.

Calculate the hydrogen ion concentration of a mixture consisting of 0.1 M acetic acid and 0.1 M ammonium chloride.

pK_a of HOAc = 4.52; pK_a of NH_4^+ = 9.24.

In this mixture we can recognize that the acetic acid is a much stronger acid than the NH_4^+, and therefore will contribute most of the $[H^+]$ in solution.

As a first approximation $[H^+] = (0.1 \times 10^{-4.52})^{1/2}$

$$= 10^{-2.76},$$

whereas the $[NH_4^+]$ alone would give only $(10^{-9.24} \times 0.1)^{\frac{1}{2}} = 10^{-5.12}$ moles/liter of hydrogen ions in solution, which even without the common ion effect may be seen to be insignificant.

Hence the concentration of hydrogen ions in solution = $10^{-2.76} = 1.7 \times 10^{-3}$M.

If the $[H^+]$ contributions from the acids in a mixture are close together, then it will be necessary to employ the expression VII-6 as shown below.

Suppose a solution contains a mixture of acids HA_1 and HA_2 whose molar concentrations are C_1 and C_2 respectively.

The equilibrium constants are given by

$$K_1 = \frac{[H^+] \times [A_1^-]}{[HA_1]} \quad \text{and } K_2 = \frac{[H^+] \times [A_2^-]}{[HA_2]} \qquad \text{(VII-1)}$$

Three other equations may be written on the basis of charge or proton balance and the mass balance.

$$[H^+] = [A_1^-] + [A_2^-] + [OH^-] \qquad \text{(VII-2)}$$

$$C_1 = [A_1^-] + [HA_1] \qquad \text{(VII-3)}$$

$$C_2 = [A_2^-] + [HA_2] \qquad \text{(VII-4)}$$

Combining VII-1 with VII-3 and VII-4, we get

$$[A_1^-] = \frac{K_1 C_1}{[H^+] + K_1} \quad \text{and} \quad [A_2^-] = \frac{K_2 C_2}{[H^+] + K_2}$$

Substituting these values of $[A_1^-]$ and $[A_2^-]$ in VII-2 we obtain the equation:

$$[H^+] = \frac{K_1 C_1}{H^+ + K_1} + \frac{K_2 C_2}{H^+ + K_2} + \frac{K_w}{H^+} \qquad \text{(VII-5)}$$

Usually $[H^+]$ is greater than either K_1 or K_2, that is the extent of dissociation of each of the weak acids is less than 5%.

Hence equation VII-5 may be simplified to give

$$[H^+] = \frac{K_1 C_1}{[H^+]} + \frac{K_2 C_2}{[H^+]} + \frac{K_w}{[H^+]}$$

or

$$[H^+]^2 = (K_1 C_1 + K_2 C_2 + K_w) \qquad \text{(VII-6)}$$

If the assumption $[H^+] > K_1$ or K_2 is not valid, the approximate value obtained from (VII-6) may be substituted in the denominators in (VII-5) to obtain a more accurate value of $[H^+]$.

The graphical approach to the problem of the calculation of pH of a mixture of acids is shown in Figure VII-1. Using the proton balance equation VII-2, the pH of the mixture is given by the point of intersection P, between the $[H^+]$ and $[A_1^-]$

lines. This solution is justified since the line segment PQ is greater than 1.3 log units (hence $[A_2^-] \ll [A_1^-]$ as is also the segment PR ($[OH^-] \ll [A^-]$). In cases where the differences in pK_1 and pK_2 are smaller or alternatively where $C_2 \gg C_1$ then PQ might be smaller than 1.3. The appropriate correction would then be carried out in the manner described earlier (p. 49).

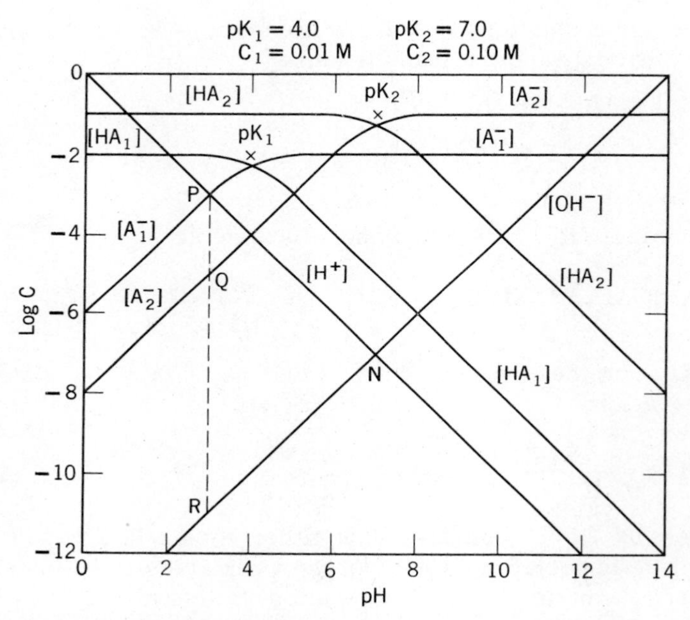

Fig. VII-1 — Logarithmic diagram for the determination of the pH of a mixture of acids

Example 3.

Calculate the hydrogen ion concentration of the solution in Example 2, to which has been added sufficient propionic acid to make the solution 0.05 M in this acid.

pK_a of propionic acid = 4.63

$pK_w = 13.76$

$$[H^+] = (10^{-4.63} \times 0.05 + 10^{-4.52} \times 0.10 + 10^{-9.24} \times 0.1 + 10^{-13.76})^{1/2} = (10^{-5.93} + 10^{-5.52} + 10^{-10.24} + 10^{-13.76})^{1/2}$$

To the nearest 5% then,

$$[H^+] = (10^{-5.93} + 10^{-5.52})^{1/2}$$
$$= 10^{-2.69}$$
$$= 2.0 \times 10^{-3}\ M$$

VII-2. MIXTURES OF ACIDS AND BASES

When we mix acids with bases the resulting chemical reactions will yield solutions which will often be recognized as previously dealt with cases. For example consider mixing equivalent amounts of a strong acid and a strong base such as HCl and NaOH. The hydrogen ion concentration of the resulting solution, containing only NaCl and H_2O can be calculated readily. In the event that either the HCl or NaOH is present in excess, the hydrogen ion concentration of the solution is obtained from the appropriate strong acid or strong base calculation.

Similarly a mixture of equivalent amounts of a strong acid and a weak base such as HCl and NH_3 will yield a solution of the conjugate acid of the weak base, namely NH_4^+. Likewise, when we mix a strong base and weak acid in equivalent amounts we will obtain a solution of the conjugate base of the weak acid. If in either of these cases the weak acid or weak base is in excess, conjugate acid-base pairs will be encountered. On the other hand, if in the mixture there is an excess of the strong acid or strong base, we will be dealing with a mixture of a strong and a weak acid or base, e.g. HCl and NH_4Cl or NaOH and NaOAc.

An important case that we have not dealt with yet is a mixture of a weak acid and a weak base such as acetic acid and ammonia. For such systems let us derive an expression that will apply equally well for mixtures containing any proportion of acid to base. Let the mixture contain C_a moles/1. of a weak acid HA and C_b moles/1. of a weak base B.

The proton balance equation for this solution will be

$$[H^+] + [BH^+] = [OH^-] + [A^-] \qquad \text{(VII-7)}$$

The equilibrium expressions for the weak base and weak acid are:

$$K_b = \frac{[BH^+][OH^-]}{[B]} \quad \text{and } K_a = \frac{[A^-]\ [H^+]}{[HA]} \tag{VII-8}$$

The mass balance equations are:

$$C_a = [HA] + [A^-] \quad \text{and} \quad C_b = [BH^+] + [B] \tag{VII-9}$$

Combining VII-8 and 9 we get:

$$[A^-] = \frac{K_a C_a}{K_a + [H^+]} \quad \text{and} \quad [BH^+] = \frac{K_b C_b}{K_b + [OH^-]} \tag{VII-10}$$

Substituting these expressions in equation (VII-7) we obtain

$$[H^+] + \frac{K_b C_b}{K_b + [OH^-]} = [OH^-] + \frac{K_a C_a}{K_a + [H^+]}$$

This equation may be rearranged to give:

$$[H^+]^2 K_b + [H^+]\{K_b(C_b + K_a) + K_w\} - \{K_a K_b (C_a - C_b) - K_w(K_a - K_b)\} - [OH^-]\{K_a(C_a + K_b) + K_w\} - [OH^-]^2 \cdot K_a = 0 \tag{VII-11}$$

Converting $[OH^-]$ to its equivalent value, $K_w/[H^+]$ and simplifying, we get:

$$[H^+]^4 \cdot K_b + [H^+]^3 \{K_b(C_b + K_a) + K_w\} - [H^+]^2 \{K_a K_b (C_a - C_b) - K_w(K_a - K_b)\} - [H^+] \cdot K_w\{K_a(C_a + K_b) + K_w\} - K_w^2 K_a = 0 \tag{VII-12}$$

This quartic equation simplifies to at least a quadratic in most cases. For example when the mixture is definitely acidic, i.e. when $[H^+] \gg [OH^-]$ only the terms in $[H^+]^4$, $[H^+]^3$ and $[H^+]^2$ are of importance. Hence equation (VII-12) becomes:

$$[H^+]^2 K_b + [H^+]\{K_b(C_b + K_a) + K_w\} - \{K_a K_b (C_a - C_b) - K_w(K_a - K_b)\} = 0 \tag{VII-13}$$

In solutions in which $[H^+]$ is not very far from that in neutral solution i.e. $C_b \gg [H^+]$ and $C_a \gg [OH^-]$, the first and last terms in equation (VII-12) may be neglected since these terms correspond to the highest order terms in $[H^+]$ and $[OH^-]$. (See also equation VII-11.) Hence we obtain:

$$[H^+]^2 \{K_b(C_b + K_a) + K_w\} - [H^+]\{K_aK_b(C_a - C_b) - K_w(K_a - K_b)\} - K_w\{K_a(C_a + K_b) + K_w\} = 0 \qquad \text{(VII-14)}$$

A very important special example that may be considered with this equation results when the mixture contains equivalent quantities of acid and base, namely

$C_a = C_b$. Equation (VII-14) then simplifies to:

$$[H^+] = \sqrt{\frac{K_wK_a(C_a + K_b)}{K_b(C_a + K_a)}} \qquad \text{(VII-15)}$$

When K_a and K_b are small compared to C_a, this simplifies further to give:

$$[H^+] = \sqrt{\frac{K_wK_a}{K_b}} \qquad \text{(VII-16)}$$

Equation VII-16 may be conveniently recast in the form

$$[H^+] = \sqrt{(K_a'K_a)} \qquad \text{(VII-17)}$$

Where K_a' will be recognized as the dissociation constant of the conjugate acid BH^+ of the base B. The student may notice the formal resemblance of (VII-17) to equation (VI-30) describing the $[H^+]$ of an ampholyte solution. (Why is this so?)

Notice as long as the concentration of the salt is greater than K_a or K_b, the pH of the solution is concentration independent.

Finally, in solutions that are definitely basic, i.e. $[OH^-] \gg [H^+]$, equation (VII-11) can be simplified to give a quadratic in $[OH^-]$ by including only those terms in this equation.

$$[OH^-]^2 K_a + [OH^-]\{K_a(C_a + K_b) + K_w\} - \{K_aK_b(C_b - C_a) - K_w(K_b - K_a)\} = 0 \qquad \text{(VII-18)}$$

The three quadratic equations just developed might seem formidable, but in actual practice reduce to expressions that are readily solved. This is illustrated in the following example.

Example 4.

Calculate the hydrogen ion concentration in a mixture containing 0.01 M weak acid HA, ($pK_a = 5.0$) and 0.005 M weak base B, ($pK_{BH+} = 9.0$).

Since HA is in excess the solution will be acidic. Hence

equation VII-13 may be used. Substituting the appropriate numbers we obtain the expression:

$$10^{-9} \times [H^+]^2 + [H^+]\{10^{-9} \times (0.005 + 10^{-5}) + 10^{-14}\} - 10^{-14}(0.005) + 10^{-14}(10^{-5} - 10^{-9}) = 0$$

This may be reduced by inspection to:

$$10^{-9}[H^+]^2 + 5 \times 10^{-12}[H^+] - 5 \times 10^{-17} = 0$$

Therefore

$$[H^+] = 1.0 \times 10^{-5} \text{ M.}$$

This problem could have been more appropriately solved by considering that the weak base reacted completely with the weak acid to give 0.005 M A^- leaving 0.005 M HA unreacted. Since we would now be dealing with an equimolar mixture of HA and A^- the hydrogen ion concentration is equal to K_a. Thus the student should always start by treating non-stoichiometric mixtures of a weak acid and a weak base as if the component present in smaller concentration were a strong, rather than a weak acid or base. This takes advantage of the fact that the component present in excess will tend to drive the reaction to completion as would be the case if the component present in smaller concentration were indeed a strong acid or base. Of course as the mixtures involve increasingly weak acids and bases and also when the deviation from stoichiometric ratio becomes smaller, the use of the above quadratic equations becomes increasingly necessary for accurate calculations.

A few illustrative examples will serve to show how various forms of the above equations are used.

Example 5.

Calculate the hydrogen ion concentration of a 0.01 M solution of ammonium acetate. $pK_{NH_4^+} = 9.24$ $pK_{HOAc} = 4.65$

Since NH_4OAc represents a mixture containing equivalent amounts of acid and base, equation VII-15 or as a suitable approximation equation VII-17 can be used.

$$[H^+] = \sqrt{(K_{a_1} K_a)}$$

$$= \sqrt{(10^{-9.24} \times 10^{-4.65})}$$

Therefore,

$$pH = 6.95$$

Example 6.

Calculate the pH of a 0.01 M solution of NH_4CN.

K_a for HCN = 9.31 K_a for NH_4^+ = 9.24

This problem may be solved in exactly the same manner as shown in example 5.

pH = 9.28.

It is useful to note however that equation VII-17 still applies even though the solution has a pH which is significantly higher than 7.

Example 7.

Calculate the pH of a 0.001 M solution of a hypothetical ammonium salt NH_4A in which the pK_a of the acid is 2.0.

Equation VII-17 may not be applied here since $C < K_a$. For this case, in which $C > K_b$ equation VII-15 can be modified to:

$$[H^+] = \sqrt{\frac{(K_w C_a)}{K_b}}$$

$$= \sqrt{\frac{10^{-14} \times 10^{-3}}{10^{-4.76}}} = 10^{-6.12}$$

pH = 6.12

As might be expected, with increasingly dilute solutions, equation VII-15 gives a limiting value of $[H^+] = (K_w)^{1/2}$.

Up to this point we have treated the 1 : 1 salts of weak acids and weak bases. It would be interesting to consider the 2 : 1 and higher salts of weak acids and weak bases. For example how would we calculate the pH of a solution of $(NH_4)_2CO_3$? Qualitatively it is useful to observe that the reaction: $NH_4^+ + CO_3^= \rightarrow HCO_3^- + NH_3$ goes almost to completion (since the K value for this reaction is

$$\frac{10^{-9.24}}{10^{-10.27}} = 10^{+1.03}).$$

Since this is so the mixture contains equimolar quantities of NH_4^+, NH_3 and HCO_3^-. Therefore the pH of this solution will be determined largely by the NH_4^+, NH_3 buffer and is equal to 9.24. Since this pH value is 1.03 units below pK_2 of H_2CO_3, the ratio of $[HCO_3^-]$ to $[CO_3^=]$ will be $10^{-1.03}$. Hence $[CO_3^=]$ will be 7% of the $[HCO_3^-]$. Of course the small additional amount of NH_4^+ resulting from the unreacted $CO_3^=$ alters the pH slightly. By using this value to make a further approximation we arrive at the pH value of 9.21.

A more exact approach that would apply to this and similar problems would be as follows. Consider a solution of C molar $(NH_4)_2CO_3$. The proton balance equation gives us:

$$[H^+] + [HCO_3^-] + 2[H_2CO_3] = [NH_3] + [OH^-]$$

Substituting appropriate equilibrium expressions we obtain: (readily derived using α values, see chap. VI.)

$$[H^+] + \frac{C[H^+]K_1}{[H^+]^2 + [H^+]K_1 + K_1K_2} + \frac{2C[H^+]^2}{[H^+]^2 + [H^+]K_1 + K_1K_2} =$$

$$\frac{2CK'}{K' + [H^+]} + \frac{K_w}{[H^+]}$$

where K_1 and K_2 are the acid dissociation constants of H_2CO_3 and K' is that of NH_4^+. This equation is fifth order in $[H^+]$ and therefore not amenable to exact solution. However in solutions that are not too dilute, namely $C > [H^+]$ or $[OH^-]$, the first and last terms disappear to give an equation that is independent of the concentration, C.

$$\frac{[H^+]K_1 + 2[H^+]^2}{[H^+]^2 + [H^+]K_1 + K_1K_2} = \frac{2K'}{K' + [H^+]}$$

Now if $K_1 > [H^+] > K_2$, then this equation simplifies to:

$$[H^+] = \frac{K' + \sqrt{(K'^2 + 8K'K_2)}}{2} = 10^{-9.17}$$

This general approach can be used profitably not only for cases of other salts of weak acids and weak bases but for all sorts of non-stoichiometric mixtures of weak acids and weak bases.

Consider for example the case of a mixture of C_1 molar NH_4HCO_3 and C_2 molar NH_3. The charge balance equation gives the following:

$$[H^+] + [NH_4^+] = [OH^-] + [HCO_3^-] + 2[CO_3^=]$$

The material balance equations are:

$$C_1 + C_2 = [NH_4^+] + [NH_3]$$

$$C_1 = [H_2CO_3] + [HCO_3^-] + [CO_3^=]$$

Substituting appropriate equilibrium expressions (α values) into the charge balance equation we have:

$$[H^+] + (C_1 + C_2)\frac{[H^+]}{K' + [H^+]} = \frac{K_w}{[H^+]} + \frac{C_1([H^+]K_1 + 2K_1K_2)}{[H^+] + [H^+]K_1 + K_1K_2}$$

Now one proceeds as above using appropriate approximations.

The value of the qualitative approach which places more stress on chemical intuition may be seen from the following analysis of the problem just considered.

The addition of NH_3 to NH_4HCO_3 will result in the predominance of the $NH_4^+ - NH_3$ buffer in fixing the pH of the solution. First calculate the pH from the ratio $C_1 : C_2$ of $NH_4^+ : NH_3$; then calculate what fraction of the HCO_3^- has been transformed to $CO_3^=$ at this pH. This will provide the basis for a second approximation as carried out in the previous example using $(NH_4)_2CO_3$.

The graphical method provides our optimum combination of the general method with chemical intuition as illustrated in the following examples.

In Fig. 1 the concentrations of the components of a 0.01 M NH_4OAc solution are graphically represented. The proton balance equation for this solution (see equation VII-7) is

$$[H^+] + [HOAc] = [NH_3] + [OH^-]$$

An examination of Fig. VII-1 shows that the HOAc and NH_3 are the two major components as shown by the fact that the lines representing these two concentrations cross first, (i.e. at the highest log C value).

It should be noticed from the graph that as the concentration of NH_4OAc decreases, the pH of the point of intersection does not change, until this point comes within 1.3 log units of either the $[H^+]$ or $[OH^-]$ line. From this concentration, on, the pH of the intersection will be smoothly displaced towards the pH of neutrality, i.e. point N.

In Fig. VII-2 we may also solve Example 6 dealing with NH_4CN. The pK values of NH_4^+ and HCN are within 0.05 log units of each other so that on the scale of Fig. VII-2 the NH_4^+, NH_3 and HCN, CN^- pairs of lines overlap. Writing the proton balance equation,

$$[H^+] + [HCN] = [NH_3] + [OH^-]$$

the important intersection is seen to be that of the HCN and NH_3 lines. This occurs at a pH just between the pK values

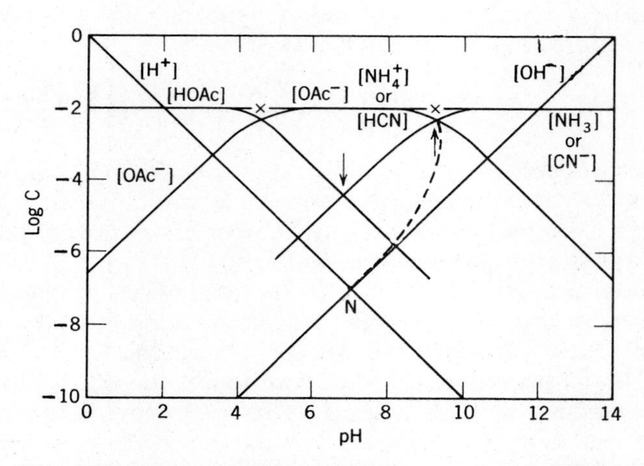

Fig. VII-2 — Logarithmic diagram for the determination of the pH of (a) 0.01 M NH_4OAc and (b) 0.01 M NH_4CN

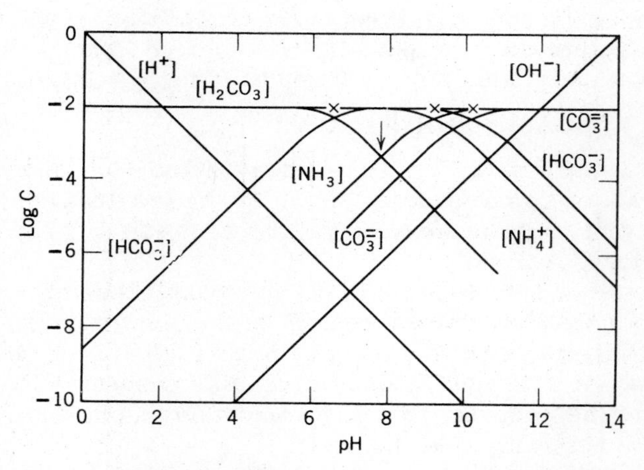

Fig. VII-3 — Logarithmic diagram for the determination of the pH of a 0.01 M NH_4HCO_3 solution.

(9.24 and 9.31) i.e. at 9.3. The dotted line indicates how the pH of the NH_4CN solution would vary with concentration.

In a similar manner the pH of a solution of 0.01 M NH_4HCO_3 may be graphically determined using Fig. VII-3. Writing the proton balance equation,

$$[H^+] + [H_2CO_3] = [CO_3^=] + [OH^-] + [NH_3]$$

we see that the important intersection is that between the NH_3 and H_2CO_3 lines at pH = 7.8.

SUGGESTIONS FOR FURTHER READING

1. T. B. Smith, Analytical Processes, Edward Arnold Publishers Ltd., London (1940), Chapter 10.
2. L. G. Sillen, Part I, Vol. I, Chapter 8, Treatise on Analytical Chemistry, Interscience, (1959), I. M. Kolthoff and P. J. Elving, Editors.

VIII-3. PROBLEMS

1. Calculate the hydrogen ion concentrations of the following solutions:
   (a) A mixture that is 0.10 M in propionic acid (pK_a = 4.83) and 0.20 M in acetic acid (pK_a = 4.72)
   (b) A mixture that is 0.20 M in ammonium chloride (pK_a = 9.24) and 0.10 M in hydrocyanic acid (pK_a = 9.1)
   (c) A mixture that is 0.05 M in phenol (pK_a = 9.85) and 0.03 M in ammonium chloride (pK_a = 9.24)
   (d) A mixture that is 0.02 M in ammonium nitrate (pK_a = 9.24) and 0.02 M in zinc nitrate. (pK_{a_1} = 9.62)
   (e) A mixture that is 1.0×10^{-3} M in acetic acid (pK_a = 4.75) and 2.5×10^{-3} M in benzoic acid (pK_a = 4.19)
2. What is the pH of a solution which is 0.01 M in acetic acid (pK_a = 4.75) and 0.10 M in H_2S (pK_1 = 7.0, pK_2 = 12.9)?
3. Calculate the pH of a solution which is obtained by mixing 500 ml of 1.0×10^{-2} M phosphoric acid and 250 ml of 1.00×10^{-3} M acetic acid (pK_a = 4.66) (pK_1 = 2.05; pK_2 = 7.00; pK_3 = 12.1 for H_3PO_4)
4. What is the pH of a solution containing 2.5 g. succinic acid (pK_1 = 3.90, pK_2 = 5.02) and 2.5 g. sulfamic acid (pK_a = 0.7) in one liter with sufficient indifferent electrolyte to give a solution an ionic strength 0.2?
5. Calculate the hydrogen ion concentration of a solution which is 1.0×10^{-3} M in pyridine (pK_a = 5.30) as well as in trimethylamine. (pK_a = 9.80, pK_w = 13.97).
6. Calculate the pH of the solutions obtained by adding 25.0 ml of 0.05 M HCl to each of the following:
   (a) 25.0 ml of 0.2 M acetic acid (pK_a = 4.62)

(b) 50.0 ml of 0.3 M nitric acid
(c) 25.0 ml of 0.05 M ammonia (pK_a = 9.24)
(d) 50.0 ml of 0.03 M sodium hydroxide (pK_w = 13.88)

7. Calculate the pH of the solutions obtained by adding 50.0 ml of 0.10 M acetic acid to each of the following:
(a) 25.0 ml of 0.20 M sodium hydroxide (pK_{HOAc} = 4.56, pK_w = 13.80)
(b) 50.0 ml of 0.10 M sodium hydroxide (pK_{HOAc} = 4.58, pK_w = 13.82)
(c) 50.0 ml of 0.10 M ammonia ($pK_{NH_4^+}$ = 9.24, pK_{HOAc} = 4.58)
(d) 25.0 ml of 0.10 M ammonia ($pK_{NH_4^+}$ = 9.24, pK_{HOAc} = 4.56)

8. What is the pH of a 0.05 M solution of the following salts? ($pK_{NH_4^+}$ in all cases is 9.24)
(a) Ammonium propionate (pK_a = 4.69)
(b) Ammonium cyanide (pK_a = 9.2)
(c) Ammonium acetate (pK_a = 4.58)
(d) Anilinium propionate ($pK_{\text{anilinium ion}}$ = 4.59, pK_a = 4.69)

9. Calculate the concentrations of all ionic and molecular species in a 0.02 M solution of ammonium sulfide. ($pK_{NH_4^+}$ = 9.24, pK_1 = 6.9, pK_2 = 12.7 for H_2S, pK_w = 13.88)

10. What is the pH of a 2.0×10^{-2} M solution of ammonium carbonate? ($pK_{NH_4^+}$ = 9.24, pK_1 = 6.18, pK_2 = 10.00 for H_2CO_3, pK_w = 13.83)

11. Calculate the hydrogen ion concentrations of the following solutions:
(a) 1.0×10^{-3} M ammonium cyanide ($pK_{NH_4^+}$ = 9.24, pK_a = 9.4, pK_w = 13.97)
(b) A mixture that is 0.02M in ammonium carbonate as well as in ammonium bicarbonate ($pK_{NH_4^+}$ = 9.24, pK_1 = 6.18, pK_2 = 10.00, pK_w = 13.83)
(c) A mixture that is 0.05 M in sodium sulfide as well as in sodium hydrogen sulfide (pK_1 = 6.8, pK_2 = 12.4, pK_w = 13.76)
(d) 0.02 M ammonium oxalate ($pK_{NH_4^+}$ = 9.26, pK_1 = 1.1, pK_2 = 3.87 for $H_2C_2O_4$, pK_w = 13.80)

12. Solve examples 1, 5, 8, 9, 10 and 11 graphically.

VIII

*Precipitation Equilibria**

VIII-I. FACTORS THAT AFFECT THE SOLUBILITY OF ELECTROLYTES

The solubility of a solid in water depends upon the difference between the energy consumed in separating the ions or molecules from the crystal lattice, (i.e., the lattice energy), and the energy released by the solvation of these ions or molecules (called the hydration energy). The lattice energy of strong electrolytes increases with the ionic charge and decreases with the ionic size. Thus CaO has a higher lattice energy than NaCl which in turn has a higher lattice energy than KCl. Similarly, hydration energies are largest for small highly charged ions. Since hydration energies and lattice energies vary in the same manner, it is difficult to predict solubility trends in various types of electrolytes from such data.

With many electrolytes, the lattice energy is somewhat greater than the hydration energy so that the dissolution of electrolytes in water is generally an endothermic process. Hence the solubility of most electrolytes increases with increasing temperature. See Equation (II-13). A very few substances such as $CaSO_4$ have negative heats of solution and therefore exhibit a decrease in solubility with increase in temperature.

---

* Any possible Bronsted or Lewis acid-base interactions of the ions of the precipitated substances will be ignored until considered in Chapter X.

The solubilities of solid electrolytes in organic solvents are generally lower than they are in water. Thus the solubility of $PbSO_4$ or $SrSO_4$ can be decreased by the addition of ethanol to water, which thereby facilitates their quantitative precipitation. The solubility decrease effected by the organic solvent probably reflects the influence of the lower dielectric constant which increases the energy required to dissociate the ions. Just what effect the change of solvent has on the solubility of molecules (or uncharged ion pairs) is far too complex to be considered here, and in any event is usually of much less importance.

Since solubility of electrolytes involves dissociation phenomena, factors affecting dissociation either directly (ionic strength) or indirectly (reactions involving ions formed) will inevitably affect solubility. These factors will now be considered in detail.

VIII-2. SOLUBILITY AND SOLUBILITY PRODUCT

The solubility of a substance at any given temperature is defined as the concentration of the substance in a saturated solution, i.e. a solution which is in equilibrium with the undissolved solid. This equilibrium may be represented as follows, using a solution of naphthalene, ($C_{10}H_8$) in water as an illustration.

$$(C_{10}H_8)_{solid} \rightleftharpoons (C_{10}H_8)_{solution}$$

with the corresponding equilibrium expression

$$\mathbf{K} = a_{C_{10}H_8} = [C_{10}H_8] \times \gamma_{C_{10}H_8}$$

The activity of the solid, being a constant, is included in the equilibrium constant, **K**. A distinction should be made between saturated and concentrated solutions since these categories are not equivalent. Here, for example, the low solubility of naphthalene in water results in a saturated solution of approximately 2×10^{-4} M which is obviously extremely dilute.

If , as in the case of naphthalene, we are dealing with solutes for which the activity coefficients are very close to unity, then the equilibrium constant, **K**, can be recognized as the molar solubility. However it should be kept in mind that the activity coefficients of even undissociated solutes will be affected to some extent by the total solution composition. Hence solubilities of non-electrolytes are not absolutely invariant at a constant temperature.

The foregoing discussion of solubility does not cover the case of the solubility of a strong electrolyte since strong electrolytes will dissociate completely in solution.* NaCl in water for example may be represented by the following equations:

$$(NaCl)_{solid} \rightleftharpoons (NaCl)_{solution}$$

However since $(NaCl)_{solution}$ is totally dissociated into Na^+ and Cl^- ions, this equation must be rewritten as:

$$(NaCl)_{solid} \rightleftharpoons Na^+ + Cl^-$$

The equilibrium constant here is:

$$\mathbf{K} = a_{Na^+} \times a_{Cl^-} = [Na^+][Cl^-]\,\gamma_{Na^+} \times \gamma_{Cl^-}$$

As before the activity of the solid NaCl is a constant and is included in the equilibrium constant. In this case, in contrast with that of the non-electrolyte, the equilibrium constant may not be equated to the solubility as may be seen from the fact that the right-hand-side of the equation contains the product of two concentration terms, rather than a single concentration term. This equilibrium constant is called the solubility product constant $\mathbf{K}_{sp}$. An added complexity arises in the illustration that we have chosen, from the fact that a saturated solution of NaCl happens to be quite concentrated. Hence the activity coefficients included in the equation may be quite different from unity, rendering attempts to describe quantitatively the solubility relationships of NaCl with the use of this equation extremely difficult. The solubility product equilibrium just described is much more useful when applied to systems of slightly soluble electrolytes.

Let us consider the case of $BaSO_4$, for example, which has a solubility of 1.0×10^{-5} moles/l. in water. Here, as with NaCl, we may write:

$$(BaSO_4)_{solid} \rightleftharpoons Ba^{++} + SO_4^{=}$$

$$\mathbf{K}_{sp} = [Ba^{++}][SO_4^{=}] \times \gamma_{Ba^{++}}\gamma_{SO_4^{=}}$$

In this dilute solution we may consider the activity coefficients of both the Ba^{++} and $SO_4^{=}$ to be unity, and the equilibrium expression may be written as:

*The concept of total dissociation of strong electrolytes in aqueous media has been modified in recent years by the admission of the existence of ion-pairs or other ion-association complexes. Such species, although differing from coordination complexes in the nature of their bonding, are described by the same type of equilibrium expressions (see Chapter X).

$$\mathbf{K}_{sp} = K_{sp} = [Ba^{++}][SO_4^{=}]$$

In cases where the activity coefficients differ from unity, the appropriate value of $K_{sp} = \mathbf{K}_{sp} / \gamma_{Ba^{++}} \times \gamma_{SO_4^{=}}$

can be calculated by methods involving the Debye-Huckel Theory described in Chapter III. In any event we will use the concentration constant expressions rather than the activity expressions hereafter. Variations in values of K_{sp} used in problems and examples, will be understood to reflect an attempt to provide appropriate values for the prevailing ionic strengths.

VIII-3. FACTORS THAT AFFECT THE SOLUBILITY PRODUCT CONSTANT

The true or thermodynamic solubility product constant $\mathbf{K}_{sp}$ will, for most substances, increase with the temperature by an amount that depends on the heat of solution. The addition of any organic solvent such as alcohol to an aqueous solution will generally result in a lower $\mathbf{K}_{sp}$ (see Table VIII-1). This may be easily understood in terms of the increased work of separation of ions in a medium of lower dielectric constant. Finally the particle size of the undissolved solid in equilibrium with the dissolved solute will affect the solubility and solubility product constant. If the particle size is sufficiently small, the surface area per mole becomes large enough to require taking the surface energy into account in describing the equilibrium. In effect the smaller the particle size, (i.e. smaller than 10^{-4} cm radius), the higher the solubility. For example the molar solubility of lead chromate having particles of 9×10^{-5} cm radius was found to be 2.1×10^{-4} in contrast to a value of 1.24×10^{-4} for particles having a radius of 3.0×10^{-3} cm or larger. In this case the $\mathbf{K}_{sp}$ has changed by 250%. There are probably cases in which the effect of particle size on $\mathbf{K}_{sp}$ is much larger, but these are exceedingly difficult to verify experimentally.

The accurate determination of K_{sp} values involves yet another difficulty, namely the slow equilibration of the solid with the solution. Thus kinetic effects may give rise to results which deviate significantly from those calculated.

In dealing with solubility equilibria it is usually assumed that the solid phase or precipitate is pure. In practice this is rarely achieved and at best only approximately so. For this reason separations that are predicted without taking such contamination into account are never fully realized.

The effect of changing ionic strength on the value of the concentration solubility product constant, K_{sp}, is even more significant than the corresponding effect on monoprotic acid dissociation constants. This is to be expected since many precipitates contain multiply charged ions. With the help of Table III-2 we can see that in a solution whose ionic strength is 0.1 the pK_{sp} for AgCl is 0.24 log units lower than the $\mathbf{pK_{sp}}$ whereas for $BaSO_4$, the difference is 4×0.24 or 0.96 log units lower. (See problem 12 in Chapter III.)

Table VIII-1

Effect of Dielectric Constant on the Solubility of $CaSO_4$

% w/w of Ethanol	Dielectric Constant	Solubility gms $CaSO_4$/l.
0.0	80	2.084
3.9	78	1.314
10.0	73	0.970
13.6	71	0.436

VIII-4. RULES FOR PRECIPITATION

The solubility product expression can be used for predicting whether or not precipitation will occur upon mixing two solutions and whether or not a precipitate will dissolve when in contact with a given solution. This represents an application of the general criteria for predicting the direction of reactions and was developed earlier (p. 12). For the purposes of this discussion it will be convenient to call the product of the concentrations of the ions each raised to the appropriate power, i.e., the right-hand-side of the solubility product expression,

$$K_{sp} = [M]^a [X]^b ,$$

the <u>ion product</u>.

The ion product would be the actual value (not the equilibrium value), assuming that no precipitation occurred. The following relations will then be seen to apply.

Ion product $< K_{sp}$: Solution is unsaturated.
No precipitate will form and precipitate present will dissolve.

Ion product > K_{sp} : Solution is supersaturated. In time a precipitate will form, and precipitate present will not dissolve.

Ion product = K_{sp} : Solution is saturated. In this equilibrium mixture no precipitate will form and precipitate present will not dissolve.

VIII-5. QUANTITATIVE RELATIONSHIP BETWEEN SOLUBILITY AND SOLUBILITY PRODUCT

It may readily be shown that the K_{sp} is not the solubility, but is of course related to it. In a saturated solution of $BaSO_4$ of 1.0×10^{-5}, molarity, the concentration of Ba^{++} and of $SO_4^=$ is each 1.0×10^{-5}, since $BaSO_4$ is completely dissociated. Hence,

$$K_{sp} = 1.0 \times 10^{-5} \times 1.0 \times 10^{-5} = 1.0 \times 10^{-10}$$

we can generalize for any slightly soluble strong electrolyte MX whose molar solubility is S,

that $K_{sp} = [M][X] = S^2$.

In the most general case where $M_a X_b$ is the formula of slightly soluble strong electrolyte, with a solubility S moles/l.,

$$(M_aX_b)_{solid} \rightleftharpoons aM + bX$$

for every mole of M_aX_b dissolved a moles of M and b moles of X are formed. Hence the solubility product expression,

$$\begin{aligned} K_{sp} &= [M]^a [X]^b \\ &= (aS)^a \times (bS)^b \\ &= a^a \times b^b \times S^{(a+b)} \end{aligned}$$

It is important to recognize that although we have multiplied S by the coefficient a and then raised the product to the power of a, that this is not equivalent to saying that the concentration of M was multiplied by a before being raised to the power of a. The concentration of M is (a S) in this case and not equal to S.

Example 1.

Derive equations relating the molar solubility S in water and the K_{sp} of the following compounds.

Ag_2CrO_4: $K_{sp} = [Ag^+]^2[CrO_4^=] = (2S)^2 \times (S) = 4S^3$

$Ce_2(C_2O_4)_3$: $K_{sp} = [Ce^{+3}]^2[C_2O_4^=]^3 = (2S)^2 \times (3S)^3 = 108.S^5$

Hg_2Cl_2: $K_{sp} = [Hg_2^{++}][Cl^-]^2 = (S) \times (2S)^2 = 4S^3$

$MgNH_4PO_4$: $K_{sp} = [Mg^{++}][NH_4^+][PO_4^{-3}] = S^3$

Note that in the various cases the K_{sp} is equated to the solubility raised to different powers, depending on the charge type of the precipitate. Hence one cannot directly compare solubilities of salts of different charge types by examination of the numerical values of the respective solubility product constants.

These expressions correctly describe the relationship between solubility product and solubility either in pure water or in solutions not containing any other source of the ions directly involved in the solubility equilibrium. These expressions must be modified when salts containing common ions are present in solution.

VIII-6. THE COMMON ION EFFECT

The common ion effect may best be illustrated by considering the solubility of M_aX_b in the presence of C_x moles/l. of anion X.

Then, $K_{sp} = (aS)^a \times (C_x + bS)^b$

In most cases that will be encountered, C_x is much greater than bs, so that this equation simplifies to:

$$K_{sp} = (aS)^a \times (C_x)^b$$

Example 2.

Calculate the solubilities of AgCl and Ag_2CrO_4 in the solutions shown.

Substance	Solution	Relation between Solubility and Solubility Product	S Solubility
AgCl	0.01M $NaNO_3$	$K_{sp} = 1.2 \times 10^{-10} = S^2$	1.1×10^{-5}
AgCl	0.01M $AgNO_3$ or 0.01M NaCl	$K_{sp} = 1.2 \times 10^{-10}$ $= S(S+0.01)$ $= 0.01S$	1.2×10^{-8}

(Table continued on following page)

Substance	Solution	Relation between Solubility and Solubility Product	S Solubility
Ag_2CrO_4	.01M $NaNO_3$	$K_{sp} = 4.5 \times 10^{-12} = 4S^3$	1.1×10^{-4}
Ag_2CrO_4	.01M $AgNO_3$	$K_{sp} = 4.5 \times 10^{-12}$	
		$= (.01 + 2S)^2 \times (S)$	
		$= (.01)^2 \times S$	4.5×10^{-8}
Ag_2CrO_4	.01M Na_2CrO_4	$K_{sp} = 7.6 \times 10^{-12}$	
		$= (2S)^2 \times (.01 + S)$	
		$= 4S^2 \times 0.01$	1.4×10^{-5}

The common ion effect may be considered to apply to any solution in which there is more than one source of the ions contained in the precipitate. In addition to the situations that have been illustrated above the common ion effect is involved in the answers to such questions as, what concentration of reagent anion would be necessary to initiate precipitation, and what is the anion concentration when the precipitation is (virtually) complete?

Example 3.

(a) What concentration of $CO_3^=$ is necessary to initiate precipitation of $CaCO_3$ from a 0.01 M $Ca(NO_3)_2$ solution? pK_{sp} for $CaCO_3$ in this solution is 7.73.

(b) What is the concentration of $CO_3^=$ when the Ca^{++} has been quantitatively precipitated, (i.e. only 0.1% Ca^{++} remained in solution)?

(a) When the solution is saturated with $CaCO_3$, we may write

$$[Ca^{++}][CO_3^=] = 10^{-7.73}$$

Since $[Ca^{++}] = 0.01$,

$$[CO_3^=] \text{ necessary for saturation } = \frac{10^{-7.73}}{0.01}$$

$$\text{i.e. } [CO_3^=] = 10^{-5.73}$$

$$= 1.9 \times 10^{-6} \text{ M.}$$

Although this is not the concentration required to initiate precipitation, it will suffice to say that slightly more than this concentration will initiate precipitation.

(b) In order to determine conditions that apply at quantitative precipitation, it is first necessary to agree on a definition of quantitative precipitation. Obviously we cannot use the point at which $[Ca^{++}] = 0$ since the concentration of $CO_3^{=}$ required would be infinite Several practical measures of quantitative precipitation have been widely used. (a) all but a small fraction of the original concentration of the ion has been precipitated. For instance 0.1 or 0.01%. (b) All but a given weight of the ion has been precipitated, e.g. 0.1 mg. The choice of one of these or similar measures should be based on its suitability to the particular problem at hand. In order to avoid ambiguity, a particular measure will be described with each problem.

When the $[Ca^{++}]$ in solution has dropped to 0.1% of the original 0.01 M, namely

$$[Ca^{++}] = 0.01 \times .001 = 1.0 \times 10^{-5}$$

$$\text{Then } [CO_3^{=}] = \frac{10^{-7.73}}{1.0 \times 10^{-5}} = 10^{-2.73}$$

$$= 1.9 \times 10^{-3} \text{ M}$$

It has been assumed that the same value of pK_{sp} applied in this solution also.

VIII-7. SELECTIVE PRECIPITATION

Up to the present we have considered the precipitation of but a single substance at a time. Of even greater interest to the analytical chemist is the use of selective precipitation procedures for separating ions.

Let us consider a mixture of cations each of which can form a precipitate with a particular reagent anion, and let us examine the changes that take place in this solution as we add this reagent. As soon as the concentration of reagent reaches a value sufficiently high for an ion product to exceed the solubility product constant of one of the substances it will precipitate. As will be seen in the following example, it does not necessarily follow that the substance having the smallest solubility product constant will be the first to precipitate. However it will be always true that the substance requiring the least amount of reagent to reach saturation will be the first to precipitate.

Example 4.

Chloride ion is added to a mixture containing 0.01 M Tl^+, 0.02 M Pb^{++} and 0.03 M Ag^+. Calculate the order in which precipitation of these metal halides occur and to what extent separations of the three metals take place. pK_{sp} of TlCl = 3.46; pK_{sp} for $PbCl_2$ = 4.08; pK_{sp} for AgCl = 9.50 (assuming that the ionic strength remains constant at 0.1 throughout).

Using the solubility product expressions for each of the metal chlorides we can calculate the $[Cl^-]$ necessary for saturating the solution with respect to each salt.

$$[Tl^+][Cl^-] = 10^{-3.46}$$

$$\therefore \quad [Cl^-] = \frac{10^{-3.46}}{.01} = 10^{-1.46}$$

$$[Pb^{2+}][Cl^-]^2 = 10^{-4.08}$$

$$\therefore \quad [Cl^-] = 10^{-1.19}$$

$$[Ag^+][Cl^-] = 10^{-9.50}$$

$$\therefore \quad [Cl^-] = 10^{-7.98}$$

From this we may conclude that the order in which the metal chlorides will precipitate is AgCl, followed by TlCl with $PbCl_2$ the last to precipitate, in keeping with the increasing concentration of $[Cl^-]$ required for saturation. This is not the order of increasing solubility product constants! (Why not?)

The next question, namely how much of each metal ion will have precipitated before the next one begins to precipitate, may be answered in the following way.

As soon as the $[Cl^-]$ has exceeded $10^{-7.98}$ moles/l., AgCl will begin to precipitate out and thereby lower $[Ag^+]$ in solution. This in turn will result in the requirement of a higher $[Cl^-]$ in order to continue the precipitation. Despite the fact that $[Cl^-]$ will increase, no TlCl will accompany the AgCl until $[Cl^-]$ reaches a value of $10^{-1.46}$. At this point the $[Ag^+]$ left in solution must be:

$$[Ag^+] = \frac{10^{-9.50}}{10^{-1.46}} = 10^{-8.04} = 9 \times 10^{-9} M$$

This represents $\frac{9 \times 10^{-9}}{0.03} \times 100\% = 3 \times 10^{-5}\%$ of the original amount present. Thus Ag^+ may be said to be quantitatively removed from the solution before either Tl^+ or Pb^{++} begin to precipitate.

In a similar fashion we can see that the $[Tl^+]$ remaining in solution when $PbCl_2$ begins to precipitate is:

$$[Tl^+] = \frac{10^{-3.46}}{10^{-1.19}} = 10^{-2.27} = 5.4 \times 10^{-3}\ M$$

Therefore % Tl^+ that remains unprecipitated at this point is $\frac{5.4 \times 10^{-3}}{.01} \times 100 = 54\%$.

From this result we may conclude that although some separation does take place, a quantitative separation of Tl^+ and Pb^{++} is not feasible.

A graphical representation of the solubility product relationships for the three chlorides can be used as the basis of the solution of the problem just considered. In Figure VIII-1 are shown

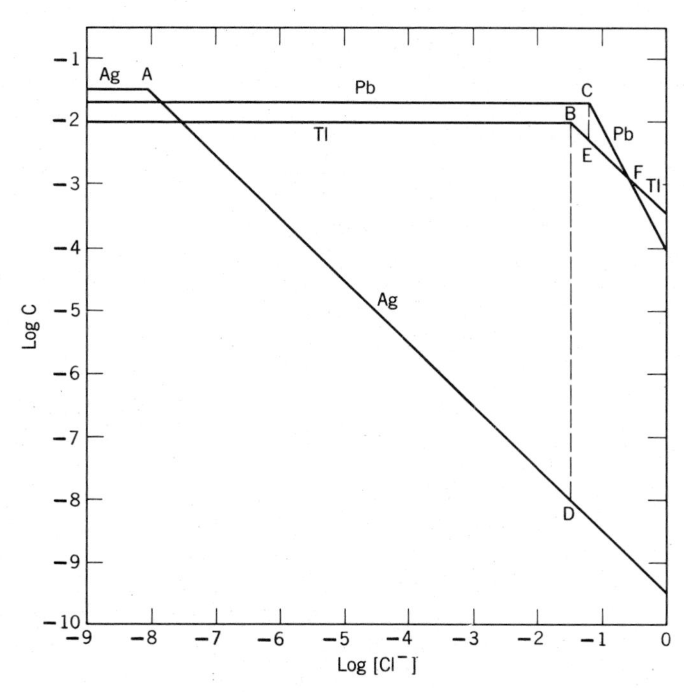

Fig. VIII-1 — Logarithmic Solubility Diagram for AgCl, TlCl and $PbCl_2$ Illustrating Example 4

horizontal lines which terminate in lines of negative slope, 1, 1 and 2 at points A, B and C respectively. These lines of negative slope are obtained by plotting the logarithmic expressions of the corresponding solubility product expressions.

$$\log [Ag^+] = -9.50 - \log [Cl^-]$$

$$\log [Tl^+] = -3.46 - \log [Cl^-]$$

$$\log [Pb^{++}] = -4.08 - 2 \log [Cl^-]$$

It is of interest to note that the difference in the nature of the dependence of the solubility upon the $[Cl^-]$ results in lines of different slope.

The significance of points A, B and C is that they represent points at which the solution is saturated with respect to each of the salts and therefore show the $[Cl^-]$ at which precipitation for each is initiated. The lines beyond these points demonstrate how the metal ion concentration decreases with increasing $[Cl^-]$.

The question as to which of the three metals will precipitate first is solved simply be comparing the points of departure from the horizontal, A, B and C. Obviously, the point occurring at the smallest $[Cl^-]$ will be that of the metal that precipitates first, in this case Ag^+.

The diagram also shows that Tl^+ is the next to precipitate. At this point B, $[Ag^+]$ may be readily obtained by dropping the perpendicular to point D as shown. Since the difference in log $[Ag^+]$ between point A (initial concentration) and point D (concentration at which TlCl begins to precipitate) is greater than 3 units (corresponding to 0.001 as the fraction of Ag^+ remaining in solution) it may be concluded that a quantitative separation of Ag^+ and Tl^+ is theoretically possible. Proceeding in a similar manner, the $[Tl^+]$ at which $PbCl_2$ begins to precipitate shown at point E is sufficiently close to the initial concentration (at point B) to predict that the quantitative separation of Tl^+ and Pb^{++} as chlorides is not feasible.

The difference in the slopes of the TlCl and $PbCl_2$ lines causes the extent of the separation that is possible, to depend on the initial metal concentration. Assuming that both Pb^{++} and Tl^+ are present in the same initial concentration, TlCl can be seen to precipitate first, so long as this concentration is higher than its value at point F where the two lines intersect. At initial concentrations below this point Pb^{++} will precipitate first.

Another approach to predicting the order of precipitation involves the use of the condition of simultaneous precipitation as the reference point. As an illustration the question of the order of the precipitation of silver salts from a solution that is 0.01 M in Cl^- and Br^-, each, will now be considered. When the solution is saturated with respect to both salts, simultaneous precipitation of AgCl and AgBr will occur. At this point both K_{sp} expressions apply.

$$[Ag^+][Cl^-] = K_{sp_{AgCl}} \text{ and } [Ag^+][Br^-] = K_{sp_{AgBr}}$$

Dividing one equation by the other, notice that the $[Ag^+]$ cancels, resulting in the following expression:

$$\frac{[Cl^-]}{[Br^-]} = \frac{K_{spAgCl}}{K_{spAgBr}}$$

From this equation we can develop rules of fractional precipitation, analogous to the rules of precipitation developed at the beginning of this chapter. Let us call the ratio of the concentrations of the two ions the ion ratio, (in analogy to the term ion product used earlier). If the ion ratio is equal to the solubility product ratio, then simultaneous precipitation will occur. If the ion ratio is greater than the solubility product ratio, then precipitation of the salt whose ion concentration is in the numerator will occur first. Conversely, if the ion ratio is smaller than the solubility product ratio, the salt whose ion concentration is in the denominator will precipitate first.

For pairs of salts of the same charge type, the ion ratio required for simultaneous precipitation is independent of the total concentration, whereas for a pair involving different charge types the ratio depends on concentration. Thus for a mixture of $CrO_4^=$ and Cl^- to which Ag^+ is added:

$$[Ag^+][Cl^-] = K_{spAgCl} \quad \text{and} \quad [Ag^+]^2[CrO_4^=] = K_{spAg_2CrO_4}$$

The $[Ag^+]$ can be eliminated between these two equations to give:

$$\frac{[Cl^-]^2}{[CrO_4^=]} = \frac{(K_{spAgCl})^2}{K_{spAg_2CrO_4}}$$

As before, comparison of the function $\frac{[Cl^-]^2}{[CrO_4^=]}$ with the value of the corresponding solubility product ratio may be used in predicting precipitation orders.

Example 5.

In a mixture containing chloride and bromide ions in a molar ratio of 1000:1, what will precipitate first upon the addition of silver ions?

pK_{sp} of AgCl = 9.52 and pK_{sp} of AgBr = 12.04.

(These constants apply to a solution whose ionic strength has to be assumed to be 0.1.)

$$[Ag^+][Cl^-] = 10^{-9.52}$$

$$[Ag^+][Br^-] = 10^{-12.04}$$

Therefore for simultaneous precipitation of AgCl and AgBr,

$$\frac{[Br^-]}{[Cl^-]} = \frac{10^{-12.04}}{10^{-9.52}} = 10^{-2.52}$$

Since the actual ratio present in solution is 10^{-3} the addition of Ag^+ will produce a precipitate of the salt whose anionic concentration appeared in the denominator, namely AgCl.

The concentration of Cl^- will decrease through the precipitation of AgCl alone, while the actual ratio reaches the value $10^{-2.52}$. In this case, the decrease in the $[Cl^-]$ is equivalent to $\frac{10^{-2.52}}{10^{-3}} = 10^{0.48} = 3.0$.

Hence in this case $\frac{2}{3}$ of the chloride will precipitate before simultaneous precipitation of AgCl and AgBr occurs.

The following example will illustrate the influence of total concentration on the order of precipitation of salts of different charge types.

Example 6.

In a mixture containing $CrO_4^=$ and Cl^- in a molar ratio of 1000:1, which salt will precipitate first on the addition of Ag^+, if the initial $[CrO_4^=]$ is (a) 1.0×10^{-4} M and (b) 1.0×10^{-2} M.

pK_{sp} of Ag_2CrO_4 = 10.71 and pK_{sp} of AgCl = 9.52.

(Assuming ionic strength = 0.1)

$$[Ag^+][Cl^-] = 10^{-9.52}$$

$$[Ag^+]^2[CrO_4^=] = 10^{-10.71}$$

$$\frac{[CrO_4^{=}]}{[Cl^-]^2} = \frac{10^{-10.71}}{(10^{-9.52})^2} = 10^{+8.33}$$

(a) Since $[CrO_4^{=}] = 1.0 \times 10^{-4}$, and $[Cl^-] = 1.0 \times 10^{-7}$

the ion ratio, $\frac{[CrO_4^{=}]}{[Cl^-]^2} = 10^{+10}$. Inasmuch as this value is greater than $10^{+8.33}$, Ag_2CrO_4 will precipitate first.

(b) Here, $[CrO_4^{=}] = 1.0 \times 10^{-2}$ and $[Cl^-] = 1.0 \times 10^{-5}$ and the ion ratio is $10^{+8.0}$ which is less than the value $10^{+8.33}$ required for simultaneous precipitation. Hence in this solution AgCl will precipitate first.

VIII-8. PURITY OF PRECIPITATES

Inasmuch as it is impossible to obtain a precipitate free of contamination in the conventional analytical procedures, it is of importance to describe the nature and extent of these impurities in order to evaluate and minimize them. Contamination by impurities from unsaturated solutions is termed coprecipitation. Coprecipitation is therefore distinguished from contamination arising through simultaneous precipitation (i.e. when for each of the substances precipitated, the ion product has exceeded the solubility product constant). Coprecipitation can occur through three general mechanisms: (a) adsorption, (b) solid solution formation and (c) ion entrapment.

VIII-8-(a). ADSORPTION

Adsorption refers to the contamination of the surface of a precipitate by foreign ions. This may occur either by molecular adsorption or by ion-exchange adsorption. In molecular adsorption, as for example the contamination of AgI by KI, an equivalent amount of both K^+ and I^- have been adsorbed. The extent of the adsorption can be described by the empirical Freundlich equation:

$$\frac{X}{m} = kC^{\frac{1}{n}}$$

where X = weight of the salt adsorbed, m = weight of the precipitate, C = concentration of the salt in the solution and k and n are constants.

Alternatively one of the lattice ions in the surface of the precipitate may exchange with a foreign ion in solution as for example the exchange of Ba^{++} in $BaSO_4$ with Pb^{++}. This type of exchange may be described by an equilibrium expression derived from the following equation:

$$BaSO_4 \text{ (solid)} + Pb^{++} \rightleftharpoons PbSO_4 \text{ (solid)} + Ba^{++}$$

from which,

$$K = \frac{[Ba^{++}][PbSO_4]_{solid}}{[Pb^{++}][BaSO_4]_{solid}}$$

In the usual equilibrium expression involving solids, no concentration terms appear for the solids on the assumption that they are perfectly pure. That is not the case here.

Obviously $\dfrac{[PbSO_4]_{solid}}{[BaSO_4]_{solid}}$ is proportional to $\dfrac{X}{m}$,

Hence, $\dfrac{X}{m} = K' \cdot \dfrac{[Pb^{++}]}{[Ba^{++}]}$

From this equation ion exchange adsorption may be seen to differ from molecular adsorption in its dependence upon the concentration of the lattice ion exchanged.

Ion exchange can also occur with precipitates whose surfaces already contain molecularly adsorbed salts. This is illustrated in the following equation:

$$AgI \cdot KI + H^+ \rightleftharpoons AgI \cdot HI + K^+$$

Such a situation would arise in the washing of an AgI precipitate with dilute HNO_3.

VIII-8-(b). SOLID SOLUTION FORMATION

If the contaminating salt crystallizes with the same lattice structure and with similar lattice constants as the precipitate, then solid solutions will form. This generally will have the effect of increasing coprecipitation. For example AgBr and AgCl form solid solutions. This results in the precipitation of AgCl from a mixture of Br^- and Cl^- before the solubility product constant of AgCl has been reached.

If the rate of equilibration in the solid crystal is sufficiently rapid then the precipitate will be homogeneous at all times and in equilibrium with the solution at all times. This situation is represented by the following equation:

$$\frac{[Br^-]_{final}}{[Cl^-]_{initial}} = K_D \cdot \frac{N_{AgBr}}{N_{AgCl}}$$

where K_D is a distribution coefficient and N represents the concentration in the solid in terms of mole fraction. However if the rate of equilibration in the solid phase is slow, then every small increment of solid formed will be in equilibrium with the solution composition at the time of its formation. Since the composition of the solution changes during the course of the precipitation, the precipitation would be heterogeneous. In this case the following equation first derived by Doerner and Hoskins applies:

$$\log \frac{[Br^-]_{final}}{[Br^-]_{initial}} = \lambda \log \frac{[Cl^-]_{final}}{[Cl^-]_{initial}}$$

where λ is identical to K_D but is used to distinguish the two types of solid solution equilibria.

VIII-8-(c). ION ENTRAPMENT

Foreign ion entrapment is a non-equilibrium process involving the rapid growth of precipitates around adsorbed ions. The nature and extent of contamination of this sort will be greatly affected by the order of mixing of the reagents. For example in the precipitation of $BaSO_4$, cation occlusion will predominate if the Ba^{++}-containing solution is added slowly to the solution containing $SO_4^=$. This follows from the fact that the precipitate is formed in the presence of an excess of $SO_4^=$. Obviously, if the reverse order of mixing reagents is employed, anion occlusion will predominate.

SUGGESTIONS FOR FURTHER READING

1. E. J. King, Qualitative Analysis and Electrolytic Solutions, Harcourt, Brace and World, Inc., New York (1959) Chapter 8.
2. H. A. Laitinen, Chemical Analysis, McGraw Hill Book Co., Inc., (1960), Chapters 6, 7, 8, 9 and 10.
3. G. Charlot, Qualitative Inorganic Analysis, Methuen and Co., Ltd., London, (1954) Chapters 9 and 10.

VIII-9. PROBLEMS*

1. Calculate the solubility of the following substances in water assuming that no side-reactions take place.
   (a) $Fe(OH)_3$ (b) CaC_2O_4 (c) $MgNH_4PO_4$ (d) Hg_2Cl_2
   (e) $Ce_2(C_2O_4)_3$
2. Arrange the following compounds in the order of decreasing solubility in water.
   AgBr, AgCl, AgI, Ag_2CrO_4, $AgIO_3$, AgSCN.
3. The solubility of lead iodate is 1.2×10^{-3} g./100 ml. Calculate its solubility product.
4. The solubilities of thallous phosphate, lead phosphate and ferric phosphate are 5.0, 0.00014 and 1.0 g. per liter. Calculate their concentration solubility products.
5. Calculate the concentrations of the metal ions in saturated solutions of (a) CuS, (b) HgS, (c) PbS, (d) $BaCrO_4$, (e) $CaCrO_4$, (f) $SrCrO_4$, (g) $PbCrO_4$.
6. Calculate the solubility of (a) AgCl, (b) $AgIO_3$, and (c) Ag_2CrO_4 in a 1.0×10^{-2} M solution of $AgNO_3$. (pK_{sp} values 9.66, 7.42, and 11.35 for (a), (b), and (c), respectively.)
7. List the following solutions in order of their effect on the solubility of $CaSO_4$ (i.e., starting from the one in which $CaSO_4$ is most soluble.)
   (a) 0.1 M $CaCl_2$ (b) 0.05 M NaCl (c) 0.1 M Na_2SO_4
   (d) 0.05 M Na_2SO_4 (e) pure water.
8. Calculate the solubility of Ag_2O in (a) 0.05 M $AgNO_3$
   (b) 0.05 M NaOH. $[Ag^+][OH^-] = K_{sp} = 10^{-7.39}$
9. Calculate the number of grams of calcium phosphate that will dissolve in 100 ml of the following solutions:
   (a) 0.02 M $Ca(NO_3)_2$ ($pK_{sp} = 25.75$) (b) 0.02 M Na_3PO_4 ($pK_{sp} = 24.87$)
10. A 0.001 M solution of $AgNO_3$ is added dropwise and with stirring to 1.0 liter of a solution which is 0.001 M in KCl as well as in K_2CrO_4. What compound precipitates first? Calculate the concentrations of all ions in solution when the second compound starts to precipitate. (pK_{sp} AgCl = 9.70, Ag_2CrO_4 = 11.49)
11. A 0.05 M solution of $AgNO_3$ is added to 500 ml of a solution containing 5.0 g. NaCl and 5.0 g. NaBr.
    (a) What is the concentration of bromide ions when AgCl starts to precipitate? (pK_{sp} AgCl = 9.52, AgBr = 12.05)

*Any possible Bronsted or Lewis acid-base interactions of the ions of the precipitated substances will be ignored in these problems.

(b) What is the concentration of chloride ions when AgBr starts to precipitate?

(c) What is the concentration of chloride ions and bromide ions when 500 ml of $AgNO_3$ have been added? (pK_{sp} AgCl = 9.48, AgBr = 12.01)

(d) Calculate the ratio, $\frac{[Cl^-]}{[Br^-]}$ when 1.0 liter of $AgNO_3$ has been added.
(pK_{sp} AgCl = 9.52, AgBr = 12.05)

(e) Calculate the ratio $\frac{[Cl^-]}{[Br^-]}$ when 2.0 liters of $AgNO_3$ has been added.
(pK_{sp} AgCl = 9.55, AgBr = 12.08)

12. Solid $AgNO_3$ is added slowly and with stirring to 1 liter of a solution which is 0.02 M in KI and 0.03 M in K_2CrO_4. Calculate the ratio $\frac{[I^-]}{[CrO_4^=]}$ when (a) 0.08 moles of $AgNO_3$ have been added, (b) 0.16 moles of $AgNO_3$ have been added.
(a) pK_{sp} AgI = 15.61, Ag_2CrO_4 = 10.96
(b) pK_{sp} AgI = 15.50, Ag_2CrO_4 = 10.63

13. Calculate the percentage of chloride that is unprecipitated when Ag_2CrO_4 starts to precipitate, on the addition of solid $AgNO_3$ to a solution that is 0.005 M in both chloride and chromate. (pK_{sp} AgCl = 9.64, Ag_2CrO_4 = 11.29)

14. Calculate the $[Ag^+]$ in a solution saturated with respect to both AgBr and AgCNS (pK_{sp} = 12.28 and 12.00, respectively.)

15. What concentration of Na_2CO_3 would be necessary to transform 1.0 millimole of $BaSO_4$ completely to $BaCO_3$ if the volume of the solution used is 100 ml? (Use the K_{sp} values tabulated in the appendix inasmuch as the ionic strength will affect the values of both K_{sp}'s to approximately the same extent.)

IX

Metal Complex Equilibria

IX-1. INTRODUCTION

The formation of metal coordination complexes can be explained in terms of the Lewis acid-base theory. A metallic cation being electron-pair deficient may be considered as a polybasic acid capable of reaction with several basic entities, whose number is related to the coordination number of the metal ion. The coordination number of a metal ion refers to the number of basic groups (ligands) that can arrange themselves around the central metal ion and depends largely on the size of the metal ion as well as the size of the coordinating groups. The basic entities are usually either neutral, as in the case of NH_3 or H_2O, or negatively charged, as with CN^-, OH^-, or halide ions.

The proton is not free but hydrated in aqueous solutions; therefore the strongest Bronsted acid in aqueous solution is the hydrated proton or hydronium ion H_3O^+ (leveling effect). Similarly the hydrated form of a metal ion will be the strongest Lewis acid of any of the complexes of this metal ion in aqueous media. For example in dealing with copper complexes the hydrated copper ion is the strongest possible Lewis acid and other copper ion complexes can be considered as weaker Lewis acids. This gives rise to a whole series of reference acids consisting of each of the hydrated metal ions in contrast to the single reference acid, the hydronium ion of the Bronsted system.

In addition to simple coordination complexes, in which metal ions combine with monofunctional (monodentate) ligands in a a number equal to their coordination number, e.g. $Ni(NH_3)_6{}^{++}$, $Fe(CN)_6^{3-}$ and $FeCl_4{}^-$, metal ions can interact with polyfunctional ligands, each capable of occupying more than one position in the coordination sphere of the metal. These complexes, called chelates, are very useful in analytical determinations and separations.

The charge of a metal complex, being the algebraic sum of the electric charges of the metal ions and the ligands, may be positive, e.g., $Zn(H_2O)_6{}^{++}$ or negative e.g. $CoCl_4^=$. Mixed ligand complexes, of course, can be neutral e.g. $ZnCl_2(H_2O)_4$. Chelates, may also be neutral since chelating agents may contain both neutral and anionic coordinating groups. For example the anion of 8-hydroxyquinoline which has one neutral and one anionic group forms neutral chelates with many metals whose coordination number is twice the charge on the metal ion.

e.g. $\left(\ldots\right)_2$Cu and $\left(\ldots\right)_3$Al

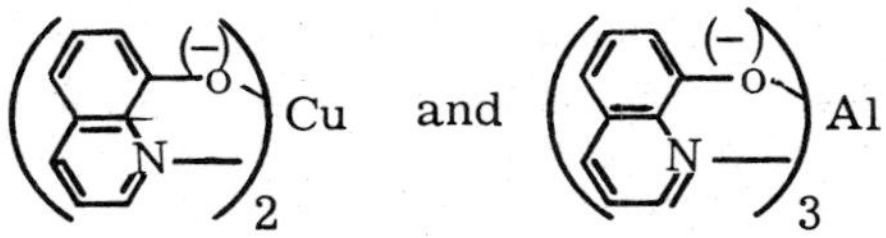

Neutral chelates are characterized by a low solubility in water and a significant solubility in organic solvents, and hence are widely used in metal separation processes such as precipitation or extraction.

Both positively and negatively charged complexes including chelates are useful as masking agents and, particularly in the case of chelates, also as titrants. Masking refers to the reduction of the concentration of a hydrated metal ion to a point at which the metal ion does not significantly participate in a reaction of interest, by virtue of formation of a sufficiently stable water-soluble complex. For example NH_3 will mask Ag^+ in the presence of Cl^- but not in the presence of I^-; CN^- is required to mask Ag^+ in the presence of I^-.

In general the overall formation constants of chelates are higher than those of similar simple coordination complexes. However there is a tremendous difference in the extent of complexation that can be best explained by reference to the equilibrium expressions for complex formation. Consider a particular metal ion M, having coordination number 4, that can react with a monodentate, ligand L, or a quadridentate chelating agent L' . Let us assume that the overall formation constant in each case is 10^{12} and that the equilibrium concentration of ligand in each case is 10^{-3} M. Then

$$\frac{[ML_4]}{[M][L]^4} = 10^{12} \text{ and } \frac{[ML']}{[M][L']} = 10^{12}$$

From these equations it may be seen that the ratio of complexed to free metal ion is unity in the case of the monodentate ligand, but 10^9 in the case of the chelate.

As mentioned in Chapter VIII, in equilibrium constants involving multiply charged species such as metal cations, etc. the effect of ionic strength and the need for activity corrections is greater than in the case of monoprotic acid-base equilibria. In each of these examples concentration constants that are appropriate for the ionic strengths, (usually 0.1) of the solutions are used.

One point of contrast between the treatment of Bronsted acid-base reactions and metal-complex reactions is that in the former, reactions and the corresponding equilibrium constants are conventionally written as dissociations. With metal complexes, however, reactions and equilibrium constants are more commonly written in the reverse manner as formations.

With the exception of this minor difference the treatment of metal-complex equilibrium calculations follows the same pattern developed for proton transfer equilibria. The difference in nomenclature or symbols used in describing metal-complex systems should not obscure this fundamental similarity. For example the notation pM, the negative logarithm of the concentration of a metal ion M^{++}, is used to denote the Lewis acidity level just as the pH is used as a measure of the Bronsted acidity level. Further, it is necessary to consider stepwise equilibrium constants in metal-complex formation analogous to the calculations involved in polybasic Bronsted acid dissociations. Successive formation constants of metal complexes are in general much closer than successive Bronsted acid dissociation constants. This results in calculations of greater difficulty. For example, although it is possible to calculate the hydrogen ion concentration in a solution of H_2S by ignoring the second dissociation constant, it is not possible to calculate the concentration of free NH_3 in a solution containing $Cu(NH_3)_4^{++}$ without taking the several stepwise formation constants into account. However these calculations can be relatively easily handled by a technique involving the definition of a set of β values that represent the fractions of the total concentration present as each complex species in a manner exactly analogous to the use of α values described in Sec. VI-4.

IX-2. GENERAL METAL-COMPLEX EQUILIBRIA

We may represent a generalized metal complex formation by the following equation from which charges have been omitted for convenience.

$$M + L \rightleftharpoons ML$$

where M is a metal ion and L is a ligand. The equilibrium expression corresponding to this reaction is:

$$k_1 = \frac{[ML]}{[M][L]}$$

If more than one ligand is bound to the metal ion the stepwise formations and their corresponding equilibrium constants can be represented similarly as follows:

$$ML + L \rightleftharpoons ML_2; \quad k_2 = \frac{[ML_2]}{[ML][L]}$$

$$ML_2 + L \rightleftharpoons ML_3; \quad k_3 = \frac{[ML_3]}{[ML_2][L]}$$

$$ML_{n-1} + L \rightleftharpoons ML_n; \quad k_n = \frac{[ML_n]}{[ML_{n-1}][L]}$$

By combining all of these equations for the reaction steps an overall metal-complex formation reaction and the overall formation constant expression may be written

$$M + nL \rightleftharpoons ML_n; \quad K_f = \frac{[ML_n]}{[M][L]^n}$$

It will be seen that, $K_f = k_1\ k_2\ k_3 \ldots\ldots k_n$ (IX-1)

A mistake that should be avoided is the incorrect interpretation of the overall formation equation as indicating that for each mole of M there will be in solution n moles of L. This is of course as absurd as the assumption that in a solution of H_2S there are 2 moles of H^+ for every mole of $S^=$. The treatment of metal-complex equilibria must be considered in the same manner as the polybasic acid equilibria.

For this purpose let us define a set of fractions β_0 to β_n which will represent the ratios of the concentrations of the metal containing species to the analytical concentration of the metal, C_M.

Thus: $\beta_0 = \frac{[M]}{C_M}$

$$\beta_1 = \frac{[ML]}{C_M}$$

$$\beta_2 = \frac{[ML_2]}{C_M}$$

$$\beta_n = \frac{[ML_n]}{C_M} \qquad \text{(IX-2)}$$

and $\beta_0 + \beta_1 + \beta_2 \text{-----} + \beta_n = 1$ (IX-3)

Incorporating the equilibrium constant expressions in (IX-2), we obtain expressions for β values that are functions of the equilibrium constants and the equilibrium ligand concentration. (Refer Sec. VI-4 for derivation.)

$$\beta_0 = \frac{1}{1 + k_1[L] + k_1 k_2[L]^2 + \text{----} + K_f[L]^n}$$

$$\beta_1 = \frac{k_1[L]}{1 + k_1[L] + k_1 k_2[L]^2 + \text{----} \; K_f[L]^n}$$

and

$$\beta_n = \frac{K_f[L]^n}{1 + k_1[L] + k_1 k_2[L]^2 + \text{----} \; K_f[L]^n} \qquad \text{(IX-4)}$$

These equations are very useful for obtaining the concentrations of all species present in solution provided that the equilibrium concentration of the ligand is known.

IX-2-(a) CASES WHERE THE EQUILIBRIUM CONCENTRATION OF LIGAND IS KNOWN

Example 1.

Calculate the concentrations of all metal containing species in Cu^{++} - NH_3 mixtures containing the following equilibrium concentrations of NH_3: 10^{-5} M, 10^{-4} M, 10^{-3}M, 10^{-2} M and 10^{-1} M. In each case the total copper ion concentration is 10^{-2} M. The stepwise formation constants for copper ammine complexes are: Log k_1 = 4.31, Log k_2 = 3.67; Log k_3 = 3.04 and Log k_4 = 2.30. (If higher concentrations of NH_3 were included it would be necessary to consider the formation of $Cu(NH_3)_5^{++}$. Log k_5 = − 0.52. The ionic strength of the solution is 0.2 using NH_4Cl which also serves the purpose of keeping the pH sufficiently low to prevent the formation of copper-hydroxy complexes in the range of low $[NH_3]$).

The answers required for this problem are obtained by substituting the appropriate values of $[NH_3]$ and the stepwise constants in equations IX-4. For example, when $[NH_3] = 10^{-3}$,

$$\beta_2 = \frac{10^{4.31} \times 10^{3.67} \times (10^{-3})^2}{1+10^{4.31}\times 10^{-3}+10^{7.98}\times 10^{-6}+10^{11.02}\times 10^{-9}+10^{13.32}\times 10^{-12}}$$

$$= 0.39.$$

Therefore, $[Cu(NH_3)_2^{++}] = \beta_2 \cdot C_M = 0.39 \times 0.01 = 3.9 \times 10^{-3}$ M. The remainder of the β values below can be obtained in a similar manner.

$[NH_3]$	β_0	β_1	β_2	β_3	β_4
10^{-5} M	0.82	0.17	8.0×10^{-3}	8.7×10^{-5}	1.7×10^{-7}
10^{-4} M	0.24	0.50	0.23	2.6×10^{-2}	5.1×10^{-4}
10^{-3} M	4.1×10^{-3}	8.4×10^{-2}	0.39	0.43	8.6×10^{-2}
10^{-2} M	3.1×10^{-6}	6.3×10^{-4}	2.9×10^{-2}	0.32	0.65
10^{-1} M	4.6×10^{-10}	9.3×10^{-7}	4.3×10^{-4}	4.8×10^{-2}	0.95

Metal complex equilibria may also be conveniently described by graphical means. In Fig. (IX-1) the variation of the concentrations of all the copper-ammonia complexes is shown as a function of the logarithm of the concentration of ammonia. The value of C_M in this diagram is 10^{-2} M. Diagrams for other total concentration values of copper will have precisely the same shape since the shape of the curve is independent of the metal ion concentration but will be displaced vertically by an appropriate amount. It is interesting to note the formal resemblance of this diagram to those of polybasic Bronsted acids such as that shown in Fig. VI-2. The construction of these diagrams is somewhat more difficult than that for most Bronsted acid-base systems in which the construction can be achieved mainly by connecting straight lines. The successive formation constants usually have values that are so close together that in a number of regions,

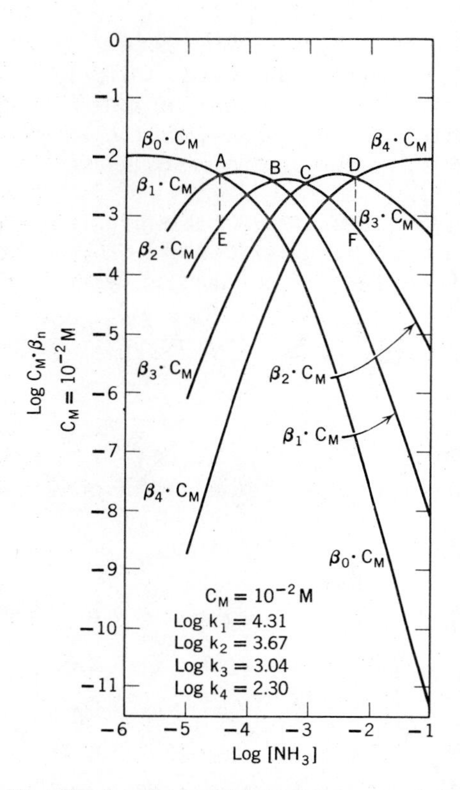

Fig. IX-1 — Logarithmic Diagram of the Copper-Ammonia System

significant concentrations of several species are involved, necessitating point by point plotting from calculated β values.

Some of the effects of closely spaced constants can be illustrated with the help of Fig. (IX-1). For example it will be noticed that the concentrations of the intermediate complexes, $Cu(NH_3)^{++}$, $Cu(NH_3)_2^{++}$, and $Cu(NH_3)_3^{++}$ never reach C_M, that is, the value of 10^{-2} M, since there are always significant fractions of the total copper, (C_M), present as other species. It can be shown that unless two successive constants are at least 3.2 log units apart the intermediate species will not reach a maximum concentration of $0.95 \times C_M$.

In analogy with the Bronsted acid-base diagrams we would expect that the cross-points A, B, C and D would correspond to values of Log $[NH_3]$ that correspond to the $-$ log k values.

This may be seen to be exactly true for all of the crosspoints.

The silver-ammonia complex system involves an unusual feature, namely the second stepwise formation constant is greater than the first. Although this does not affect the nature of the calculations described above for the copper system, it is interesting to see how the graphical representation of this system differs from the copper system. The curves in Fig. IX-2 were obtained in a manner exactly analogous to that described

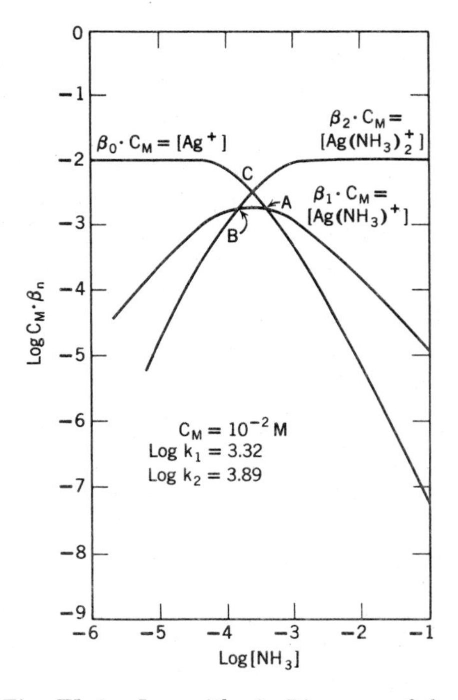

Fig. IX-2 — Logarithmic Diagram of the Silver-Ammonia System

above in a solution containing a total silver concentration of 0.01M using the following stepwise formation constants: $\log k_1 = 3.32$ and $\log k_2 = 3.89$.

As a result of $\log k_1$ being less than $\log k_2$, it can be seen from Fig. IX-2 that there is no ligand concentration at which the intermediate complex, $Ag(NH_3)^+$, predominates. Further, it can be seen that cross-points A and B correspond to pNH_3 values equal to $\log k_1$ and $\log k_2$ respectively and that point C at which $[Ag^+] = [Ag(NH_3)_2^+]$ is at a pNH_3 equal to the average of Log k_1 and Log k_2.

The diagram also clearly shows that it is in a rather narrow range of free ammonia concentration (from approximately 10^{-3} M to $10^{-4.5}$ M) that it is necessary to consider more than one species of stoichiometric significance. If the equilibrium ammonia concentration is greater than 10^{-3} M, this species is $Ag(NH_3)_2^+$ whereas in solutions containing less than $10^{-4.5}$ M ammonia, it is Ag^+.

In the region where the $Ag(NH_3)_2^+$ is the only one of stoichiometric significance, it can be readily shown that the slope of the variation of $\log [Ag(NH_3)^+]$, and $\log [Ag^+]$ with $\log [NH_3]$ is -1 and -2 respectively. Similarly in the region where $[Ag^+]$ predominates, the corresponding slopes for $Ag(NH_3)^+$ and $Ag(NH_3)_2^+$ are +1 and +2 respectively.

Having considered two real cases in which stepwise formation constants increase (a rare occurrence), or decrease, let us now examine a hypothetical case in which the two stepwise constants are identical. As shown in Fig. IX-3 the intermediate complex never predominates but rises to a maximum β value equal to 1/3 at point A, at the same time the other two species do. Point A, it will be noticed, has a pL concentration (where L is the ligand) equal to $\log k_1$.

From Figs. IX-1 and IX-2 we may conclude that the curve for ML in Fig. IX-3 would reach a maximum at a point above point A if $\log k_1 > \log k_2$, and would have reached its maximum below point A if $\log k_1 < \log k_2$. In either of these two situations, the intersections of the ML^+ curve with those of M^{++} and ML_2 correspond to $pL = \log k_1$ and $\log k_2$ respectively.

Up to this point we have considered cases of monodentate ligands. However as far as the calculations are concerned, precisely the same methods may be used for calculating β values and obtaining graphs for metal chelate systems such as the copper-ethylenediamine system shown in Fig. IX-4.

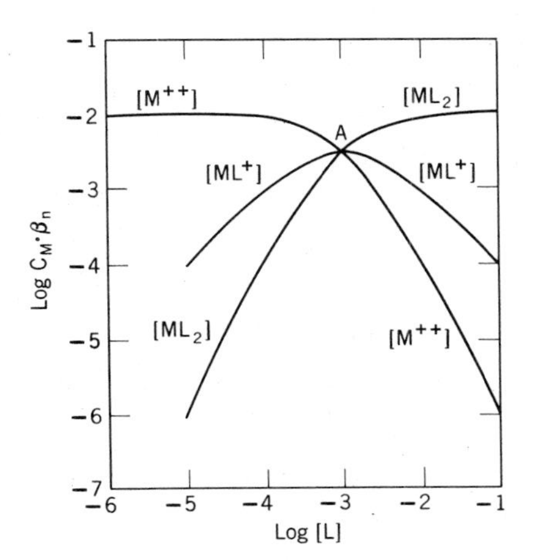

Fig. IX-3 — Logarithmic Diagram of a Metal Ligand System. $C_M = 10^{-2}$ M, Log k_1 = Log k_2 = 3.0.

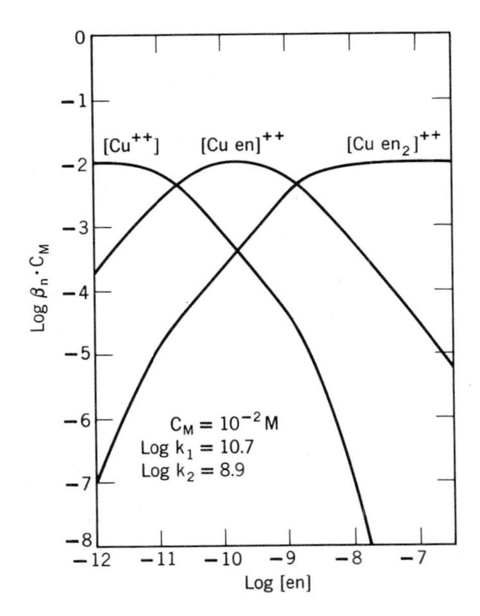

Fig. IX-4 — Logarithmic Diagram of the Copper-ethylenediamine System

It is instructive to compare the copper-ammonia and the copper-ethylenediamine systms. The greater stability of chelates over simple coordination compounds is dramatically seen in the much lower ligand concentration, 10^{-11} M, as compared to 10^{-5} M for NH_3, at which the free copper ion concentration begins to drop. It will also be observed that the greater separation of the stepwise constants than is observed with simple coordination complexes results in a simplified diagram that clearly resembles Fig. VI-1 for a dibasic acid. Of course Figs. IX-1, 2 and 3 also resemble those corresponding to di- and polybasic Bronsted acids having overlapping pK_a values. In this connection see Fig. VI-2. The improvement in the stability of copper complexes achieved in going from ammonia, a monodentate ligand, to ethylenediamine, a bidentate ligand, is further increased with the quadridentate ligand "trien," trimethylenetetramine, (NH_2 $CH_2CH_2NHCH_2CH_2NHCH_2CH_2CH_2NH_2$). In this system, Fig. IX-5 the concentration of free copper ion begins to drop at a free ligand concentration below 10^{-20} M. Since only the 1:1 complex is formed in this case, the diagram is considerably simplified. Metal-EDTA complexes have diagrams similar to this.

IX-2-(b). CASES WHERE THE EQUILIBRIUM CONCENTRATION OF LIGAND IS NOT KNOWN

In general we do not immediately know the equilibrium concentration of a ligand, but we must calculate this from the data usually available, namely the analytical composition of the system. In most cases of analytical interest, metal complexes are formed in the presence of an excess of the ligand which simplifies the calculation considerably. In such situations it is possible to assume that the metal ion is in the most highly complexed form and further that the free ligand concentration is equal to the excess ligand concentration. The validity of these assumptions can be illustrated in the following example.

Example 2.

Calculate $[Ag^+]$ in a solution obtained by mixing 1.0×10^{-2} moles $AgNO_3$ and 0.10 moles NH_3 in a manner to give one liter of solution. ($\log k_1 = 3.32$ and $\log k_2 = 3.89$ for the stepwise formation constants of $Ag(NH_3)_2^+$).

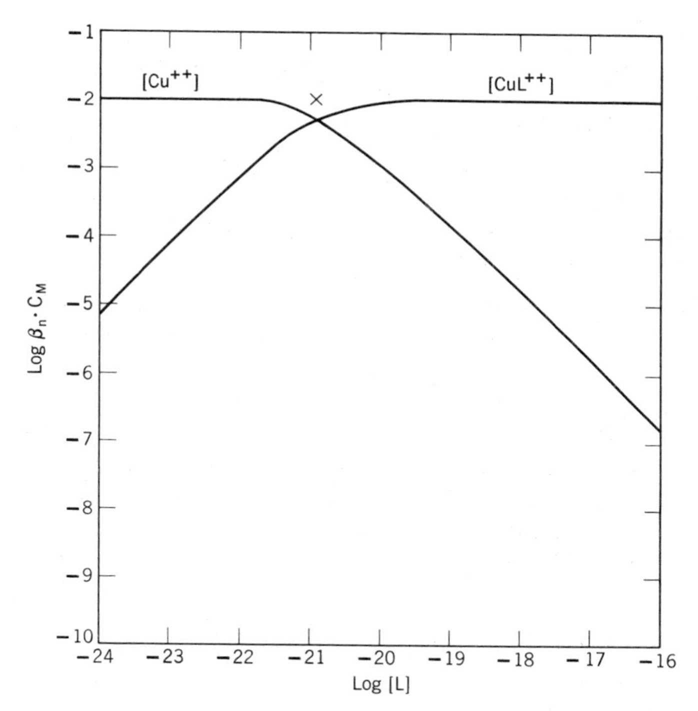

Fig. IX-5 — Logarithmic Diagram of the Copper-trien System. $C_M = 10^{-2}$ M, $\log k_1 = 20.4$

Since NH_3 is present in considerable excess we will assume that essentially all of the Ag^+ is present as $Ag(NH_3)_2^+$. Hence the free NH_3 concentration is $0.10 - 2 \times 1.0 \times 10^{-2} = 0.08$ M; $(-\text{Log}[NH_3] = 1.10)$.

Since in this case we already have a graphical representation of the silver-ammonia system available (Fig. IX-2), let us first solve the problem graphically. As a matter of fact an examination of Fig. IX-2 serves to confirm the validity of our assumption that the concentrations of Ag^+ and $Ag(NH_3)^+$ are much smaller than the concentration of $Ag(NH_3)_2^+$.

The value of $[Ag^+]$ when $-\text{Log}[NH_3] = 1.10$, is seen from Fig. IX-2 to be $10^{-7.00}$ M.

Alternatively $[Ag^+]$ in this problem can be calculated in the following manner:

Mass Balance for NH_3:

$$0.10 = [NH_3] + [NH_4^+] + [Ag(NH_3)^+] + 2[Ag(NH_3)_2^+]$$

Mass Balance for Ag:

$$0.01 = [Ag^+] + [Ag(NH_3)^+] + [Ag(NH_3)_2^+]$$

Since $[NH_3]$ is in excess, $[Ag(NH_3)_2^+] \gg [Ag(NH_3)^+]$ and $[Ag^+]$, $0.01 = [Ag(NH_3)_2^+]$.

From this and the assumption that $[NH_4^+]$ is negligible, (the extent of dissociation of NH_3 is under 5% at this concentration),

$$[NH_3] = 0.10 - 2 \times 0.01$$
$$= 0.08 M.$$

Since $[NH_3]$ and $[Ag(NH_3)_2^+]$ are known we can use the overall formation expression for $Ag(NH_3)_2^+$.

$$K_f = k_1 k_2 = \frac{[Ag(NH_3)_2^+]}{[Ag^+][NH_3]^2}$$

$$\text{i.e., } 10^{3.32} \times 10^{3.89} = \frac{0.01}{[Ag^+] \times (0.08)^2}$$

$$\therefore \quad [Ag^+] = 10^{-7.01} M$$

The slight discrepancy between the graphical solution and the numerical solution is due to the error in reading the graph.

If the ligand is not present in considerable excess the simplifications employed above cannot be used and it becomes necessary to solve exact equations. Although these equations may be readily derived in the usual manner, their solutions are obtained with difficulty. Hence approximate solutions obtained graphically are of interest.

Example 3(a).

Calculate $[Cu^{++}]$ in a solution obtained by mixing 1.0×10^{-2} moles $Cu(NO_3)_2$ and (a) 1.0×10^{-2} moles, (b) 2.0×10^{-2} moles and (c) 4.0×10^{-2} moles NH_3, so that one liter of solution is obtained in each case.

In problems of this sort, equations which express a ligand balance are very helpful. The ligand balance is derived in a manner exactly analogous to the proton balance. One side of the equation will have the concentrations of substances which release ligand upon formation from the substance considered to be originally present, and the other side will have the ligand concentration and those of substances which consume ligand upon formation. Thus for a solution of $Cu(NH_3)^{++}$, we may write:

$$[Cu^{++}] = [NH_3] + [Cu(NH_3)_2{}^{++}] + 2[Cu(NH_3)_3{}^{++}] + 3[Cu(NH_3)_4{}^{++}] \quad \text{(IX-5)}$$

Or for a solution of $Cu(NH_3)_2{}^{++}$:

$$2[Cu^{++}] + [Cu(NH_3)^{++}] = [NH_3] + [Cu(NH_3)_3^{++}] + 2[Cu(NH_3)_4{}^{++}] \quad \text{(IX-6)}$$

and for a solution of $Cu(NH_3)_4{}^{++}$:

$$4[Cu^{++}] + 3[Cu(NH_3)^{++}] + 2[Cu(NH_3)_2{}^{++}] + [Cu(NH_3)_3{}^{++}] = [NH_3] \quad \text{(IX-7)}$$

These relationships can be verified if desired from the two mass balance equations for copper and ammonia.

(a) This mixture is equivalent to a 0.01 M solution of $Cu(NH_3)^{++}$. The appropriate ligand balance equation IX-5 simplifies to:

$$[Cu^{++}] = [Cu(NH_3)_2{}^{++}] + 2[Cu(NH_3)_3{}^{++}]$$

i.e. Fig. IX-1 shows that the concentrations of $Cu(NH_3)_4{}^{++}$ and NH_3 are considerably lower than those of the substances on the right-hand-side of the equation at the point at which $[Cu^{++}] = [Cu(NH_3)_2{}^{++}] = 10^{-2.60}$ M, the approximate solution. At this point, $[Cu(NH_3)_3{}^{++}] = 10^{-3.60}$ M. As seen from the ligand balance equation, we must multiply $[Cu(NH_3)_3{}^{++}]$ by 2; this effectively raises the $Cu(NH_3)_3{}^{++}$ line by 0.30 and therefore to within 0.70 of the $Cu(NH_3)_2^{++}$ line. A new line is drawn parallel to the $Cu(NH_3)_2{}^{++}$ line and $(1 + 10^{-.70})$ or 0.08 units higher. Now the point of intersection with the copper line, and the solution to the problem, is:

$[Cu^{++}] = 10^{-2.55}$ M. Note also that $[NH_3] = 10^{-4.10}$ M.

(b) 0.01 M $Cu(NH_3)_2{}^{++}$.

The approximate solution of the ligand balance equation IX-6 is:

$$[Cu(NH_3)^{++}] \approx [Cu(NH_3)_3{}^{++}]$$

At this point $[Cu^{++}] = 10^{-3.65}$ M.

Since this is also the approximate concentration of $Cu(NH_3)_4^{++}$, these two terms balance each other out in Equation IX-6. At the point of approximate solution, $[NH_3]$ is 0.45 units below. Hence a new line $(1 + 10^{-0.45})$ or 0.15 units above the $Cu(NH_3)_3^{++}$ line is drawn. This new line meets the $Cu(NH_3)^{++}$ line at $[NH_3] = 10^{-3.40}$ at which $[Cu^{++}] = 10^{-3.55}$ M.

(c) .01 M $Cu(NH_3)_4^{++}$.

The approximate solution of equation (IX-7) is

$[Cu(NH_3)_3^{++}] = [NH_3] = 10^{-2.30}$ M.

However, $2[Cu(NH_3)_2^{++}]$ is within 0.70 units of the $Cu(NH_3)_3^{++}$ line. A new line $(1 + 10^{-.70})$ or 0.08 units higher must be drawn. This intersects the NH_3 line at $10^{-2.23}$, at which $[Cu^{++}] = 10^{-6.90}$ M.

IX-3. EFFECT OF pH ON METAL-COMPLEX EQUILIBRIA

Many metal complexing agents (Lewis bases) are also Bronsted bases and will therefore be affected by changes in pH. For example, NH_3, CN^-, $NH_2CH_2CH_2NH_2$ and the EDTA anion will accept protons and hence the fraction of their total concentrations which are available for metal complexation, vary with the pH. The calculation of these fractions, which has been previously described, (Chapter VI, Sec. 4), can be readily incorporated into metal complex equilibrium calculations. This approach can be illustrated with reference to the silver-ammonia system. The variation of $[NH_3]$ with pH can be expressed as:

$$[NH_3] = \alpha_1 \cdot C_L$$

where C_L is the total uncomplexed NH_3 concentration assumed to be sufficiently large to permit neglect of any intermediate complex formation.

$$\alpha_1 = \frac{K_a}{K_a + [H^+]} \qquad \text{(See Example 7, Chapter VI)}$$

Substituting $C_L\alpha_1$ for $[NH_3]$ in the expression for K_f we obtain,

$$K_f = \frac{[Ag(NH_3)_2^+]}{[Ag^+](\alpha_1 \cdot C_L)^2}$$

This equation may be rearranged to give:

$$K_f' = K_f \cdot \alpha_1^2 = \frac{[Ag(NH_3)_2^+]}{[Ag^+]\ C_L^2}$$

Here K_f' is called a conditional formation constant, whose value depends on the pH of the solution but naturally does not depend on the concentration of NH_3 and Ag^+.

An illustration of a more general case can be made using the Cu^{++} - NH_3 system. Here, employing equations IX-2, and 4

$$[Cu^{++}] = \beta_0\, C_M \quad (C_M = \text{total copper concentration}).$$

$$= \frac{C_M}{1 + k_1[NH_3] + k_1k_2[NH_3]^2 + k_1k_2k_3[NH_3]^3 + k_1k_2k_3k_4[NH_3]^4 + k_1k_2k_3k_4k_5[NH_3]^5}$$

As before we may substitute $\alpha_1 C_L$ wherever $[NH_3]$ appears. Hence

$$[Cu^{++}] = \frac{C_M}{1 + (k_1\alpha_1)C_L + (k_1k_2\alpha_1^2)C_L^2 + (k_1k_2k_3\alpha_1^3)C_L^3 + (k_1k_2k_3k_4\alpha_1^4)C_L^4 + (k_1k_2k_3k_4k_5\alpha_1^5)C_L^5}$$

It should be noticed that this expression can be obtained by replacing every stepwise constant, k_i, in equation IX-4 by $k_i \times \alpha_n$, the conditional constant. The presence of higher powers of α in the equation arises in those terms having more than one stepwise formation constant. Thus, for $k_1k_2k_3$ substitute $k_1\alpha_n k_2\alpha_n k_3\alpha_n$ or $k_1k_2k_3\alpha_n^3$. The symbol α_n refers to the appropriate α value; when the ligand is a polyacidic Bronsted base, e.g. for $C_2O_4^=$ complexes, α_2 is used, and for EDTA complexes, α_4 is used. (Fig. IX-6).

Example 3 (b).

Calculate $[Zn^{++}]$ in a solution containing 10^{-4} M zinc nitrate and 0.028 M total uncomplexed ammonia and having a pH of 9.0. The stepwise formation constants for zinc-ammonia complexes are $10^{2.27}$, $10^{2.34}$, $10^{2.40}$ and $10^{2.05}$; pK_a for NH_4^+ is 9.26.

$$[Zn^{++}] = \beta_0 \cdot C_M;\ \alpha_1 = \frac{10^{-9.26}}{10^{-9.26} + 10^{-9.0}} = 0.36$$

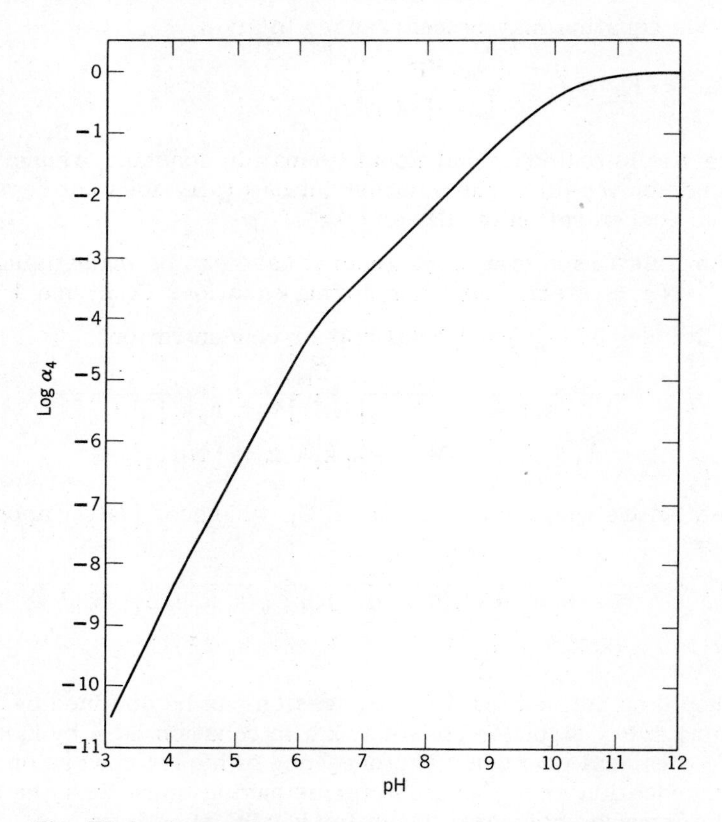

Fig. IX-6 — Logarithmic Diagram for the Variation of α_4 for EDTA with pH

$$\beta_0 = \frac{1}{1 + 10^{2.27}(0.36\times 0.028) + 10^{2.27}\times 10^{2.34}(0.36\times 0.028)^2 + 10^{2.27}\times 10^{2.34}\times 10^{2.40}(0.36\times 0.028)^3 + 10^{2.27}\times 10^{2.34}\times 10^{2.40}\times 10^{2.05}(0.36\times 0.028)^4}$$

$$\beta = \frac{1}{28.1} = 3.56\times 10^{-2}$$

Hence $[Zn^{++}] = 3.56\times 10^{-6}$ M.

Example 4.

Calculate the concentration of $[Ni^{++}]$ in a solution containing a total uncomplexed EDTA concentration, C_{H_4Y}, of 10^{-2} M and a total metal ion concentration of 10^{-4} M, at pH 6.0. Log K_f for the Ni^{++}-EDTA complex, ($NiY^=$) is 18.62.

Since EDTA, (H_4Y) is a polydentate ligand and forms a 1:1 chelate with most metal ions, the equilibrium expression is simply:

$$K_f = \frac{[NiY^=]}{[Ni^{++}][Y^{-4}]}$$

For $[Y^{-4}]$ we may write: $\alpha_4 \cdot C_{H_4Y}$, substituting this in the formation constant expression we can obtain in an expression involving the conditional formation constant, K_f .

$$K_f' = K_f \cdot \alpha_4 = \frac{[NiY^=]}{[Ni^{++}] \cdot C_{H_4Y}}$$

The mass balance for nickel,

$$10^{-4} = [Ni^{++}] + [NiY^=]$$

can be simplified to $10^{-4} = [NiY^=]$
since in this example K_f' will be seen to be large and further we have a reasonably large excess of ligand.

$$K_f' = 10^{18.62} \times 10^{-4.66} = 10^{13.96}$$

In this case it was convenient to evaluate α_4 using Fig. IX-6.

$$\text{Hence } [Ni^{++}] = \frac{[NiY^=]}{K_f' \cdot C_{H_4Y}}$$

$$= \frac{10^{-4}}{10^{13.96} \times 10^{-2}}$$

$$= 10^{-15.96}$$

Since values of K_f for a whole series of metal complexes with EDTA will differ in the appropriate K_f values (which are pH independent), it is possible to use Fig. IX-6 as a template for the convenient construction of an entire series of conditional constants of these complexes. Such a set of curves (Fig. IX-7) is useful if many calculations are involved.

One of the ways in which the usefulness of metal chelating agents can be compared is by examination of their respective K_f values. A more revealing comparison of these reagents, most of which are involved in acid-base reactions, is made by using the respective conditional constants, K_f'. For example a comparison of the reagents 'trien' and EDTA reveals that the former gives a more stable copper chelate ($K_f = 10^{20.5}$) than does EDTA ($K_f = 10^{18.80}$) . The variation of the respective α_4

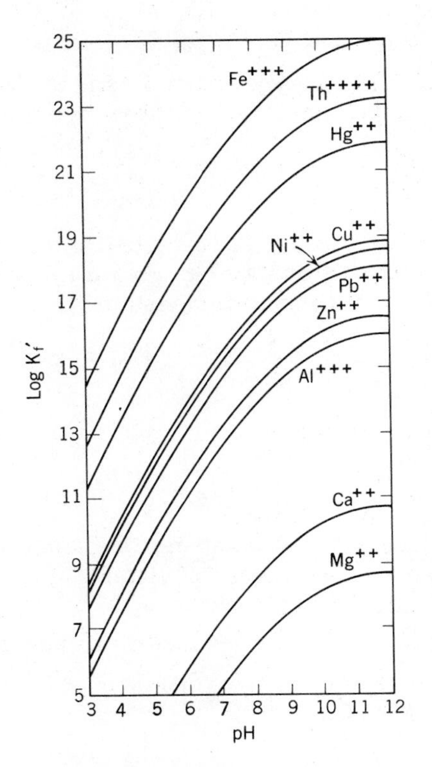

Fig. IX-7 — Conditional Constants for EDTA-Metal Complexes

values for these two ligands is such as to reverse the order of conditional stability constants at pH values below 7. This may be clearly seen from Fig. IX-8 in which the respective conditional constants (K_f') are plotted as a function of pH.

IX-4. METAL COMPLEX EQUILIBRIA INVOLVING MORE THAN ONE COMPLEXING AGENT

The hydrated metal cation can of course react with any substance present in the solution which behaves as a Lewis base. Such substances may include the hydroxide ion or buffer components (e.g. NH_3). In some instances auxiliary complexes are intentionally added to mask certain metal ions and thereby increase the

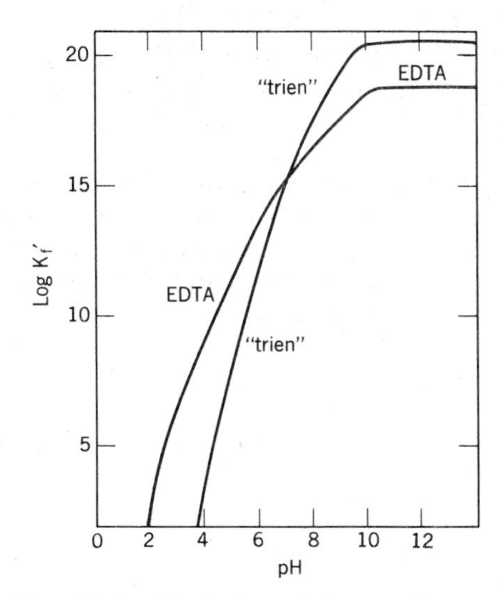

Fig. IX-8 — Variation of the Conditional Stability Constants of EDTA and "trien" with pH

selectivity of the complexation reaction. For example the addition of CN^- to a mixture of Ni^{++} and Ca^{++} will permit the selective reaction of EDTA with Ca^{++} by forming a sufficiently strong complex with Ni^{++} to prevent any appreciable reaction of this ion with EDTA. The role of masking in precipitation reactions will be considered in Chapter X, Sec. 2-(b).

The presence of auxiliary complexing reagents can be most readily taken into account in metal-complex equilibria calculations by use of the conditional equilibrium constant in a manner similar to that employed in the previous section. Just as before, where the total ligand concentration was corrected by means of an α value representing the fraction of the ligand that reacts, now β values will be used to correct the available metal ion concentration.

Example 5.

Calculate the total concentration of Ni^{++} unreacted with EDTA in a solution containing a total uncomplexed EDTA concentration (C_{H_4Y}) of .005 M, * a total nickel concentration of 10^{-2} M and 10^{-1} M NH_3 having enough NH_4Cl to give a pH of 9.0. (The logarithms of the stepwise constants for the nickel-ammine complexes are 2.75, 2.20, 1.69, 1.15, 0.71, and -0.01. Log K_f for $NiY^{=}$ = 18.62.)

This problem differs from Example 4 in the need to consider the Ni^{++}-NH_3 complexes in addition to the simple hydrated nickel ion. This use of β, the ratio of the hydrated nickel ion concentration to the total concentration of nickel not complexed with EDTA, simplifies the calculation.

$$\beta_0 = \frac{[Ni^{++}]}{C_{Ni} - [NiY^{=}]}$$

Incorporating this expression for β into the form developed earlier, we have:

$$K_f = \frac{[NiY^{=}]}{\beta_0 \{C_{Ni} - [NiY^{=}]\} \alpha_4 C_{H_4Y}}$$

or

$$K_f' = K_f \alpha_4 \beta_0 = \frac{[NiY^{=}]}{(C_{Ni} - [NiY^{=}])C_{H_4Y}}$$

The reason the term (C_{Ni} − $[NiY^{=}]$) is used in place of the total nickel concentration, C_{Ni}, is that the EDTA-Ni complex is so stable (K_f' is very large here), in comparison to the NH_3-Ni complexes, that we may assume that essentially all of the nickel will be in the form of $NiY^{=}$, i.e. $C_{Ni} \approx [NiY^{=}]$. This assumption should always be checked from the final answer.

From Fig. IX-6, at pH 9.0, $\alpha_4 = 10^{-1.30}$.

β_0 may be calculated in a manner similar to that shown in example 3. Here,

$$\beta_0 = \frac{1}{1 + 56 + 492 + 4350 + 6160 + 3170 + 320}$$
$$= 10^{-4.17}$$

* Alternatively the total EDTA concentration, 1.5×10^{-2} M may be given, in which case it would be necessary to subtract from this value the EDTA consumed in the reaction with nickel. The student should always be careful to distinguish between total and equilibrium concentrations.

Hence $K_f' = 10^{18.62} \times 10^{-1.30} \times 10^{-4.17}$

$= 10^{13.15}$

$$\text{Therefore } 10^{13.15} = \frac{NiY^{=}}{\{C_{Ni} - [NiY^{=}]\}\, C_{H_4Y}} = \frac{10^{-2}}{\{C_{Ni} - [NiY^{=}]\} \times .005}$$

and the total concentration of nickel unreacted with EDTA, $C_{Ni} - [NiY^{=}] = 10^{-12.85}\ M = 1.4 \times 10^{-13}\ M$.

Suppose however we wish to know the $[Ni^{++}]$. This is given by

$$[Ni^{++}] = \beta_0\ \{C_{Ni^{++}} - [NiY^{=}]\}$$
$$= 10^{-4.17} \times 10^{-12.85}$$
$$= 10^{-17.02}\ M$$

It should be noticed that the value for $[Ni^{++}]$ is independent of the value of β_0 since the quantity $(C_{Ni^{++}} - [NiY^{=}])$ is inversely proportional to K_f' which in turn is proportional to β_0. This implies that the pNi of a mixture of $NiY^{=}$ and excess EDTA does not change upon the addition of NH_3, (provided of course that the pH which affects α_4, is kept constant). Hence such a mixture is said to act as a metal buffer (see Chapter XV).

SUGGESTIONS FOR FURTHER READING

1. A. Ringbom, in Treatise on Analytical Chemistry, Editors I. M. Kolthoff and P. J. Elving. Interscience Publishers (1959), Part I, Vol. I, Chapter 14.
2. S. Chaberek and A. E. Martell, Organic Sequestering Agents, John Wiley and Sons, Inc. (1959).
3. H. A. Laitinen, Chemical Analysis, McGraw Hill Book Co. Inc. (1960), Chapter 13.
4. L. G. Sillen, in Treatise on Analytical Chemistry, Editors I. M. Kolthoff and P. J. Elving, Interscience Publishers (1959), Part I, Vol. I, Chapter 8.

IX-5. PROBLEMS

1. What is the concentration of cadmium ion in a solution that contains 0.02M $Cd(NH_3)_4(NO_3)_2$ and 3.0M NH_3 ? ($\log k_1 = 2.51$, $\log k_2 = 1.96$, $\log k_3 = 1.30$, $\log k_4 = 0.79$)

2. (a) Calculate the silver ion concentration in a solution of 0.01M $AgNO_3$ which contains (1) 0.1M NH_3 (2) 0.5M NH_3 (3) 1.0M NH_3 ($\log k_1 = 3.32$ and $\log k_2 = 3.89$)
   (b) Plot pAg vs log [NH_3] and explain the slope of the line. Would the slope be the same at low [NH_3] ?
3. (a) Calculate the cadmium ion concentration in a solution of 0.05M $Cd(NO_3)_2$ which contains (1) 0.3M KCN (2) 0.5M KCN (3) 0.7M KCN and (4) 1.0M KCN.
   Assume a constant ionic strength of 3 throughout.
   ($\log k_1 = 5.48$ $\log k_2 = 5.14$ $\log k_3 = 4.56$
   $\log k_4 = 3.58$
   (b) Plot pCd vs log [CN^-] and explain the slope of the line obtained.
4. What is the nickel ion concentration in a solution that is 0.10M in $Ni(en)_3(NO_3)_2$ and 0.01M in ethylenediamine (en)? ($\log k_1 = 7.60$, $\log k_2 = 6.48$ and $\log k_3 = 5.03$).
5. (a) Derive expressions for the β values of all the metal complexes present in solutions of the following metal-ligand pairs. Assume a constant ionic strength of 1.0 in all cases.

Metal	Ligand	$\log k_1$	$\log k_2$	$\log k_3$	$\log k_4$	$\log k_5$	$\log k_6$
Ni(II)	NH_3	2.68	2.12	1.60	1.07	0.52	−0.08
	ethylene-diamine	7.60	6.48	5.03	-	-	-
Co(II)	NH_3	2.11	1.63	1.05	0.76	0.18	−0.62
	ethylene-diamine	5.93	4.73	3.30	-	-	-
Zn(II)	NH_3	2.18	2.25	2.31	1.96	-	-
	Cl^-	−0.5	−0.5	1.0	−1.0		
Cd(II)	NH_3	2.51	1.96	1.30	0.79	-	-
	Cl^-	1.95	0.54	−0.15	−0.70	-	-
Hg(II)	Cl^-	6.74	6.48	0.85	1.00	-	-
	Br^-	9.05	8.28	2.41	1.26	-	-
	I^-	12.87	10.95	3.78	2.23	-	-

(b) Calculate β values for each of the above systems at a series of seven appropriate equilibrium ligand concentrations, each varying by a factor of 10.

(c) Use the values obtained in part (b) to draw log β vs log (ligand concentration) diagrams for each of the above systems.

6. Calculate $[Cu^{++}]$ in a solution obtained by mixing 2.0×10^{-3} moles of $CuSO_4$ and (a) 4.0×10^{-3} moles, (b) 0.05 moles of ethylenediamine and diluting to 1.0 liter in each case. ($\log k_1 = 10.75$, $\log k_2 = 9.28$).

7. Calculate the concentrations of all species in a solution made by mixing 1.0×10^{-3} moles of $Hg(NO_3)_2$ and
(a) 1.0×10^{-3} moles KI (b) 2.0×10^{-3} moles KI
(c) 4.0×10^{-3} moles KI (d) 8×10^{-3} moles KI and
diluting to 1.0 liter in each case. (Use log k values given in problem 5.)

8. One hundred ml of a solution that is initially 0.01 M in $Cu(NO_3)_2$ and 0.10 M in NH_4NO_3 is titrated with 0.01 M NaOH. Calculate and plot pCu values as a function of pH. (Use log k values given in appendix.)

9. Calculate $[Cd^{++}]$ in a solution containing a total uncomplexed EDTA concentration of 1.0×10^{-2} M and a total metal ion concentration of 1.0×10^{-3} M at pH values 3.0, 5.0, 7.0 and 9.0. ($\log K_f = 16.46$)

10. (a) Calculate the concentrations of each of the metal containing species in a solution containing a total uncomplexed EDTA concentration of 0.005 M, total zinc concentration of 1.0×10^{-3} M and 0.10 M ethylenediamine at pH values of 7.0, 8.0, 9.0 and 10.0.
($\log K_f\ ZnY^{=} = 16.50$; $\log k_1 = 6.00$, $\log k_2 = 5.08$ for zinc-ethylenediamine complexes, and pK_a of $enH_2^{++} = 7.42$, $enH^{+} = 10.14$).

(b) What is the conditional formation constant of $ZnY^{=}$ at each of these pH values ?

X

Effect of Acid-Base Interaction on Precipitation Equilibria

In chapter VIII precipitation processes were considered to take place without the added complexity of other reactions involving either the cation or the anion of the precipitated substance. Since in many cases the cation is a Bronsted or Lewis acid and the anion may also be a Bronsted or Lewis base, consideration of such acid-base interactions must be included in a description of the precipitation process. From this point of view, the effect of pH as well as the concentration of ligand upon the solubility of a precipitate will be examined here.

X-1. BRONSTED ACID-BASE INTERACTIONS

X-1-(a). THE ANION IS A BRONSTED BASE

If the anion of a precipitated substance is a Bronsted base, its concentration will depend upon the pH of the solution. Therefore the solubility will also depend on pH. For example the $[CO_3^{=}]$ in a saturated solution of $CaCO_3$ is described by the following equation: (See Sec. VI-4)

$$[CO_3^{=}] = \alpha_2 S = \frac{S\,K_1 K_2}{[H^+]^2 + [H^+]K_1 + K_1 K_2}$$

where S is the solubility of $CaCO_3$ in the solution. The need for the use of α_2 arises from the fact that the $CO_3^{=}$ is transformed in part to HCO_3^- and H_2CO_3.

$$\text{i.e. } S = [Ca^{++}] = [H_2CO_3] + [HCO_3^-] +]CO_3^{=}]$$

Since the solubility product of $CaCO_3$ is given by:

$$K_{sp} = [Ca^{++}][CO_3^{=}]$$

then, $$K_{sp} = S \times \alpha_2 S$$

and, $$K_{sp} = \alpha_2 \cdot S^2 \qquad \text{(X-1)}$$

Equation (X-1) differs from what would obtain if the $CO_3^{=}$ were not a base, (i.e., $K_{sp} = S^2$). In order to make a direct comparison of the effect of pH on the solubility, S, equation (X-1) can be transformed as follows:

$$\frac{K_{sp}}{\alpha_2} = S^2 = K'_{sp} \qquad \text{(X-2)}$$

K'_{sp} which has been termed a conditional solubility product, can be seen to be a function of pH.

Example 1.

Calculate the solubility of $Ca_3(PO_4)_2$ in a solution whose pH = 5.0. (Assume that the ionic strength is 0.025 and therefore $pK_{sp} = 26.66$ and for H_3PO_4, $pK_1 = 2.01$, $pK_2 = 6.93$ and $pK_3 = 11.99$).

$$K_{sp} = [Ca^{++}]^3[PO_4^{\equiv}]^2$$

If S = solubility of $Ca_3(PO_4)_2$,

$$[Ca^{++}] = 3S \text{ and } [PO_4^{\equiv}] = 2\alpha_3 S$$

Therefore $$K'_{sp} = \frac{K_{sp}}{\alpha_3^2} = 108\, S^5$$

$$\alpha_3 = \frac{K_1 K_2 K_3}{[H^+]^3 + K_1[H^+]^2 + K_1 K_2 [H^+] + K_1 K_2 K_3}$$

$$\alpha_3 = \frac{10^{-20.93}}{10^{-15} + 10^{-12.01} + 10^{-13.94} + 10^{-20.93}}$$

$$\alpha_3 = 10^{-8.92}$$

$$\text{Therefore} \quad K'_{sp} = \frac{10^{-26.66}}{10^{-17.84}} = 10^{-8.82} = 108\ S^5$$

$$\therefore \quad S = 10^{-2.17} = 6.8 \times 10^{-3}\,M$$

When the pH of the solution is not given or when the pH of the solution is altered by the dissolution of the precipitate, the calculations cannot be carried out without the prior determination of the equilibrium pH.

Example 2.

(a) What is the solubility of silver acetate as a function of the hydrogen ion concentration?

(b) What is the solubility of silver acetate in water? (Assume a constant ionic strength of 0.05.)

pK_{sp} for silver acetate = 2.46 : pK_a for acetic acid = 4.58. pK_w for H_2O = 13.82.

(a) Using equation (X-2), the following may be written:

$$\frac{K_{sp}}{\alpha_1} = S^2$$

$$\text{where} \quad \alpha_1 = \frac{K_a}{[H^+] + K_a}$$

(The interaction of Ag^+ with H_2O has been neglected.)

Therefore,

$$S = \left(\frac{K_{sp}}{\alpha_1}\right)^{\frac{1}{2}} = \left(\frac{K_{sp}([H^+] + K_a)}{K_a}\right)^{\frac{1}{2}}$$

As long as the equilibrium hydrogen ion concentration is known or is much less than K_a, the solubility may be calculated readily from the above equation.

(b) The solution of silver acetate must have a pH greater than that of pure water since the acetate ion is a base. Hence the condition that $[H^+] \ll K_a$, results in the simplification of the equation derived in (a) to:

$$= K_{sp}^{\frac{1}{2}} = 10^{-1.23}$$

The pH of this solution may be readily calculated to be 8.59. (See equation V-22 for a weak base.)

Let us consider an example where the hydrogen ion concentration is neither given nor less than the pertinent dissociation constant.

Example 3.

Calculate the solubility of MnS in water. Assume that K_{sp} for MnS = 1.0×10^{-11}; $K_1 = 1.0 \times 10^{-7}$ and $K_2 = 1.3 \times 10^{-13}$ for H_2S and $K_w = 1.0 \times 10^{-14}$.

$$K_{sp} = [Mn^{++}][S^{=}] \text{ and}$$

$$K_{sp} = \alpha_2 S^2 \text{ where } \alpha_2 = \frac{K_1 K_2}{[H^+]^2 + [H^+]K_1 + K_1 K_2}$$

Assuming that enough MnS dissolves to raise the pH significantly above 7 (by the reaction $S^{=} + H_2O \rightleftharpoons SH^- + OH^-$)

then, $K_1 > [H^+]$

Hence $\alpha_2 = \dfrac{K_2}{[H^+] + K_2}$

The solubility of MnS would have to be exceptionally high for the $[H^+]$ to be reduced to a value as small as K_2. Since this is not the case,

$$\alpha_2 = \frac{K_2}{[H^+]}$$

For every mole of MnS that dissolves, one mole of OH^- is formed according to the equation,

$$MnS + H_2O \rightleftharpoons Mn^{++} + SH^- + OH^-$$

Therefore, $[OH^-] = [Mn^{++}] = [SH^-] = S$

and $\therefore\ \alpha_2 = \dfrac{K_2 \times S}{K_w}$

Combining this with the solubility product expression, then we obtain

$$\frac{K_{sp} \times K_w}{K_2} = S^3$$

i.e., $\dfrac{1.0 \times 10^{-11} \times 10^{-14}}{1.3 \times 10^{-13}} = S^3$

Therefore, $S = 9.1 \times 10^{-5}$ M

and the pH of the solution is 9.95.

In all of the above examples the general effect of acids and bases upon solubility has been discussed. Suppose, however, that the acid in question contains the precipitating anion; for instance what is the solubility of calcium phosphate in phosphoric acid or the solubility of silver acetate in acetic acid? To illustrate this effect, the dependence of the solubility of $BaSO_4$ upon the concentration of H_2SO_4, C_a, will be considered as a typical case.

$$K_{sp} = [Ba^{++}][SO_4^{=}]$$

if S = solubility of $BaSO_4$ and C_a the concentration of H_2SO_4,

$$[Ba^{++}] = S \text{ and } [SO_4^{=}] = \alpha_2(C_a + S)$$

Since the solubility of $BaSO_4$ in even the most dilute H_2SO_4 solutions is very small, $C_a \gg S$, and the above equation simplifies to:

$$K_{sp} = S \cdot (\alpha_2 C_a)$$

and $$\alpha_2 = \frac{K_2}{[H^+] + K_2}$$

Hence, $$S = \frac{K_{sp}([H^+] + K_2)}{K_2 \cdot C_a} \qquad \text{(X-3)}$$

Once the hydrogen ion concentration is expressed as a function of C_a, the solubility may be calculated.

Mass Balance: $C_a = [HSO_4^-] + [SO_4^{=}]$

(Unless concentrated H_2SO_4 is involved, $[H_2SO_4]$ is negligible.)

Charge Balance: $[H^+] = 2[SO_4^{=}] + [HSO_4^-]$

(Since $[Ba^{++}]$ is negligibly small, due to the low solubility of $BaSO_4$ in H_2SO_4, it may be omitted.)

Substituting for $[HSO_4^-]$ and $[SO_4^{=}]$ in the charge balance equation,

$$[H^+] = \frac{2 \cdot K_2 \cdot C_a}{[H^+] + K_2} + \frac{[H^+] \cdot C_a}{[H^+] + K_2}$$

i.e. $$[H^+] = C_a \frac{[H^+] + 2K_2}{[H^+] + K_2}$$

and solving explicitly for $[H^+]$,

$$[H^+] = \frac{(C_a - K_2) + \sqrt{C_a^2 + 6C_a K_2 + K_2^2}}{2}$$

At concentrations where $C_a \gg K_2$, $[H^+]$ approaches C_a. When this happens, the solubility of $BaSO_4$ reaches a constant value, as seen from equation (X-3),

$$S = \frac{K_{sp}}{K_2}$$

This conclusion presupposes that the ionic strength is essentially constant.

Why doesn't the solubility of silver acetate in acetic acid become constant with increasing acetic acid concentration?

Finally it should be remarked that in many cases the precipitating agent is not a Bronsted base itself but rather the conjugate acid. For example H_2S is used in precipitating sulfides. Such precipitation reactions will result in the release of hydrogen ions.

$$Zn^{++} + H_2S \rightarrow ZnS + 2H^+$$

The hydrogen ions released must be taken into account in calculating the solubility of ZnS. In the design of suitable buffers for precipitations, adequate buffer capacity to handle the hydrogen ion released, must be employed.

X-1-(b). THE CATION IS A BRONSTED ACID

Many hydrated metal ions act as polyprotic Bronsted acids, e.g.

$$Al(H_2O)_6^{+++} \rightleftharpoons Al(H_2O)_5(OH)^{++} + H^+$$

Reactions of this type will naturally result in the pH dependence of the hydrated cation which is the conjugate acid form.

In analogy to the term α applied to the fraction of the conjugate base used earlier, the term β, signifying the fraction of the fully hydrated cation is useful here. Thus in a solution containing Al^{+++}, $Al(OH)^{++}$, $Al(OH)_2^{++}$, etc. (the water of hydration has been omitted for convenience),

$$\beta = \frac{[Al^{+++}]}{C_{Al^{+++}}}$$

or in general,

$$\beta = \frac{[M^{n+}]}{C_M} \qquad (X\text{-}4)$$

Since M^{n+} is a polyprotic acid, β is readily obtained from equation VI-16 as:

$$\beta = \frac{[H^+]^n}{[H^+]^n + K_1[H^+]^{n-1} + K_1K_2[H^+]^{n-2} + \text{---}\ K_1K_2\ K_n} \qquad (X\text{-}5)$$

Where K_1, K_2 etc. are the successive acid dissociation constants of the hydrated metal ion.

The solubility of precipitates involving cations which are Bronsted acids will depend on pH. The solubility, S, of MX, where M is a Bronsted acid is given by:

$$K_{sp} = [M][X] = \beta S \times S.$$

or $$\frac{K_{sp}}{\beta} = S^2 = K'_{sp} \qquad (X\text{-}6)$$

Equation (X-6) may be compared directly with equation (X-2). Of course if X is a Bronsted base, then equation (X-6) becomes:

$$K_{sp} = \beta S \cdot \alpha S$$

or $$\frac{K_{sp}}{\alpha\beta} = S^2 = K'_{sp} \qquad (X\text{-}7)$$

Example 4.

Calculate the solubility of CuS at pH 9.0.

pK_{sp} of CuS = 35.1 pK_1 = 7.0 and pK_2 = 12.9 for H_2S:

$$K_{a_1} \text{ for } Cu^{++} = 10^{-8.30} = \frac{[CuOH^+][H^+]}{[Cu^{++}]}$$

If S = solubility of CuS, then

$$\frac{K_{sp}}{\beta\,\alpha_2} = S^2$$

$$\alpha_2 = \frac{K_1 K_2}{[H^+]^2 + [H^+] K_1 + K_1 K_2} = \frac{10^{-1.99}}{10^{-18} + 10^{-16} + 10^{-19.9}}$$

$$= 10^{-3.9}$$

$$\beta = \frac{[H^+]}{K_{a_1} + [H^+]} = \frac{10^{-9}}{10^{-8.30} + 10^{-9}}$$

$$= 10^{-0.78}$$

(β has been calculated here on the assumption that only the first acid dissociation step is of importance.)

Therefore, $$S^2 = \frac{10^{-35.1}}{10^{-3.9} \times 10^{-0.78}}$$

$$S = 10^{-15.2}$$

$$= 6.3 \times 10^{-16} \text{ M}$$

In cases where the conditional solubility product, K'_{sp}, involves both α and β, the effects of pH on the solubility caused by each of these factors tend to counteract each other, since α increases and β decreases with increasing pH.

Problems in which the pH is not specified fall into two general categories; very slightly soluble salts in which the pH may be considered to be that of pure water such as CuS and those salts whose solubilities are sufficiently large to affect the pH. In this latter case, the pH of the solution must be calculated as in Example 3.

Example 5.

What is the solubility of CuS in water?

This problem may be solved exactly as Example 4 was solved keeping in mind that, because of the small solubility of CuS, the pH of a saturated solution is equal to 7.0.

At this pH, $\alpha = 10^{-6.20}$ and $\beta = 1$.

Hence, $$S^2 = \frac{10^{-35.1}}{10^{-6.20} \times 1}$$

$$S = 10^{-14.5}$$

$$= 3.2 \times 10^{-15} \text{ M}$$

X-2. LEWIS ACID-BASE INTERACTIONS

In all the cases to be considered the cation acts as a Lewis acid, and the Lewis base or ligand can either be the precipitating anion or another complexing agent.

X-2-(a). THE PRECIPITATING ANION IS A LEWIS BASE

A number of slightly soluble metal salts interact with excess anion to form soluble complexes. In general a metal ion is capable of forming a whole series of coordination or ion-association complexes with an anion, of which the least soluble is the uncharged species. For example, Pb^{++} and Cl^- will form $PbCl^+$, $PbCl_2$ (which is slightly soluble in water), $PbCl_3^-$ and $PbCl_4^=$. The solubility of lead in chloride solutions will go through a minimum when the opposing effects of the common ion and formation of anionic complexes counterbalance each other.

Consider a saturated solution of $PbCl_2$. This dissolves to give undissociated $PbCl_2$,

$$PbCl_{2\,(solid)} \rightleftharpoons PbCl_{2\,(solution)}$$

which may either react with chloride ion in solution,

$$PbCl_2 + Cl^- \rightleftharpoons PbCl_3^-$$

and $$PbCl_3^- + Cl^- \rightleftharpoons PbCl_4^=$$

or dissociate,

$$PbCl_2 \rightleftharpoons PbCl^+ + Cl^-$$

and $$PbCl^+ \rightleftharpoons Pb^{++} + Cl^-$$

Thus the overall solubility of $PbCl_2$ in a series of solutions containing chloride ions, may be quantitatively described in terms of the equilibrium expressions corresponding to the four stepwise complex formation constants and the intrinsic solubility of $PbCl_2$. The latter corresponds to the equilibrium constant, $S^0 = [PbCl_2]_{(solution)}$, of the first of the series of equations above.

Since the solubility of $PbCl_2$, S, is given by the sum of the concentrations of all species containing Pb^{++},

$$S = [Pb^{++}] + [PbCl^+] + [PbCl_2] + [PbCl_3^-] + [PbCl_4^=]$$

i.e. $$S = K_{sp}\left[\frac{1}{[Cl^-]^2} + \frac{k_1}{[Cl^-]} + k_1 k_2 + k_1 k_2 k_3 [Cl^-] + k_1 k_2 k_3 k_4 [Cl^-]^2\right]' \qquad (X\text{-}8)$$

It will be of interest to note that in equation (X-8) S^0 has been replaced by its equivalent, $k_1 k_2 \cdot K_{sp}$.

i.e.
$$K_{sp} = \frac{S^0}{k_1 k_2} \qquad \text{(X-9)}$$

This demonstrates that the solubility product depends not only on the complex formation constants but upon the intrinsic solubility as well.

Example 6.

Calculate the solubility of $PbCl_2$ in solutions containing the following concentrations of chloride ions:

0.010 M; 0.10 M; 1.0 M and 2.0 M.

Assume that the stepwise formation constants of $PbCl_4^{=}$ are 40, 1.5, 0.83 and 0.50 respectively, and K_{sp} for $PbCl_2$ is 2.6×10^{-5}.

Substituting the appropriate $[Cl^-]$ in equation (X-8), the following solubilities are obtained:

$[Cl^-]$	Solubility (M)
1.0×10^{-2}	0.36
0.10	1.5×10^{-2}
1.0	4.6×10^{-3}
2.0	7.2×10^{-3}

Note that there is a minimum in the solubility somewhere between 0.10 and 1.0 M chloride ion concentration.

X-2-(b). OTHER LEWIS BASES

The presence of Lewis bases which form complexes with the metal ion will naturally affect the solubility of slightly soluble metal salts. For example, ammonia will increase the solubility of silver halides and transition metal sulfides and hydroxides. Other complexing agents which are of interest in this connection include cyanide, tartrate, citrate, and EDTA ions.

Formation of charged metal complexes generally tends to increase the solubility of metal ions in solution, because these complexes do not form compounds which are as insoluble as those of the free or hydrated metal ion. That is to say, whereas the hydrated copper ion forms an insoluble hydroxide, the $Cu(NH_3)_4^{++}$ complex does not. Hence the decrease in the concentration of the hydrated copper ion, upon the addition of ammonia will permit a higher copper concentration in solution at a given hydroxide ion concentration.

In general if a metal ion, M, forms a series of complexes with a ligand, L, ML, ML_2, ML_3, etc.,

$$C_M = [M] + [ML] + [ML_2] + [ML_3] + \cdots$$
$$= [M](1 + k_1[L] + k_1k_2[L]^2 + k_1k_2k_3[L]^3 + \cdots)$$

where k_1, k_2, k_3, etc. are the stepwise formation constants of the metal complexes.

Defining β in the usual way (Equation X-4), as the fraction of the total metal ion concentration, C_M, corresponding to the uncomplexed metal ion concentration, [M], we may write:

$$\frac{1}{\beta} = \frac{C_M}{[M]} = 1 + k_1[L] + k_1k_2[L]^2 + k_1k_2k_3[L]^3 + \cdots \qquad \text{(X-10)}$$

Now we may write for the solubility of MX, in the presence of L,

$$K_{sp} = [M][X]$$
$$= \beta C_M[X]$$

The solubility of MX is equal to C_M and also to [X],

so that $$S = \sqrt{\frac{K_{sp}}{\beta}} \qquad \text{X-11)}$$

This equation is identical with equation (X-6) which was developed for cases where the metal ion acted as a Bronsted acid.

Example 7.

Calculate the solubility of AgCl in 0.01 M NH_3. K_{sp} for AgCl = 2.0×10^{-10}. The stepwise formation constants for $Ag(NH_3)_2^+$ are $10^{3.2}$ and $10^{3.8}$.

From equation (X-10),

$$\frac{1}{\beta} = 1 + 10^{3.2} \times 10^{-2} + 10^{7.0} \times 10^{-4}$$

$$\beta = 10^{-3.0}$$

Therefore $S = \sqrt{\dfrac{2.0 \times 10^{-10}}{10^{-3.0}}}$

$= 4.5 \times 10^{-4}$ M.

It may be noted that the implicit assumption that the initial concentration of ammonia would not be significantly reduced by either

(a) the amount consumed in complexation with Ag^+ or

(b) The amount transformed to NH_4^+,

is justified, since neither (a) nor (b) involves as much as 5% of the initial concentration.

As the concentration of NH_3 is increased, the solubility of AgCl likewise increases to the point where it might be necessary to correct for the amount of NH_3 consumed in the complexation of Ag^+. A compensating simplification however results from the fact that with increasing NH_3 concentration, the contribution of intermediate complex formation becomes less important and may be ignored.

In cases where the precipitating anion complex formation also contributes to the overall situation, then a combined approach is necessary as shown in the following example.

Example 8.

What is the solubility of AgCl in a buffer mixture containing 0.01 M NH_3 and 0.10 M NH_4Cl ? Assume that the same constants in example 7 apply. The stepwise formation constants for the chloro complexes of Ag^+ are 3×10^3 and 57.

The solubility of AgCl is given by the total concentration of all forms of silver.

i.e. $$S = [Ag^+] + [AgCl] + [AgCl_2^-] + [Ag(NH_3)^+] + [Ag(NH_3)_2^+]$$

$$= [Ag^+](1 + k_1[NH_3] + k_1 k_2 [NH_3]^2) + [AgCl] + [AgCl_2^-]$$

$$= [Ag^+] \times \frac{1}{\beta} + [AgCl] + [AgCl_2^-]$$

$$S = K_{sp}\left(\frac{1}{\beta \cdot [Cl^-]} + k_1' + k_1' k_2' [Cl^-]\right) \qquad \text{(X-12)}$$

where k' and k_2' represent the stepwise formation constants of the chloro complexes of Ag^+.

Therefore,

$$S = 2 \times 10^{-10} \left\{ \frac{1}{10^{-3} \times 0.1} + 3 \times 10^{3} + 1.7 \times 10^{5} \times 0.1 \right\}$$

$$= 6.0 \times 10^{-6} \text{ M.}$$

From equation (X-12) it may be seen that the $[Cl^-]$ in which the AgCl has a minimum solubility (calculated by differentiating S with respect to $[Cl^-]$ will change with the value of β. The minimum solubility can be shown to be given by

$$S_{minimum} = K_{sp} \left\{ 2 \sqrt{\frac{k_1' k_2'}{\beta}} + k_1' \right\}$$

and $$[Cl^-]_{minimum} = \sqrt{\frac{1}{k_1' k_2' \beta}}$$

Suppose that the concentration of the complexing agent as well as that of the precipitating anion is pH dependent, as is not uncommon in precipitation processes of analytical interest. Then the situation will be formally analogous to Example 4.

Example 9.

Derive an expression for the solubility of CuS in 0.01 M EDTA as a function of pH.

Calculate the solubility at pH 9.0.

pK_{sp} of CuS = 35.1; Log K_f of Cu-EDTA = 18.80

$pK_1 = 7.0$ and $pK_2 = 12.9$ for H_2S

$pK_1 = 2.0$, $pK_2 = 2.67$, $pK_3 = 6.16$ and

$pK_4 = 10.26$ for EDTA.

If S = solubility of CuS, then

$$S^2 = \frac{K_{sp}}{\beta \alpha_2}$$

where,

$$\alpha_2 = \frac{[S^=]}{S} = \frac{K_1 K_2}{[H^+]^2 + [H^+]K_1 + K_1 K_2}$$

and,

$$\frac{1}{\beta} = \frac{C_M}{[Cu^{++}]} = \frac{S}{[Cu^{++}]} = 1 + K_f [Y^{4-}]$$

$$= 1 + K_f \alpha_4 C_y$$

where,

C_M = total copper concentration

C_y = total EDTA concentration = 0.01

$\alpha_4 = \dfrac{[Y^{4-}]}{C_y}$ = fraction of EDTA that is present as the quadruply charged anion.

$$\alpha_4 = \frac{K_1 K_2 K_3 K_4}{[H^+]^4 + K_1[H^+]^3 + K_1 K_2 [H^+]^2 + K_1 K_2 K_3 [H^+] + + K_1 K_2 K_3 K_4}$$

Hence the solubility, S, of CuS in EDTA, as a function of pH is given by:

$$S^2 = \frac{K_{sp}(1 + K_f \alpha_4 C_y)}{\alpha_2} \qquad \text{(X-13)}$$

At pH 9.0

$$\alpha_2 = 10^{-3.9}$$

$$\alpha_4 = 10^{-1.29}$$

Then

$$S^2 = \frac{10^{-35.1}(1 + 10^{18.8} \times 10^{-1.29} \times 10^{-2})}{10^{-3.9}}$$

$$= 10^{-15.7}$$

and

$$S = 10^{-7.9} = 1.3 \times 10^{-8}\ M$$

Although the solubility of CuS in EDTA solution is much greater than it was in water (see example 4), as might be expected from the high stability of the Cu-EDTA complex, it is still fairly insoluble

The action of Cu^{++} as a Bronsted acid in this solution is negligible, unlike the situation described in Example 4, because the Cu-EDTA complex is much more stable than the $CuOH^+$ complex and furthermore the $[OH^-]$ is much smaller than the $[Y^{4-}]$.

In this particular system for most of the pH range $K_f \alpha_4 C_y$ is sufficiently greater than 1 to permit the simplification of the expression, X-13, to

$$S^2 = K_{sp} K_f \alpha_4 C_y / \alpha_2$$

the logarithmic equivalent of which may be written as

$$\log S = \tfrac{1}{2}(\log K_{sp} + \log K_f + \log C_y) + \tfrac{1}{2}(\log \alpha_4 - \log \alpha_2)$$

This logarithmic relation is convenient for the calculation of log S as a function of pH inasmuch as it indicates that only values of $\log \alpha_4$ and $\log \alpha_2$ at various pH values need be calculated (or graphically evaluated.) If the function $(\log \alpha_4 - \log \alpha_2)$ is plotted against pH it will be noted that there are regions in which the function is constant (why?), and hence the solubility remains constant in these pH regions.

Another frequently encountered solubility problem is one in which the excess concentration of the precipitant is kept constant.

Example 10.

How does the total copper concentration in a solution in which $C_{H_4Y} = 0.01$ M and $C_{H_2S} = 0.10$ M vary with the pH?

The equilibria that need to be considered here are

$$[Cu^{++}][S^{=}] = K_{sp} = [Cu^{++}]C_{H_2S}\,\alpha_2$$

and

$$\frac{[CuY^{=}]}{[Cu^{++}][Y^{-4}]} = K_f = \frac{[CuY^{=}]}{[Cu^{++}]C_{H_4Y}\cdot\alpha_4}$$

Eliminating $[Cu^{++}]$ to solve for $[CuY^{=}]$ we obtain:

$$[CuY^{=}] = \frac{K_{sp}\cdot K_f\cdot C_{H_4Y}}{C_{H_2S}}\cdot\frac{\alpha_4}{\alpha_2}$$

Inasmuch as $[CuY^{=}]$ is much greater than $[Cu^{++}]$ it is considered equivalent to the total dissolved copper concentration. Note that this varies with pH according to the value of the ratio α_4/α_2. The logarithmic form of the solution:

$$\log[CuY^{=}] = (\log K_{sp} + \log K_f + \log C_{H_4Y} - \log C_{H_2S}) + (\log \alpha_4 - \log \alpha_2)$$

differs from that obtained in example 9 by the absence of the coefficient $\tfrac{1}{2}$ in the right-hand side of the equation.

The variation of the solubility of copper over a wide pH range can be conveniently visualized by a graphical solution to this problem, as shown in Fig. X-1. The Cu^{++} line is constructed from the equation

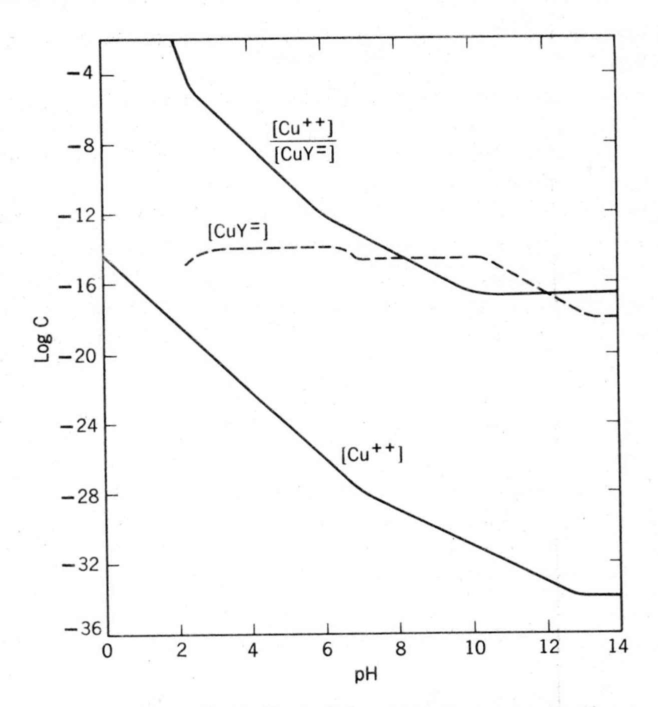

Fig. X-1 — Variation of the copper concentration with pH in a solution containing 0.01 M EDTA (H_4Y) and 0.10 M H_2S.

$$\log [Cu^{++}] = \log K_{sp} - \log C_{H_2S} - \log \alpha_2$$

Note that the line at high pH is horizontal down to pH 12.9 (i.e. to pK_2 for H_2S), has a slope of −1 from this point to pH 7 (pK_1) and below this pH a slope of 2; these changes being required by the variation of α_2 with pH. There is also plotted the line:

$$\log \frac{[Cu^{++}]}{[CuY^{=}]} = -\log K_f - \log C_{H_4Y} - \log \alpha_4 .$$

The slope of this line changes at appropriate pH values dictated by the variation of α_4 with pH. The difference between the ordinates of these two lines is equal to log $[CuY^{=}]$ and hence represents the solution to the problem at any pH.

$$\log [Cu^{++}] - \log \frac{[Cu^{++}]}{[CuY^{=}]} = \log [CuY^{=}]$$

The difference so obtained is plotted in the Fig. (X-1) as the line labelled $[CuY^{=}]$.

In systems where the difference between the values of these two lines decreases, it is necessary to take into account the total amount of copper in order to evaluate the pH at which the precipitate totally dissolves. This is most readily done by examining the line corresponding to the difference in comparison to the maximum possible copper concentration calculated on the basis of total solution (which should be drawn as a horizontal line across the diagram). The intersection of this line with the "difference" line marks the pH at which total solution occurs.

SUGGESTIONS FOR FURTHER READING

1. H. A. Laitinen, Chemical Analysis, McGraw Hill Book Co. Inc. (1960), Chapter 6.
2. G. Charlot, Qualitative Inorganic Analysis, John Wiley and Sons, Inc. (1954), Chapters 11, 12, and 14.

X-3. PROBLEMS

1. Between what pH values will it be possible to quantitatively separate zinc from iron(II) by saturating the solution with H_2S if the initial concentrations of Zn^{++} and Fe^{++} are each 1.0×10^{-2} M? (Consider that 99.9% precipitation of zinc is a quantitative removal.) Assume $\mu = 0.3$ throughout.

   What concentrations of Na_2SO_4 and $NaHSO_4$ should be used in order that the final pH is sufficiently high to ensure the quantitative ZnS precipitation and that the pH of the solution will not have changed by more than 0.3 during the precipitation? (The solubility of H_2S is 0.07 M) pK_{sp} of FeS is 15.9, for ZnS is 22.7. pK_1 is 6.6 and $pK_2 = 12.2$ for H_2S; pK_2 for HSO_4^- is 1.28.
2. How many moles of KI must be added to 100 ml of a solution containing 0.05 M $Hg(NO_3)_2$ and having a pH of 0.03 to just prevent the precipitation of HgS when the solution is saturated with H_2S? (The solubility of H_2S is 0.07 M. Assume $\mu = 0.03$) pK_{sp} for HgS = 50.1; pK_1 is 6.6, $pK_2 = 12.2$ for H_2S; the overall formation constant of $HgI_4^{=}$ is $10^{31.0}$.
3. The solubility of $BiPO_4$ is 0.75 mg/l. in a solution that is 0.10 M in H_3PO_4 and 0.20 M in HNO_3. Calculate the solubility

product of $BiPO_4$. (pK_1 = 1.84, pK_2 = 6.58 and pK_3 = 11.5 for H_3PO_4.)

4. Calculate the solubility of CaF_2 in 0.20 M $HClO_4$. (pK_{sp} of CaF_2 is 9.37 and pK_a of HF is 2.86.
5. What is the solubility of $CaCO_3$ in
   (a) 1.0×10^{-3} M HCl (b) 0.0015 M NaCl
   pK_{sp} for $CaCO_3$ = 8.17, pK_1 = 6.31 and pK_2 = 10.25 for H_2CO_3.
6. (a) Calculate the pH of a solution in which the solubility of $Ca_3(PO_4)_2$ is 2.0×10^{-2} M. (pK_1 = 1.80, pK_2 = 6.48, pK_3 = 11.3 for H_3PO_4 and pK_w = 13.64)
   (b) What would be the solubility of $Ca_3(PO_4)_2$ in a solution whose pH is 3.0? (pK_{sp} 28.25; pK_1 = 2.12, pK_2 = 7.14, pK_3 = 12.3, pK_w = 13.97)
   (c) Show graphically the manner in which the solubility of $Ca_3(PO_4)_2$ varies with pH, by plotting pCa against pH.
7. (a) How does the solubility of $BaSO_4$ in H_2SO_4 vary with the concentration of the H_2SO_4?
   (b) At what H_2SO_4 concentration is the solubility of $BaSO_4$ a minimum?
8. (a) Plot the negative logarithm of the solubility of (i) $BaSO_4$ (ii) $BaCO_3$, against pH.
   (b) Explain the difference in the two solubility curves in the pH region 4 to 7.
9. (a) Calculate the pH of a saturated solution of CaC_2O_4. (pK_{sp} = 8.68; pK_1 = 1.27 and pK_2 = 4.27 for $H_2C_2O_4$)
   (b) What is the solubility of CaC_2O_4 in a solution that is 1.0×10^{-3} M in $H_2C_2O_4$? (pK_{sp} = 8.56; pK_1 = 1.24 and pK_2 = 4.21 for $H_2C_2O_4$)
10. Calculate the solubility of lead fluoride in
    (a) pure water
    (b) 0.15 M $HClO_4$ (pK_{sp} = 6.73 and pK = 2.89 for HF)
11. Calculate the solubility of (a) CuS (b) Ag_2O in H_2O. (K_{sp} for $Ag_2O = [Ag^+][OH^-]$)
12. Calculate the concentrations of all ionic and molecular species in a solution which is saturated with $Pb(OH)_2$. (Formation constant of $Pb(OH)^+$ is $10^{7.51}$. pK_{sp} = 14.40)
13. (a) Plot the manner in which the solubility of AgCl varies with the concentration of added HCl.
    (b) Calculate the minimum solubility of AgCl in HCl.
14. Calculate the solubility of AgCN in
    (a) 2.5×10^{-3} M NaCN (pK_{sp} = 15.65, log $k_1 k_2$ for formation of $Ag(CN)_2^-$ is 19.8)

(b) 1.0×10^{-5} M HCN ($\log k_1 k_2$ for formation of $Ag(CN)_2^-$ is 19.9, $pK_{sp} = 15.7$ pK_a for HCN = 9.4)

15. Calculate the solubility of the following substances in both 0.5 M NH_3 and 1.0×10^{-3} M NH_3. In both these ammonia solutions 0.10 M NH_4Cl is present.
    (a) AgCl ($pK_{sp} = 9.76$ $\log k_1 = 3.32$ and $\log k_2 = 3.92$ for the formation of the silver ammine complexes.
    (b) $AgIO_3$ ($pK_{sp} = 7.27$)
    (c) AgBr ($pK_{sp} = 12.04$)
16. What is the concentration of silver ion in a solution which is 0.3 M in NH_3 and saturated with respect to AgBr and AgSCN? (pK_{sp} AgBr = 12.28, pK_{sp} AgSCN = 12.00)
17. Calculate the solubility of ZnS in (a) pure H_2O ($pK_{sp} = 21.52$, $K_1 = 7.0$ and $K_2 = 12.9$ for H_2S)
    (b) 0.3 M $HClO_4$ ($pK_{sp} = 20.10$; $K_1 = 6.6$ and $K_2 = 12.2$ for H_2S)
    (c) 0.3 M NaOH (Use constants in (b) along with the overall formation constant for $Zn(OH)_4^{--}$ $\log k_1 k_2 k_3 k_4 = 16.2$.)
18. Calculate the intrinsic solubility of (a) AgCl (b) AgI in H_2O.
19. Plot the solubility of ZnS in a 0.10 M solution of EDTA as a function of pH.
20. What is the solubility of zinc 8-hydroxyquinolinate at pH 2.0, 3.0, 4.0, 5.0, 6.0 and 7.0? The acid dissociation constants of 8-hydroxyquinoline are $10^{-5.13}$ and $10^{-9.89}$. The formation constants of the zinc chelate are $10^{8.5}$ and $10^{7.5}$. The solubility product of the chelate is $10^{-27.13}$.

XI

Oxidation Reduction Equilibria

Oxidation-reduction reactions involve electron transfer and therefore can be used to produce electrical work. This is accomplished by suitably separating the reaction components so that electron transfer must occur through an external circuit. Such a system is called a galvanic cell. The electromotive force of the cell is a measure of the driving force of the chemical reaction involved. The equilibrium constant of the chemical reaction is another measure of this driving force. The e.m.f. of the cell and the equilibrium constant of the chemical reaction involved are therefore related as will be shown below. As a matter of fact oxidation-reduction equilibria are more often characterized in terms of e.m.f. values rather than equilibrium constants.

In redox reactions, to an extent greater than found in other types of reactions, systems which are not reversible or are not in equilibrium are encountered. The behavior of irreversible systems will deviate more or less from the predictions made below with the Nernst equation. An accurate description of such systems is beyond the scope of this book.

XI-1. THE NERNST EQUATION

If a cell is operated under reversible conditions, the e.m.f., E, observed is related to the free energy change of the reaction, ΔF.

i.e. $\Delta F = -n\mathfrak{F}E.$ (XI-1)

where n is the number of Faradays of electricity transferred, $\mathfrak{F}$ is the Faraday, (96500 coulombs).

The free energy change of a reaction is also related to the activities of the reactants and products (Chapter II Sec. 3.)

$$\Delta F = \Delta F^\circ + RT \ \ln \frac{a_P^p \cdot a_Q^q \cdot a_R^r \ldots}{a_A^a \cdot a_B^b \cdot a_C^c \ldots}$$

Substituting for ΔF°, its equivalent, $-RT\ln\mathbf{K}$, (XI-2) and combining equations XI-1 and 2, we obtain the Nernst equation:

$$E = \frac{RT}{n\mathfrak{F}} \ln\mathbf{K} - \frac{RT}{n\mathfrak{F}} \ln \frac{a_P^p \cdot a_Q^q \cdot a_R^r}{a_A^a \cdot a_B^b \cdot a_C^c} \qquad \text{(XI-3)}$$

In equation (XI-3) the term $(RT/n\mathfrak{F})\ \ln K$ can be replaced by E°, which is known as the standard e.m.f. of the cell. E° is seen to be the value of the cell e.m.f. that would be obtained if all of the substances in the reaction were at unit activity. On the other hand if the system were at equilibrium then the value of E as well as ΔF would be zero, and in this case equation XI-3 reduces to the usual equilibrium expression. (Chapter II Sec. 3.)

XI-2. HALF-CELL REACTIONS

Any oxidation-reduction reaction may be considered to be a combination of two half-reactions, each involving an oxidation-reduction couple. For example in the reaction of Fe^{+3} with Sn^{+2},

$$2Fe^{+3} + Sn^{+2} \rightleftharpoons 2Fe^{+2} + Sn^{+4}$$

the two half reactions are:

$$Fe^{+3} + e^- \rightleftharpoons Fe^{+2}$$

$$Sn^{+2} \rightleftharpoons Sn^{+4} + 2e^-$$

Half-cell reactions bear a striking resemblance to the Bronsted acid-base equilibrium:

B^-	$+ p^+$	$\rightleftharpoons$	BH	;	B^-/BH
Base	proton		acid		conjugate acid-base pair
O_x	$+ ne^-$	$\rightleftharpoons$	Red	;	O_x/Red_1
oxidized form	elec- trons		reduced form		Redox couple

In the generalized half-cell equation, n represents the number of electrons necessary to transform a species to the next stable lower oxidation state. In contrast to proton transfer reactions which occur in steps of one proton at a time, many redox reactions involve the simultaneous transfer of several electrons.

Just as a redox equation can be considered as a combination of two half-cell equations, the cell e.m.f. can be considered to be the difference between the potentials of the two half-cells. Values of the absolute potential of any single redox couple cannot be measured, but for all practical purposes the relative values are sufficient.

XI-3. ELECTRODE POTENTIALS

The electrode potential of any single redox couple is defined as the e.m.f. of a cell consisting of the standard hydrogen electrode and the electrode in question, written in the following manner:

$$\text{Pt, } H_2 \mid H^+(a = 1) \parallel Ox_1, Red_1 \mid \text{Pt}$$

In this diagrammatic representation of the cell, a single vertical line stands for a phase boundary at which a potential difference is taken into account. The double vertical line represents a liquid junction whose potential difference is considered to be small enough to be ignored.

This cell diagram implies that the overall chemical equation is written with the hydrogen gas acting as a reducing agent. If the cell diagram is written:

$$\text{Pt} \mid Red_1, Ox_1 \parallel H^+ (a = 1) \mid H_2, \text{Pt}$$

then the chemical equation is reversed. This cell e.m.f. is not called the electrode potential of the electrode in question.

This diagrammatic representation illustrates a general method of describing galvanic cells. In general the cell reaction corresponding to the diagram is written to correspond with the passage of positive electricity through the cell from left to right. A positive cell e.m.f. signifies that the reaction occurs spontaneously as written, (i.e. ΔF is negative).

Thus the diagram:

$$\text{Pt} \mid Red_1, Ox_1 \parallel Ox_2, Red_2 \mid \text{Pt}$$

corresponds to the reaction:

$$n_2 Red_1 + n_1 Ox_2 \rightleftharpoons n_1 Red_2 + n_2 Ox_1$$

The e.m.f. of this cell, E, is given by

$$E = E_{right} - E_{left} = E_2 - E_1 \qquad \text{(XI-3)}$$

The relationship between the electrode potential and the Nernst Equation can be developed by considering the e.m.f. of the cell:

$$Pt, H_2 \mid H^+ (a = 1) \parallel Ox_i, Red_i \mid Pt$$

The half-cell reactions involved are

$$2H^+ + 2e^- \rightleftharpoons H_2 \qquad\qquad Ox_i + ne^- \rightleftharpoons Red_i$$

These two equations combine to give:

$$nH_2 + 2Ox_i \rightleftharpoons 2nH^+ + 2\,Red_i$$

in which 2n Faradays of electricity have been transferred. The e.m.f. of this cell, $E = E_i - E_H = E_i$ since $E_H = 0$, is given by the Nernst Equation XI-3.

$$E_i = E_i^\circ - \frac{RT}{2n\mathfrak{F}} \ln \frac{a_{H^+}^{2n} \cdot a_{Red_i}^{2}}{a_{H_2}^{n} \cdot a_{Ox_i}^{2}}$$

Since in this cell $a_{H_2} = a_{H^+} = 1$, the equation reduces to:

$$E_i = E_i^\circ - \frac{RT}{2n\mathfrak{F}} \ln \frac{a_{Red_i}^2}{a_{Ox_i}^2}$$

and finally, $$E_i = E_i^\circ + \frac{RT}{n\mathfrak{F}} \ln \frac{a_{Ox_i}}{a_{Red_i}} \qquad \text{(XI-4)}$$

It is more practical to use logarithms to the base 10. Substituting numerical values for R, T, and $\mathfrak{F}$ at 25°C we have:

$$E_i = E_i^\circ + \frac{0.059}{n} \log \frac{a_{ox_i}}{a_{Red_i}} \qquad \text{(XI-5)}$$

In this expression E_i° is the standard electrode potential, i.e. the value of E_i when the activities of both the reduced and oxidized forms are unity. Tables of E° values (see Appendix) are useful for the selection of reductants of suitable strength. It will be noticed that for redox couples having positive E° values, the oxidized form is superior to the hydrogen ion as an oxidant. The more effective reducing agents are the reduced forms of couples having negative E° values. Not all E° values are obtained from galvanic cell measurements. Difficulties in the form of slow rates of equilibration and the presence of reaction intermediates or extremely reactive components prevent the direct determination of E° values in such systems as $Cr_2O_7^{=}$, Cr^{+++} and Na^+, Na. E° values in such systems may be calculated from appropriate thermodynamic data.

A common error in using the Nernst equation is the reversal of the ratio of activities. This can be avoided by remembering that increasing the activity of the oxidized form should result in an increase of E_i since this represents an increase in the effectiveness of the couple as an oxidant. Hence in Equations such as XI-5, $a_{ox\,i}$, must be in the numerator.

XI-4. TYPES OF HALF-CELLS

The wide variety of electrode systems or half-cells of analytical importance can be classified in terms of the nature of the components of the redox couple. Of particular importance is the metal that is used as an electron conductor.

If the redox couple involves a metal then the half-cell consists of a metal immersed in a solution containing the metal ions.

e.g. $$Zn^{++} + 2e^- \rightleftharpoons Zn$$

$$Ag^+ + e^- \rightleftharpoons Ag$$

The form of equation XI-5 that applies in these cases is:

$$E_{Zn^{++},Zn} = E^\circ_{Zn^{++},Zn} + \frac{0.059}{2} \log a_{Zn^{++}}$$

and $$E_{Ag^+,Ag} = E^\circ_{Ag^+,Ag} + 0.059 \log a_{Ag^+}$$

The activity of the metal, Zn or Ag, is unity as will be that of every solid component (as well as H_2O) of a half-cell reaction. (Chapter III)

A useful variant of this type of half-cell is obtained by coating the metal with a slightly soluble salt or oxide.

$$AgCl_{solid} + e^- \rightleftharpoons Ag + Cl^-$$

$$PbO_{2\,solid} + 4H^+ + 4e^- \rightleftharpoons Pb + 2H_2O$$

The Nernst equations for these two electrode systems take the form:

$$E_{AgCl,\,Ag} = E^\circ_{AgCl,\,Ag} + 0.059 \log \frac{1}{a_{Cl^-}}$$

or

$$E_{AgCl,\,Ag} = E^\circ_{AgCl,\,Ag} - 0.059 \log a_{Cl^-}$$

and $$E_{PbO_2, Pb} = E^\circ_{PbO_2, Pb} + \frac{0.059}{4} \log a^4_{H^+}$$

or $$E_{PbO_2, Pb} = E^\circ_{PbO_2, Pb} + 0.059 \log a_{H^+}$$

These equations illustrate that the activities of each component in the half-cell equation appear in the Nernst equation whether or not the substances undergo a change in oxidation state. In the examples above neither Cl^- nor H^+ is oxidized or reduced.

The presence of the slightly soluble substance changes the electrode system from one whose potential responds to changes in the activity of metal ion to one whose potential depends on the activity of another ion. It is of analytical importance to be able to follow changes in a_{H^+} (for example in pH measurements) or a_{Cl^-} by means of these or similar electrode systems.

Finally, if the redox couple does not include a metal as one of its components, then an inert metal such as platinum must be used as the necessary electron conductor. In such systems the platinum metal does not appear either in the half-cell or in the Nernst equation.

$$Ce^{+4} + e^- \rightleftharpoons Ce^{+3}$$

$$2H^+ + 2e^- \rightleftharpoons H_2 \text{ (1 atmosphere)}$$

$$Cr_2O_7^= + 14H^+ + 6e^- \rightleftharpoons 2Cr^{+++} + 7H_2O$$

The corresponding expressions for the electrode potentials are:

$$E_{Ce^{+4}, Ce^{+3}} = E^\circ_{Ce^{+4}, Ce^{+3}} + 0.059 \log \frac{a_{Ce^{+4}}}{a_{Ce^{+3}}}$$

$$E_{H^+, H_2} = E^\circ_{H^+, H_2} + \frac{0.059}{2} \log \frac{a^2_{H^+}}{a_{H_2}}$$

or $E_{H^+, H_2} = 0.059 \log a_{H^+}$

(since $E^\circ_{H^+, H_2} = 0$ and $a_{H_2} = 1$, at one atmosphere pressure).

$$E_{Cr_2O_7^=, Cr^{+++}} = E^\circ_{Cr_2O_7^=, Cr^{+++}} + \frac{.059}{6} \log \frac{a_{Cr_2O_7^=}\, a^{14}_{H^+}}{a^2_{Cr^{+++}}}$$

XI-5. ELECTRODE POTENTIAL SIGN CONVENTIONS

A great deal of confusion exists in the literature concerning sign conventions. Since it is possible to write half-cell reactions either as reductions (electrons on the left-hand side of the equation) or as oxidations (electrons on the right) and further to show cell diagrams with the standard hydrogen reference electrode on either the right or left hand side, there exists in principle at least four distinctly different sign conventions. Lively controversies culminated in a decision by the International Union of Pure and Applied Chemistry (IUPAC) in 1953, to write all half-cell reactions as reductions and with a sign that corresponds to the electrostatic charge on the metal. (Thus metals more active than H_2 acquire a negative charge with respect to the hydrogen electrode and therefore are given negative values for their electrode potentials).

The sign convention for electrode potentials that will be used in this book is described above and is based on the IUPAC convention. However a number of useful reference works make use of other sign conventions. To avoid confusion when obtaining data from such sources, it is imperative to ascertain what sign convention is employed.

For example, in the text, Oxidation Potentials, by W. M. Latimer, the half cell equation is written in a manner exactly opposite to that described above, namely as an oxidation reaction. Thus in Latimer's text, as well as in many physical chemistry texts, a positive value for E° means that the reduced form of the couple is a better reducing agent than H_2. Such E° values can be readily transformed into E° values conforming to the IUPAC convention by merely changing their sign. Of course in texts using the Latimer convention of oxidation potentials the signs employed in the Nernst equation are different. Thus for the Cd, Cd^{++} electrode system:

IUPAC Convention: $Cd^{++} + 2e^- \rightleftharpoons Cd(E^\circ_{Cd^{++}, Cd} = -.40\,v)$

$$E_{Cd^{++}, Cd} = -0.40 + \frac{.059}{2} \log a_{Cd^{++}}$$

Latimer Convention: $Cd \rightleftharpoons Cd^{++} + 2e^- (E^\circ_{Cd, Cd^{++}} = +0.40\,v.)$

$$E_{Cd, Cd^{++}} = +0.40 - \frac{0.059}{2} \log a_{Cd^{++}}$$

XI-6. FACTORS THAT AFFECT ELECTRODE POTENTIALS

The electrode potential of any redox couple E_i will vary with the temperature, which affects the value of both E_i° as well as the coefficient of the logarithmic term, and with the activities of the oxidized and reduced forms of the couple (Equation XI-4). The temperature variation of E° like its counterpart K will depend on the heat of reaction. (See Chapter II-V-b)

Values of the electrode potentials, E_i, may vary with (1) changes in the analytical concentrations of the reaction components, (2) the ionic strength of the solution which will affect values of the activity coefficients, (3) the pH of the solution where hydrolysis or hydroxy complex formation is involved, and (4) the presence of complexing agents other than hydroxide ion which affect the concentrations of uncomplexed oxidized or reduced forms when metal ions are involved.

If the solvent is modified as for example by the addition of ethanol to the aqueous solution, the value of E_i° is likely to change significantly. Modifying the nature of the solvent will affect not only E_i° but also the activities of the components of the redox couple by altering activity coefficients and extents of complexation reactions.

XI-6-(a). EFFECT OF CONCENTRATION ON ELECTRODE POTENTIALS

It is obvious from the Nernst equation XI-5 that the value of the electrode potential of a redox couple will depend on the values of the activities of the reaction components. This means of course that changes in the concentrations of any of the components will likewise affect the value of the electrode potential. The precise way in which the concentration affects the potential is related to the type of half-cell reaction involved, i.e., in the half-cell reactions

$$Ag^+ + e^- \rightleftharpoons Ag$$

and

$$Hg^{++} + 2e^- \rightleftharpoons Hg$$

the electrode potentials vary with the concentrations of the metal cations. In the following half-cell reactions however:

$$Fe^{+3} + e^- \rightleftharpoons Fe^{+2}$$

$$Sn^{+4} + 2e^- \rightleftharpoons Sn^{+2}$$

the values of the electrode potentials are independent of the concentration of the individual metal ion, but depend on the ratio of the concentrations of the oxidized and reduced forms.

Another type of concentration dependence occurs in the half-cell reactions:

$$Cr_2O_7^{=} + 14H^+ + 6e^- \rightleftharpoons 2Cr^{+3} + 7H_2O$$

$$I_3^- + 2e^- \rightleftharpoons 3I^-$$

where the coefficients of the oxidized and reduced forms are not the same. For this reason the concentration factors in the Nernst equation will be raised to different powers. Hence in such cases the electrode potential values will vary with the absolute values of the concentrations as well as concentration ratios.

It is useful to visualize the way in which the electrode potential of a redox couple varies with the concentrations of the reaction components. A logarithmic diagram analogous to the type used in acid-base systems can be constructed in the following manner.

Let us rewrite the Nernst Equation XI-5 as:

$$\frac{E_i}{0.059} = \frac{E_i^\circ}{0.059} + \frac{1}{n} \log \frac{a_{Ox}}{a_{red}}$$

and define $\frac{E_i}{0.059}$ as η. The use of η which is dimensionless is more convenient than the use of E_i since every tenfold change in the activity ratio will cause a unit change in η. Thus the quantity η is seen to be analogous to pH.

$$\eta = \eta^\circ + \frac{1}{n} \log \frac{a_{Ox}}{a_{red}} \qquad \text{(XI-6)}$$

$$pH = pK_a + \log \frac{C_B}{C_{HB}}$$

In Fig. XI-1 is shown a redox diagram for iron representing a variation of the activity of the reaction components as a function of η.

The half-cell reactions diagrammatically represented in Fig. XI-1 are:

$$Fe^{+3} + e^- \rightleftharpoons Fe^{+2}$$

$$Fe^{+2} + 2e^- \rightleftharpoons Fe$$

with

$$\eta = \eta^\circ_{Fe^{+3}, Fe^{+2}} + \log a_{Fe^{+3}} - \log a_{Fe^{+2}} \qquad \text{(XI-7)}$$

$$\eta = \eta^\circ_{Fe^{+2}, Fe} + \frac{1}{2} \log a_{Fe^{+2}} \qquad \text{(XI-8)}$$

From Equation XI-7 we can construct the right hand portion of the Fig. XI-1 which looks quite similar to Fig. V-1. The activities of Fe^{+3} and Fe^{+2} are equal at $\eta = \eta^\circ_{Fe^{+3}, Fe^{+2}}$

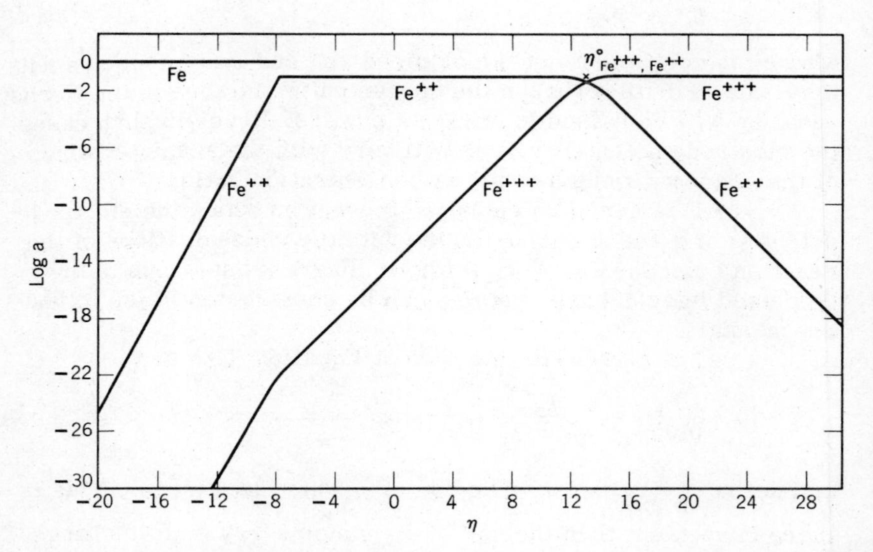

Fig. XI-1 — Redox Diagram for Iron. The sum of the activities of Fe^{++} and Fe^{+++} is 0.1 except in the region where solid Fe exists in equilibrium.

(intersection of Fe^{+3} and Fe^{+2} line). At values of η greater than $\eta°$, $a_{Fe^{+3}}$ becomes constant and essentially equal to the total iron activity. Similarly at values of η lower than $\eta°$, Fe^{+2} is the predominant species. This situation continues with decreasing η values down to $\eta = \eta°_{Fe^{+2}, Fe} + \frac{1}{2} \log 0.10$. Below this value the total activity of Fe^{+2} cannot exceed what corresponds to equilibrium with pure Fe, ($a_{Fe} = 1$ for pure metallic iron).

Fig. XI-2 is a redox diagram for copper. This diagram presents a contrast to Fig. XI-1 where there is a range of η values where each oxidation state predominates. In the case of copper, the concentration of Cu(I) is always small (unless suitable complexing agents are present).

XI-6-(b). EFFECT OF COMPLEX FORMATION ON ELECTRODE POTENTIALS

In this section we will consider the effect of complex formation on E_i values of two types of half-cells which will be illustrated by:

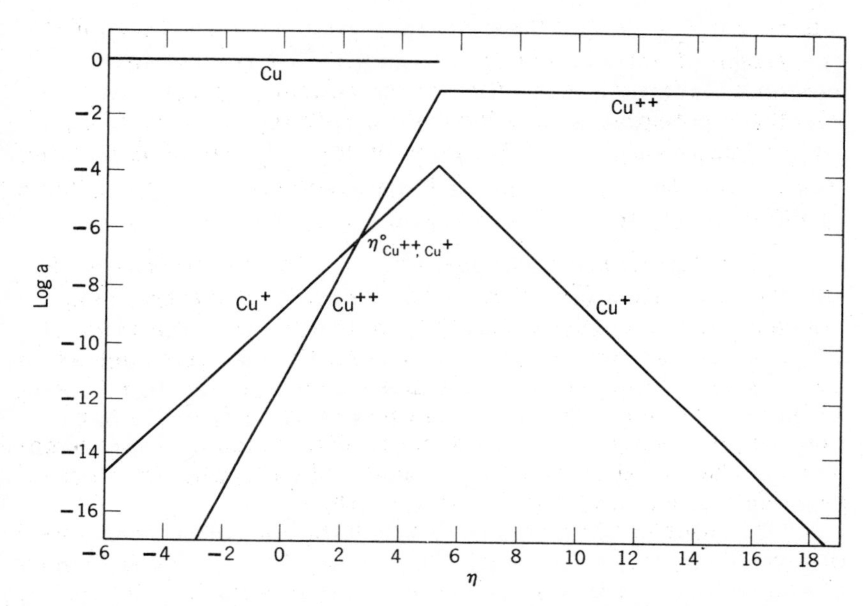

Fig. XI-2 — Redox Diagram for Copper. The sum of the activities of Cu^+ and Cu^{++} is 0.1 except where solid Cu exists at equilibrium.

$$Ag^+ + e^- \rightleftharpoons Ag$$

and $$Fe^{+3} + e^- \rightleftharpoons Fe^{++}$$

Let us consider the effect that a complexing agent such as NH_3 will have on the electrode potential of the Ag^+/Ag half-cell.

$$E_{Ag^+, Ag} = E^\circ_{Ag^+, Ag} + .059 \log a_{Ag^+}$$

If the total silver concentration in solution is C_{Ag} and the total uncomplexed ammonia is $[NH_3]$, then

$$a_{Ag} = [Ag^+]\,\gamma_{Ag^+} \text{ and } [Ag^+] = \beta_0 C_{Ag}$$

where $$\beta_0 = \frac{1}{1 + k_1[NH_3] + k_1 k_2 [NH_3]^2}$$

Hence:

$$E_{Ag^+, Ag} = E^\circ_{Ag^+, Ag} + .059 \log \gamma_{Ag^+} + .059 \log \beta_0 + 0.059 \log C_{Ag}$$

or $$E_{Ag^+, Ag} = E^{\circ\prime}_{Ag^+, Ag} + .059 \log C_{Ag}$$

where $E^{\circ\prime}_{Ag^+, Ag}$, called the formal potential (see below), includes the effect of activity coefficient as well as the extent of complex formation. The effect of the complexing agent on the electrode potential is thus incorporated in the value of $E^{\circ\prime}_{Ag^+, Ag}$ which changes with $[NH_3]$. Values of $E^{\circ\prime}_{Ag^+, Ag}$ can be calculated since values of γ_{Ag^+} as well as the stepwise formation constants of the complex $Ag(NH_3)_2^+$ are known.

The formal potential, $E^{\circ\prime}_{Ag^+, Ag}$, is seen to change 59 millivolts for each factor of 10 change in β_0. If the ammonia concentration increases tenfold from $[NH_3] = 1.0$, the value of $E^{\circ\prime}_{Ag^+, Ag}$ is lowered by 2×59 or 118 millivolts. Inasmuch as in this region of C_{NH_3} the predominant species is $Ag(NH_3)_2^+$, (see problem 2, Chapter IX), β_0 varies inversely as $[NH_3]^2$. Such considerations are helpful in deriving diagrammatic representations of the variation of $E^{\circ\prime}_{Ag^+, Ag}$ and $E^{\circ\prime}$ values of other systems as a function of ligand concentrations.

One of the methods by which the formation constants of metal complexes can be determined involves the measurement of electrode potentials of half cells of the type, $M^{n+} + ne^- \rightleftharpoons M$, in a series of solutions of varying ligand concentrations.

In a redox system involving two metal ions such as the Fe^{+3}/Fe^{+2} couple, a ligand is likely to complex both ions although to different extents. Consider for example a solution containing $C_{Fe(III)}$ moles/ l of Fe III, $C_{Fe(II)}$ moles/l of Fe (II) and $[Cl^-]$ moles/l of uncomplexed Cl^-. The electrode potential of the Fe^{+3}/Fe^{+2} couple is given by

$$E_{Fe^{+3}, Fe^{+2}} = E^{\circ}_{Fe^{+3}, Fe^{+2}} + 0.059 \log \frac{a_{Fe^{+3}}}{a_{Fe^{+2}}}$$

$$a_{Fe^{+3}} = [Fe^{+3}]\, \gamma_{Fe^{+3}} \text{ and } a_{Fe^{+2}} = [Fe^{+2}]\, \gamma_{Fe^{+2}}$$

Further, $[Fe^{+3}] = \beta_{o(III)}\ C_{Fe(III)}$ and $[Fe^{+2}] = \beta_{o(II)}\ C_{Fe(II)}$

where the β_0 values refer to the fraction of metal ions not complexed with Cl^- in each case

$$\beta_{0_{(III)}} = \frac{1}{1 + k_1[Cl^-] + k_1 k_2 [Cl^-]^2 + k_1 k_2 k_3 [Cl^-]^3 + k_1 k_2 k_3 k_4 [Cl^-]^4}$$

A similar expression can be written for $\beta_{0(II)}$. Substituting these expressions in the Nernst equation, the relation of the total concentrations of Fe(III) and Fe(II) to the electrode potential, is obtained.

$$E_{Fe^{+3}, Fe^{+2}} = E^{\circ}_{Fe^{+3}, Fe^{+2}} + 0.059 \log \frac{\beta_{o(III)}}{\beta_{o(II)}} + 0.059 \log \frac{\gamma_{Fe^{3+}}}{\gamma_{Fe^{2+}}} + 0.059 \log \frac{C_{Fe(III)}}{C_{Fe(II)}}$$

or $$E_{Fe^{+3}, Fe^{+2}} = E^{\circ\prime}_{Fe^{+3}, Fe^{+2}} + 0.059 \log \frac{C_{Fe(III)}}{C_{Fe(II)}}$$

where the formal potential $E^{\circ\prime}_{Fe^{+3}, Fe^{+2}}$ varies with the ratio of activity coefficients of Fe^{+3} and Fe^{+2} and with the difference in the extent of complexation of Fe^{+3} and Fe^{+2} (i.e. with the β_0 ratio).

Example I

What is the effect of a tenfold increase in the $[Cl^-]$ on the electrode potential of Fe^{+3}/Fe^{+2} in a solution such that the bulk of the Fe(III) is in the form of $FeCl_4^-$ and that of Fe(II) is present as $FeCl^+$?

$\beta_{o(III)}$ varies inversely as $[Cl^-]^4$ and $\beta_{o(II)}$ varies inversely as $[Cl^-]$.

Hence the ratio $\frac{\beta_{o(III)}}{\beta_{o(II)}}$ is inversely proportional to to $[Cl^-]^3$ so that the drop in $E^{\circ\prime}_{Fe^{+3}, Fe^{+2}}$ and hence in $E_{Fe^{+3}, Fe^{+2}}$ is 3×0.059 or 0.177 v.

XI-6-(c). EFFECT OF pH ON ELECTRODE POTENTIALS

As mentioned in Section XI-4, when hydrogen ions (or hydroxyl ions) appear in the half-cell reaction the corresponding electrode potentials are seen to vary with a_{H^+}. In addition to such systems those involving metal ions capable of forming hydroxyl complexes will also have pH-dependent electrode potentials. As a matter of fact this situation is very much like that dealt with above describing the effect of Cl^- on the Fe^{+3}/Fe^{+2} system. It is merely necessary to replace $[Cl^-]$ by $K_w/[H^+]$ (the equivalent of $[OH^-]$) and of course the appropriate k values.

XI-7. FORMAL POTENTIALS

In view of the complications arising from the formation of hydroxy and other complexes as well as from the variation of activity coefficients, the use of standard potentials (E°) is not nearly as practical as that of formal potentials (E°′). Values of formal potentials which have been experimentally determined for a large number of systems under well-defined conditions of acidity and ionic strength, will be used wherever possible. Such values of formal potentials are sometimes tabulated or, alternatively, diagramatically represented and will be illustrated now for a few typical examples. These diagrams differ from Figs. XI-1 and 2 where the formal potentials were considered to be invariable while the activities were varying. In redox diagrams such as Fig. XI-3 and others, the activities of all the components are considered to be unity on each of the lines, and will show the variation of E°′ with pH and complexing agents.

The areas between the lines represent regions in which one species predominates and are labeled accordingly. The line of demarcation between any two areas represents the variation of E°′ of a couple composed of the two species, as a function of pH.

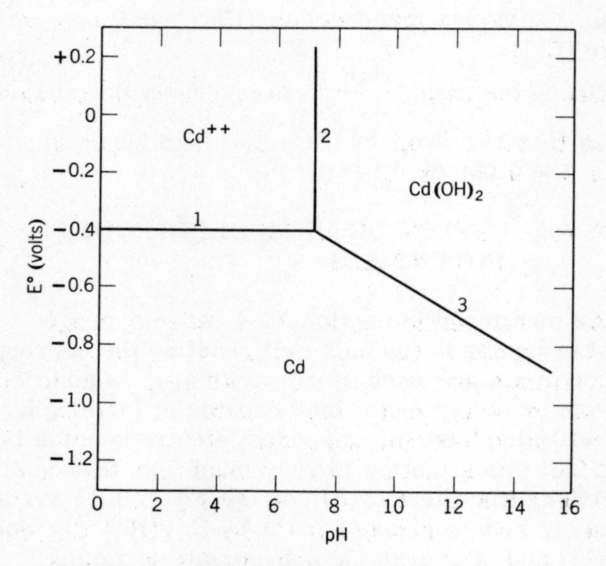

Fig. XI-3 — Formal Potentials for Cadmium

In addition to affording a concise presentation of a great deal of data, these diagrams are useful for depicting the redox and acid-base behavior of an element.

To illustrate the construction of a potential-pH redox diagram, let us use the system Cd, Cd^{+2}.

Line 1. $Cd^{+2} + 2e^{-} \rightleftharpoons Cd$; $E°' = -0.40$ v.

This value is independent of pH; hence Line 1 is horizontal.

Line 2. $Cd(OH)_2 \rightleftharpoons Cd^{+2} + 2OH^{-}$; $\log K_{sp} = -13.5$;

At $a_{Cd^{+2}} = 1$, pOH = 6.8; pH = 7.2.

This value is independent of E and hence line 2 is vertical.

Line 3. $Cd(OH)_2 + 2H^{+} + 2e^{-} \rightleftharpoons Cd + 2H_2O$; at pH 7.2, $E°' = 0.02 - 0.059$ pH.

A question might arise as to why line 3 does not continue into the region between lines 1 and 2, i.e. into the region of pH > 7.2. Since line 3 represents the formal potential of a redox couple involving $Cd(OH)_2$, the line must stop at pH 7.2, since below this value, no $Cd(OH)_2$ exists. A more complex diagram is shown in Fig. XI-4.

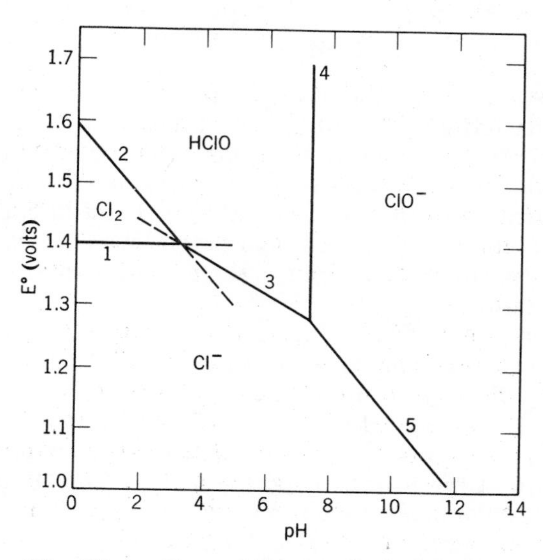

Fig. XI-4 — Formal Potentials for Chlorine

Line 1. $Cl_2 + 2e^- \rightleftharpoons 2Cl^-$; $E°' = 1.4$ V.

This value is independent of pH; hence line 1 is horizontal.

Line 2. $HClO + H^+ + e^- \rightleftharpoons \frac{1}{2} Cl_2 + H_2O$; $E°' = 1.6 - 0.059$ pH

This line crosses line 1 at pH 3.3. At higher pH values HClO is a poorer oxidant than Cl_2. Hence line 2 stops at pH 3.3, i.e. the dashed portion of line 2 is imaginary. Similarly the dashed portion of line 1 is imaginary since Cl_2 disproportionates to HClO and Cl^- beyond pH 3.3.

Line 3. $HClO + H^+ + 2e^- \rightleftharpoons Cl^- + H_2O$; $E°' = 1.5 - \frac{0.059}{2}$ pH

The lower pH limit for the existence of line 3 is pH 3.3. The dashed portion of this line which extends into the area where Cl_2 predominates lies below line 2. This means that HClO in the reaction for line 2 is a better oxidant than it is in the reaction for line 3. Hence reaction for line 2 will take precedence in this region. The upper pH limit of line 3 occurs at pH 7.3 when HClO is transformed to ClO^-.

Line 4. $HClO \rightleftharpoons H^+ + ClO^-$; pK = 7.3

This value is independent of E and results in a vertical line.

Line 5. $ClO^- + 2H^+ + 2e^- \rightleftharpoons Cl^- + H_2O$;

$$E°' = 1.7 - 0.059 \text{ pH}$$

Notice that line 5 has the same slope as line 2 even though the half cell reactions have different numbers of H^+ and e^-; the ratio of H^+ to e^-, which is the factor that determines the slopes of these lines, is the same for both lines.

Redox diagrams of this sort can also be used to represent the effect of complexation on formal potentials ($E°'$ values). Fig. XI-5 is the redox diagram for the copper system in a chloride medium.

The vertical lines in this diagram (for processes that are independent of E) are analogous to those in Figs. XI-3 and 4 and represent the logs of the formation constants. The vertical line that would correspond to solid CuCl would appear at pCl = 6.7, if it were not for the fact that Cu(I) disproportionates to Cu^{++} and Cu at the lower pCl value of 3.9. The horizontal lines involve redox processes for which the half-cell reactions contain the same number of Cl^- on both sides.

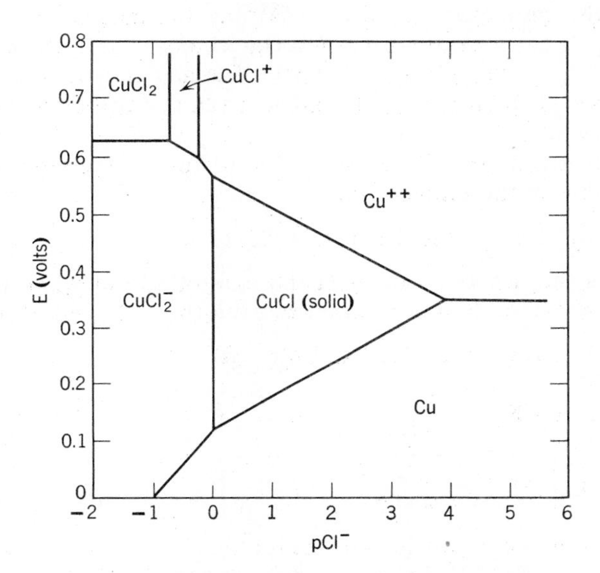

Fig. XI-5 — Formal Potentials for Copper in Chloride Solutions.

XI-8. CELL REACTIONS

Up to this point we have been considering the half-cell reaction of a particular redox couple and its electrode potential which is the e.m.f. of a cell composed of the standard hydrogen electrode coupled with this half-cell. The e.m.f.'s of cells consisting of any two redox couples may be obtained by taking the algebraic difference between the electrode potentials. The cell reaction, however, is obtained by the subtraction of the two half-cell reactions in such a way as to cancel the number of electrons.

$$Ag^+ + e^- \rightleftharpoons Ag \qquad E^\circ_{Ag^+,Ag} = 0.80 \text{ v.}$$

$$Zn^{++} + 2e^- \rightleftharpoons Zn \qquad E^\circ_{Zn^{++},Zn} = -0.76 \text{ v.}$$

Multiplying the Ag^+/Ag half-cell equation by 2 and subtracting the Zn^{++}/Zn half-cell equation, we have:

$$2\,Ag^+ + Zn \rightleftharpoons 2\,Ag + Zn^{++}$$

$$E^\circ_{Cell} = 0.80 - (-0.76) = 1.56 \text{ v.}$$

It is important to note that even though the Ag^+/Ag half-cell equation was multiplied by 2 in obtaining the balanced cell reaction, the $E°_{Ag^+, Ag}$ value remained the same. The reason for this is that a Nernst equation written for any multiple of a half-cell reaction is identical with that written for the half-cell reaction itself.

The cell e.m.f. is related to the equilibrium constant of the cell reaction by the expression

$$\Delta F° = -n\mathfrak{F}E°_{Cell} \quad \text{(See equation XI-1)}$$

where n is the number of electrons cancelled out when the two half-cell reactions are combined to give the cell reaction.

Also $-\Delta F° = RT \ln K$ (See Chapter II)

Therefore $\log K = \frac{n\mathfrak{F}}{2.3\,RT} \cdot E°_{Cell}$

or at 25°C $\log K = \frac{nE°_{Cell}}{0.059}$

For the cell reaction just described:

$$\log K = \frac{2 \times 1.56}{0.059} = 52.8$$

If we had reversed the two half-cell reactions to give the reverse cell reaction:

$$2\,Ag + Zn^{++} \rightleftharpoons Zn + 2Ag^+$$

this would have resulted in a negative $E°_{Cell}$ and

$\log K = -52.8$.

Thus the sign as well as the magnitude of the $E°_{Cell}$ indicates the direction and extent of a reaction.

On occasion the need arises to calculate an E° value of a half-cell reaction which may only be obtained by combining two known half-cell reactions. For example the half-cell reaction:

$$Fe^{+3} + 3e^- \rightleftharpoons Fe$$

is the sum of the half-cell reactions:

$$Fe^{+3} + e^- \rightleftharpoons Fe^{+2}$$

$$Fe^{+2} + 2e^- \rightleftharpoons Fe$$

The calculation of $E^\circ_{Fe^{+3},Fe}$ is analogous to the calculation of the equilibrium constant K_3 for the reaction resulting from the combination of two other equilibrium reactions:

$$K_3 = K_1 \times K_2 \text{ or } \log K_3 = \log K_1 + \log K_2$$

Since the E° for each half-cell may be considered the e.m.f. of a cell involving the standard hydrogen electrode, we may write for each half-cell:

$$\log K = \frac{n \cdot E^\circ}{0.059}$$

From this it will be seen that

$$3 \cdot \frac{E^\circ_{Fe^{+3},Fe}}{0.059} = \frac{E^\circ_{Fe^{+3},Fe^{+2}} + 2E^\circ_{Fe^{+2},Fe}}{0.059}$$

$$E^\circ_{Fe^{+3},Fe} = \frac{0.77 + 2 \times (-0.44)}{3} = -0.04 \text{ v.}$$

SUGGESTIONS FOR FURTHER READING

1. H. A. Laitinen, Chemical Analysis, McGraw Hill Book Co., Inc., (1960), Chapter 17.
2. G. Charlot, Qualitative Inorganic Analysis, Methuen and Co., Ltd., London,)1954), Chapters 3, 7, 8, 13 and 14.
3. L. G. Sillen, in Treatise on Analytical Chemistry, Editors I. M. Kolthoff and P. J. Elving, Interscience Publishers (1959), Part I, Vol. I, Chapter 8.

XI-9. PROBLEMS*

1. Write expressions for the electrode potentials of the following redox systems.
   (a) $Zn^{++} + 2e \rightleftharpoons Zn$
   (b) $AgBr + e \rightleftharpoons Ag + Br^-$
   (c) $Tl^{+3} + 2e \rightleftharpoons Tl^+$
   (d) $O_2 + 2H^+ + 2e \rightleftharpoons H_2O_2$
   (e) $Sb_2O_5 + 6H^+ + 4e \rightleftharpoons 2\,SbO^+ + 3\,H_2O$

*In this and subsequent problem sets, the student is expected to calculate his own values of concentration constants and potentials or, if Chap. III has been omitted, to use the values tabulated in the appendix.

(f) $MnO_4^- + 8H^+ + 5e \rightleftharpoons Mn^{++} + 4H_2O$
(g) $MnO_4^- + 4H^+ + 3e \rightleftharpoons MnO_2 + 2H_2O$

2. Calculate the e.m.f's of the following cells:
   (a) Pt | H_2 (0.5 atm.), HCl (0.01 M) || H^+(a = 1), H_2 (1 atm.) | Pt
   (b) $Ag_{(solid)}$, $AgCl_{(solid)}$ | KCl (0.03 M) || H^+(a=1), H_2 (1 atm.)| Pt
   (c) Pt | Fe^{+3} (0.05 M), Fe^{+2} (0.001 M) || H^+(a=1), H_2 (1 atm.)| Pt
   (d) Zn | $ZnSO_4$ (0.10 M) || H^+(a=1), H_2 (1 atm.) | Pt
   (e) Ag | $AgNO_3$ (0.25 M) || H^+(a=1), H_2 (1 atm.) | Pt
   (f) Pt | Tl^+ (0.10 M), Tl^{+3} || (0.03 M) H^+(a=1), H_2 (1 atm.) | Pt
   (g) Tl | Tl^+ (0.004 M) || H^+(a=1), H_2 (1 atm.) | Pt
   (h) Cd | $Cd(NO_3)_2$ (0.059 M) || H^+(a=1), H_2 (1 atm.) | Pt
3. Calculate the potential difference between a platinum electrode immersed in solution (i) and a platinum electrode immersed in solution (ii), electrical contact between solutions (i) and (ii) being maintained by means of a salt bridge which contributes no additional potential difference.

	Solution (i)	Solution (ii)
(a)	0.03M Fe^{+2} and 0.05M Fe^{+3}	0.10M Fe^{+2} and 0.05M Fe^{+3}
(b)	0.05M Sn^{+4} and 0.02M Sn^{+2}	0.005M Sn^{+4} and 0.003M Sn^{+2}
(c)	0.10M Ce^{+4} and 0.02M Ce^{+3}	1.0×10^{-4} M Ce^{+3} and 0.03M Ce^{+4}
(d)	0.01M Ce^{4+} and 0.02M Ce^{3+}	1×10^{-4} M Ce^{4+} and 2×10^{-4} M Ce^{3+}

4. Calculate the e.m.f. of the following cell when the concentration, C, of KCl is (a) 0.01 M (b) 0.1 M and (c) when the solution is saturated with respect to KCl.
   Pt | H_2 (1 atm), H^+ (0.10 M) || C M KCl, Hg_2Cl_2 (solid) | Hg
5. Calculate the thermodynamic equilibrium constants for the following reactions:
   (a) $Fe^{+2} + Cu^{+2} \rightleftharpoons Fe^{+3} + Cu^+$
   (b) $Ce^{+4} + Fe^{+2} \rightleftharpoons Fe^{+3} + Ce^{+3}$
   (c) $SO_2 + 2H_2O + I_2 \rightleftharpoons 2I^- + 3H^+ + HSO_4^-$
   (d) $Ag^+ + Fe^{+2} \rightleftharpoons Fe^{+3} + Ag$
6. Construct a potential - log C diagram of Ce^{4+} - Ce^{3+} system similar to Fig. XI-1. On the same diagram construct the Fe^{3+} - Fe^{2+} system. How can this combined diagram be used to calculate the equilibrium constant for the reaction
   $Fe^{2+} + Ce^{4+} \rightleftharpoons Ce^{3+} + Fe^{3+}$.
   Compare answer with that obtained in previous problem.
7. (a) Calculate the solubility product constants of AgI and CuI, and the formation constant of $Ag(S_2O_3)_2^{-3}$ from the following data:

   $I_2 + 2e \rightleftharpoons 2I^-$ $E° = +0.536$ v.
   CuI(solid) + e $\rightleftharpoons$ Cu° + I^- $E° = -0.185$ v.

$Cu^{++} + e \rightleftharpoons Cu^{+}$ $E° = .15$ v.
$Cu^{++} + I^{-} + e \rightleftharpoons Cu\,(solid) + \frac{1}{2} I_2$ $E° = +0.086$ v.
$Ag^{+} + e \rightleftharpoons Ag\,(solid)$ $E° = +0.80$ v.
$AgI\,(solid) + e \rightleftharpoons Ag\,(solid) + I^{-}$ $E° = -0.15$ v.
$Ag(S_2O_3)_2^{-3} + e \rightleftharpoons Ag\,(solid) + 2S_2O_3^{=}$ $E° = -0.01$ v.

(b) Calculate the following constants with the aid of tabulations of standard electrode potentials: overall formation constants of $Cd(CN)_4^{=}$, $Ag(NH_3)_2^{+}$, $Ag(CN)_2^{-}$, $Cd(NH_3)_4^{2+}$, the solubility product constants of AgCl, AgBr, Ag_2S, $Cd(OH)_2$ and CdS.

8. From the equilibrium constant of the reaction calculated in question 5-(b) determine the composition of a solution at equilibrium when 50 ml of 0.05 M Ce^{+4} is mixed with 250 ml of 0.10 M Fe^{+2} .
9. Calculate the concentrations of the metal ions present at equilibrium when an excess of the Tl metal is added to a solution which is 0.050 M in Cd^{++} and 0.003 M in Tl^{+}.
10. Calculate the change in the electrode potential of the system: $Ag_{(solid)}$ | $AgNO_3$ (0.15 M) vs. standard hydrogen electrode. when sufficient KCl is added to the solution until the equilibrium concentration of Cl^{-} is 0.15 M.
11. On the assumption that reversible electrode potentials are applicable here, calculate the electrode potential of a system containing 0.05 M Mn^{++} and 0.10 M MnO_4^{-} at a pH of (a) 1.50 (b) 0.0 (c) −1.0
12. (a) Calculate the change in the electrode potential of a system which has 0.1 M Cu^{2+} in contact with copper metal on the addition of 0.3 moles Na_2 EDTA per liter if the pH after addition of EDTA is maintained at 5.00. How does the potential vary with pH?

    (b) Calculate the change in the electrode potential of a system which is 0.10 M in Fe^{+2} and 0.10 M in Fe^{+3} on the addition of 0.4 moles Na_2 EDTA per liter if the pH after addition of EDTA is maintained at 5.00. How does the potential vary with pH?
13. Construct a formal potential diagram to represent
    (a) dependence of the potential of the bromine system on pH (0-14)
    (b) dependence of the potential of the As(V)-As(III) system on pH(0-14)
    (c) dependence of the potential of the cadmium system on pNH_3 (from −1 to +6).
    (d) dependence of the potential of the mercury system on pCl (from −1 to +6).

XII

Liquid-Liquid Extraction and Ion Exchange Equilibria

Liquid-liquid extraction (solvent extraction) procedures have proven very useful in analytical separations. These compare favorably with precipitation procedures because of (a) the virtual absence of phenomena resembling coprecipitation, (b) the applicability to the separation of traces of substances and (c) rapidity of separation.

Like precipitation, solvent extraction involves the removal of a substance from one phase (usually aqueous) into another (usually organic). A study of how the extent of extraction varies with experimental conditions is based on equilibrium considerations.

XII-1. THE DISTRIBUTION LAW

The equilibrium distribution of a substance such as ethanol between two essentially immiscible phases such as water and benzene is described by the expression:

$$K_D = \frac{[C_2H_5OH]_o}{[C_2H_5OH]}$$

where the subscript $_o$ represents concentration in the organic phase. This expression applies to all substances such as ethanol which do not enter into any chemical reaction in either the aqueous or organic phase. Thus for any solute A:

$$K_D = \frac{[A]_0}{[A]} \qquad \text{(XII-1)}$$

K_D, called the distribution coefficient is not strictly speaking a constant because concentrations rather than activities are used. This represents only a minor problem since variation in K_D values over a range of analytically useful conditions is usually small.

If the solute is involved in chemical interactions in either phase, then the simple distribution expression, (XII-1) although still valid, is not adequate to describe the system. For such cases an expression involving all reaction equilibria must be developed. In this connection the quantity D, the distribution ratio, describing the ratio of total concentration of the solute present in whatever form is useful. This stoichiometric ratio is defined by:

$$D = \frac{\text{Analytical concentration of A in the organic phase}}{\text{Analytical concentration of A in the aqueous phase}} = \frac{C_{A_0}}{C_A} \qquad \text{(XII-2)}$$

We will derive expressions showing how D varies with experimental conditions in terms of the appropriate equilibria concerned.

Of practical interest to the analytical chemist in describing the completeness of extraction, is the quantity % E, the percentage of solute in the organic phase. This quantity may be easily shown to be the following function of D

$$\%\,E = \frac{100\,D}{D + (V/V_0)} \qquad \text{(XII-3)}$$

where V and V_0 are the volumes of the aqueous and organic phases, respectively. From this expression we could see the advantage of using a large volume ratio of organic to aqueous phase in increasing the % E. With a given volume of organic solvent, however, a larger total amount of solute may be extracted by multiple extractions each employing a fraction of the total volume of organic solvent.

The distribution of I_2 between water and CCl_4 may be described in terms of the simple distribution law.

i.e. $$D = \frac{(C_{I_2})_0}{C_{I_2}} = K_D = \frac{[I_2]_0}{[I_2]}$$

In the presence of KI in the aqueous phase, however, iodine is involved in the reaction:

$$I_2 + I^- \rightleftharpoons I_3^- \; ; \; K = \frac{[I_3^-]}{[I_2][I^-]}$$

Now $C_{I_2} = [I_2] + [I_3^-]$

$$\text{Therefore} \quad D = \frac{[I_2]_0}{[I_2] + [I_3^-]}$$

Dividing both numerator and denominator by $[I_2]$ and substituting K_D and $K[I^-]$ for the two ratios we obtain

$$D = \frac{K_D}{1 + K[I^-]} \qquad \text{(XII-4)}$$

This expression shows how the extraction of iodine varies with the concentration of KI in the aqueous phase.

A system in which a solute is involved in both phases is illustrated with the distribution of acetic acid between benzene and water. This system is diagrammatically represented in Fig. XII-1.

$$2\,HOAc \overset{K_p}{\rightleftharpoons} (HOAc)_2 \qquad \text{Dimerisation} \quad K_p = \frac{[(HOAc)_2]_0}{[HOAc]_0^2}$$

K_D (between phases) — Benzene / water — K_a

$$HOAc \rightleftharpoons H^+ + OAc^- \qquad \text{Dissociation:} \quad K_a = \frac{[H^+][OAc^-]}{[HOAc]}$$

Fig. XII-1

$$D = \frac{[HOAc]_0 + 2[(HOAc)_2]_0}{[HOAc] + [OAc^-]}$$

Dividing both numerator and denominator by [HOAc], and making the appropriate substitutions:

$$D = \frac{K_D\,[1 + 2K_p\,[HOAc]_0]}{1 + K_a/[H^+]}$$

From this expression it may be seen that the distribution ratio of acetic acid depends on the pH of the aqueous phase and on the absolute value of the concentration of HOAc in the organic phase and therefore on the total concentration of HOAc present. In general when the degree of association of a solute in one phase is different from that of the other phase the value of D will vary with the total concentration of the solute present.

XII-2. METAL EXTRACTION SYSTEMS

One of the most important analytical uses of solvent extraction is in the separation of metal ions. There are two general types of metal extraction systems; chelate systems and ion association systems.

XII-2 (a). CHELATE EXTRACTION EQUILIBRIA

A large number of chelating agents which may be represented generally as HL (weak monoprotic acids) form uncharged metal chelates which are soluble in organic solvents and are therefore extractable. Such reagents, which include acetylacetone, dimethylglyoxime and diphenylthiocarbazone, are generally dissolved in organic solvents immiscible with water.

Fig. XII-2 illustrates the equilibria involved in a typical chelate extraction.

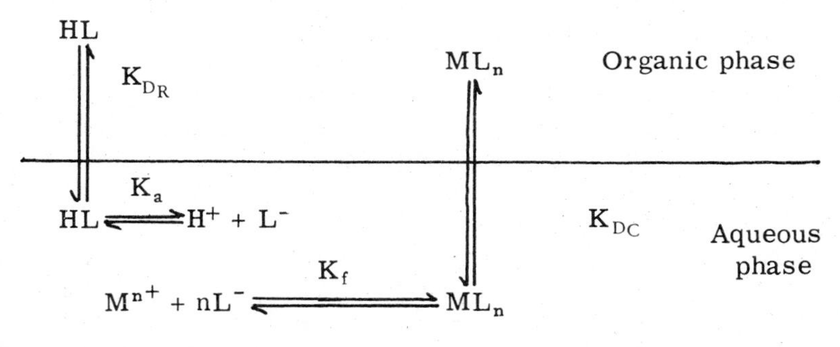

Fig. XIII-2.

With the simplifying assumptions that (a) the only metal-containing species present in the organic phase is ML_n, (b) that intermediate complexes such as ML^+_{n-1} etc. are absent, and (c) that no hydroxy or other anion complexes of the metal ion are present in the aqueous phase, the expression for D becomes:

$$D = \frac{[ML_n]_o}{[ML_n] + [M^{n+}]} \qquad \text{(XII-5)}$$

With appropriate substitutions, then

$$D = \frac{K_{D_C}}{1 + \frac{K_{D_R}^n [H^+]^n}{K_f\ K_a^n [HL]_0^n}}$$

At values $D < K_{D_C}$

$$D = \frac{K_{D_C}\ K_f\ K_a^n [HL]_0^n}{K_{D_R}^n\ [H^+]^n} \tag{XII-6}$$

Equation (XII-7) shows, by the absence of any metal concentration terms, that the extraction of a metal is independent of the total amount of metal present and hence may be used for trace and macro levels alike. For a given metal chelate system the extent of extraction increases with the concentration of the ligand in the organic phase and with the pH of the aqueous phase.

The course of solvent extraction of a series of metal ions with a given chelating agent can be conveniently represented in Fig. XII-3 where the variation of log D with pH for three metal ions at constant reagent concentration is shown. Each of the plots is seen to have a linear portion whose slope is equal to the charge of the metal ion (strictly speaking the slope is equal to the number of hydrogen ions released per metal ion in the overall extraction equation) and a horizontal portion of maximum log D which is reached at $D = K_{D_C}$. In this horizontal portion of the curve the intrinsic solubility characteristics of the chelate (as reflected by K_{D_C}) rather than the formation equilibria limits the extent of extraction. From the curves may be seen that the more highly charged the metal cation, the narrower the pH range required for complete extraction. For most analytical purposes extraction may be considered to be complete at 99% extraction. This corresponds to a value of log D = 2 (assuming a phase volume ratio of unity). Since we can usually ignore extractions of less than 1% the practical limits of extraction of a particular metal are:

$$-2 < \log D < +2.$$

By examining extraction data for various chelating agents represented in the form of curves such as Fig. XII-4 it is possible to predict which metals may be quantitatively separated from each other by one extraction as well as the pH range in which this separation should be carried out. In order to calculate the minimum separation of two curves let us consider

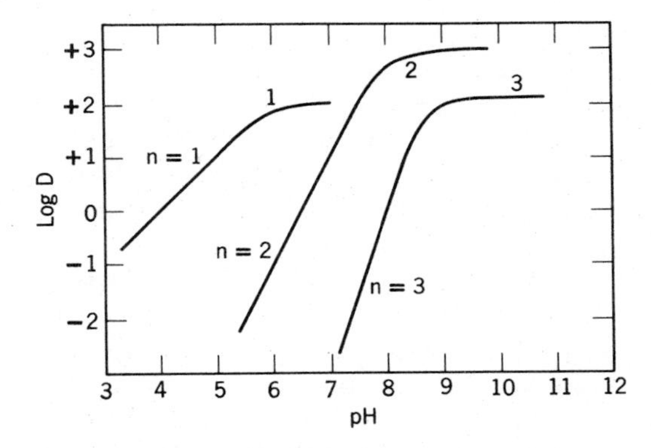

Fig. XII-3. The Solvent Extraction of a Series of Metal Ions with a Chelating Agent.
n = number of hydrogen ions released per metal ion.

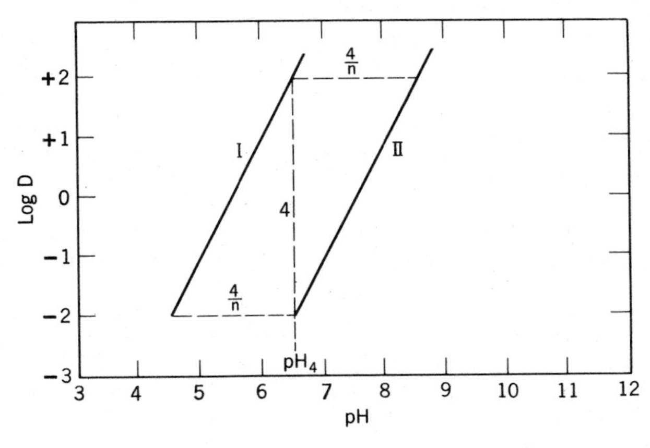

Fig. XII-4. Extraction Curves of Two Metals Having the Same Valence, n.

Fig. XII-4 in which the extraction curves for two metals I and II having the same valence n are shown. Complete separation will be possible if there is a pH, pH_A where log D = −2 for metal II when log D = +2 for metal I. From Fig. XII-4 it is obvious that the minimum separation between the two curves must be 4/n. This relationship does not hold if the two metals under consideration have extraction curves whose slopes are not the same. In general for a series of extraction curves that may have different slopes the distance separating each curve has to be described in terms of some arbitrarily chosen value of D as a reference. Usually the value log D = 0 which corresponds to half-extraction is chosen for reference and the pH at which this occurs is designated as $pH_{1/2}$.

Although not as useful in analytical separations as the log D-pH curves, an analogous set of curves of log D vs log (reagent concentration) may be derived from equation XII-6. Here again the curves will consist of linear portions whose slopes give the number of reagent molecules per metal ion in the extracted species, and horizontal portions corresponding to $D = K_{DC}$.

Now that we have described the course of chelate extraction in its simplest terms using equations XII-6 and 7, we can examine the effects of factors ignored up to now and arrive at a more general expression. The main limitation of the use of Equation XII-6 is the failure to take into account the effect of other reactions involving the metal ion such as:

(a) Formation of intermediate complexes with the reagent $ML^{(n-1)+}$ ------------ ML^{+}_{n-1} ; let $\Sigma[ML_i]$ be the sum of the concentration of all the complexes of the metal with the reagent.

(b) Formation of hydroxy complexes, $M(OH)^{(n-1)+}$ etc; let $\sum_j[M(OH)_j]$ be the sum of the concentrations of all the hydroxy complexes.

(c) Formation of complexes with other anions present in solution e.g. Cl^- or $SO_4^=$. $MCl^{(n-1)+}$, $MCl_2^{(n-2)+}$; let $\sum_K [MX_k]$ be the sum of the concentrations of all such complexes.

(d) Formation of complexes with masking agents e.g. CN^- or EDTA e.g. $MY^{(4-n)-}$; let $\Sigma[MY_m]$ be the sum of the concentrations of all such complexes.

If these additional factors are taken into account the expression for D changes from that given in Equation (XII-5) to:

$$D = \frac{[ML_n]_o}{[M^{n+}] + \sum_i[ML_i] + \sum_j[M(OH)_j] + \sum_k[MX_k] + \sum_m[MY_m]} \quad \text{(XII-8)}$$

As an illustration of the usefulness of this equation (XII-8) let us develop an expression applicable to the extraction of a metal in the presence of a masking agent which is a weak acid present in total concentration C_y. If the effect of reactions (a), (b) and (c) described above may be neglected, then the expression for D is:

$$D = \frac{[ML_n]_o}{[M^{n+}] + [ML_n] + \Sigma[MY_m]} \qquad \text{(XII-9)}$$

From Equation (IX-2), $[M^{n+}] + \sum_m [MY_m] = \frac{[M^{n+}]}{\beta_o}$ and

$$\beta_o = \frac{1}{1 + k_1' C_y + k_1' k_2' C_y^2 + \text{---} K_f' C_y^n} \qquad \text{(XII-10)}$$

where k_1', k_2' etc. are conditional formation constants whose values are pH dependent (see Sec. IX-3).

Substituting the expression for β_o in equation XII-9 we obtain:

$$D = \frac{[ML_n]_o}{[ML_n] + [M^n]/\beta_o} \qquad \text{(XII-11)}$$

Equation (XII-11) is seen to be quite similar in form to equation (XII-5). This permits us to proceed to an expression similar to (XII-6) and (XII-7).

$$D = \frac{K_{DC}}{1 + \frac{K_{DR}^n [H^+]^n}{K_f \cdot K_a^n \beta_o [HL]_o^n}} \qquad \text{(XII-12)}$$

at values $D < K_{DC}$

$$D = \frac{K_{DC} \cdot K_f \cdot K_a^n}{K_{DR}^n} \cdot \frac{\beta_o [HL]_o^n}{[H^+]^n} \qquad \text{(XII-13)}$$

This equation does not directly represent the change of D with pH with masking agents such as CN^- or EDTA which are weak acids since β_o is a function of pH.

From the foregoing discussion it is easy to see how solvent extraction measurements can be used to study various solution equilibria involving metal ions. By determining the variation of D over a variety of reaction conditions, the equilibrium constants of chelate formation, of hydroxy and other complex formation of quite a number of metal ions have been evaluated.

XII-2-b. ION-ASSOCIATION EXTRACTION SYSTEMS

Ion-pair formation can also result in an uncharged species having solubility characteristics suitable for their extraction into organic solvents. Reactions involving ion-pair formation are widely used for the extraction of metal ions. The metal may be incorporated into a very large ion containing bulky organic groups or it may pair (or associate) with an oppositely charged ion which is large. For example, $[Fe\,(phenanthroline)_3^{++}, 2ClO_4^-]$, $[(C_4H_9)_4N^+, FeCl_4^-)$, $[(C_6H_5)_4As^+, MnO_4^-]$[3] represent extractable ion pairs of analytical utility.

Since many ion-association extractions take place from aqueous solutions of relatively large ionic strengths, a region in which great differences exist between concentrations and activities, it is very difficult to describe quantitatively the behavior of such systems in terms of simple equilibrium expressions. Allowing for the uncertainty in the values of the appropriate conditional formation constants however, it is possible to derive expressions which are at least qualitatively useful. As an illustration let us consider the extraction of Zn^{++} from an HCl solution using a benzene solution of a high molecular weight amine such as tribenzylamine, $(C_6H_5CH_2)_3N$, symbolized by R_3N.

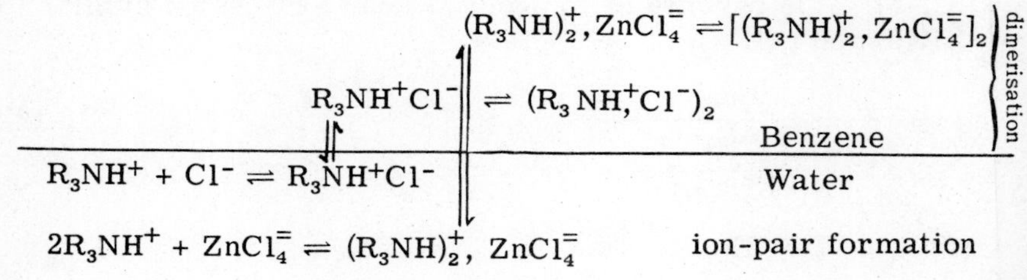

Benzene:

$$(R_3NH)_2^+, ZnCl_4^= \rightleftharpoons [(R_3NH)_2^+, ZnCl_4^=]_2 \quad \text{(dimerisation)}$$

$$R_3NH^+Cl^- \rightleftharpoons (R_3NH^+Cl^-)_2 \quad \text{(dimerisation)}$$

Water:

$$R_3NH^+ + Cl^- \rightleftharpoons R_3NH^+Cl^-$$

$$2R_3NH^+ + ZnCl_4^= \rightleftharpoons (R_3NH)_2^+, ZnCl_4^= \quad \text{ion-pair formation}$$

Fig. XII-5

The scheme in Fig. XII-5 shows that, unlike chelate complexes, ion-pair complexes are capable of dimerizing or associating to an even greater extent in the organic phase. It has been assumed that sufficient HCl is present in the aqueous phase to permit the neglect of unprotonated tribenzylamine.

For this system the distribution ratio of zinc is given by:

$$D = \frac{[(R_3NH)_2^+,\ ZnCl_4^=]_o + 2[((R_3NH)_2^+,\ ZnCl_4^=)_2]_o}{\sum_{i=2}^{4}[ZnCl_i^{(2-i)+}] + [(R_3NH)_2^+, ZnCl_4^=]}$$

which formally resembles the expression shown earlier for the distribution of acetic acid between water and benzene.

By the incorporation of the various equilibrium expressions as indicated in Fig. XII-5, the distribution ratio D may be expressed in terms of the total zinc concentration present, the total tribenzylamine concentration present and the chloride ion concentration.

XII-3. ION EXCHANGE EQUILIBRIA

In the last 25 years synthetic resins which are capable of exchanging cations or anions have been developed. The process consists essentially of passing a solution of the sample through a column of finely divided resin granules, each of which plays a part in the ion-exchange process. Thus the technique of ion exchange offers the advantage of a multistage separation process and ease of phase separation. These resins fall into two categories, cationic and anionic exchangers In each type, the polymeric molecules of the resin contain functional groups, e.g. a sulfonate anion in the cation exchanger or a quaternary ammonium cation in the anion exchanger, which are associated with oppositely charged ions that can exchange with others present in an aqueous phase Ion exchange resins to be active, imbibe a certain amount of water and indeed can be considered as possessing the properties of a gel. The ion exchange takes place between the ions which diffuse back and forth between the gel and aqueous phases. In this sense ion exchange is quite analogous to liquid-liquid extraction The exchange reaction can be described in terms of the equations:

Cation exchange: $RSO_3^-, H^+ + Na^+ \rightleftharpoons RSO_3^-, Na^+ + H^+$

Anion exchange: $R_4N^+, OH^- + Cl^- \rightleftharpoons R_4N^+, Cl^- + OH^-$

Equilibrium expressions corresponding to these reactions may be written as follows:

$$K_{ex._1} = \frac{(a_{Na^+})_{resin} \times a_{H^+}}{a_{Na^+} \times (a_{H^+})_{resin}} \qquad \text{(XII-14)}$$

and

$$K_{ex._2} = \frac{(a_{Cl^-})_{resin} \times a_{OH^-}}{a_{Cl^-} \times (a_{OH^-})_{resin}}$$

In terms of concentration $K'_{ex._1}$ for the Na^+-H^+ exchange becomes:

$$K_{ex._1} = K'_{ex._1} \frac{(\gamma_{H^+})_{resin}}{(\gamma_{Na^+})_{resin}} = \frac{[Na^+]\gamma_{Na^+} X_H}{[H^+]\gamma_{H^+} \cdot X_{Na}} \qquad \text{(XII-16)}$$

where X refers to the mole fraction of the cation in the resin phase, and $K'_{ex._1}$ is the apparent equilibrium constant which can be determined experimentally. The constancy of $K'_{ex._1}$ depends on the assumption that the ratio of the activity coefficients of the cations in the resin phase is a constant.

Values of $K'_{ex.}$ for various cation exchange reactions can be shown to be related to the ratio of activity coefficients in the resin phase by means of the Donnan theory developed by describing membrane equilibria. The Donnan theory makes use of the fact that the activity product of a salt on either side of a membrane to which it is permeable, is a constant. In this view we can consider the resin phase as being separated from the aqueous phase just as if there were a membrane which permitted the free passage of all ions except of course the ion sites of the polymeric resin molecule itself. Applying this to a system consisting of a resin in the hydrogen form in equilibrium with a solution of NaCl, we may write:

$$(a_{Na^+} \times a_{Cl^-})_{resin} = a_{Na^+} \times a_{Cl^-}$$

$$(a_{H^+} \times a_{Cl^-})_{resin} = a_{H^+} \times a_{Cl^-}$$

which can be combined to give:

$$\frac{(a_{Na^+} \times a_{Cl^-})_{resin}}{(a_{H^+} \times a_{Cl^-})_{resin}} = \frac{a_{Na^+}}{a_{H^+}}$$

When this equation is rearranged one obtains equation (XII-14) in which $\mathbf{K}_{ex._1}$ is seen to have a value of unity.
Hence we may rewrite equation XII-16 as:

$$K'_{ex._1} = \frac{(\gamma_{H^+})_{resin}}{(\gamma_{Na^+})_{resin}} \qquad \text{(XII-17)}$$

One can interpret equation (XII-17) as signifying that ion exchange equilibria depend on the differences in the properties of the ions themselves, i.e. their activity coefficients in the resin phase rather than on any specific interaction with the resin. Even though it is not possible to determine the coefficients of the cations in the resin phase directly, the resemblance of the resin phase to a highly concentrated electrolyte solution permits the comparison of experimentally determined $K'_{ex.}$ values with the activity coefficient ratios of the ions in solutions of high ionic strengths. In a series of exchanges of monovalent cations, the observed order of $K'_{ex.}$ values was found to be:

$$Cs^+ > Rb^+ > K^+ > Na^+ > H^+ > Li^+$$

Except for the reversal of H^+ and Li^+, this is the order that is expected on the basis of the activity coefficients of these ions.

$K'_{ex.}$ for a general reaction,

$$nRM_1 + M_2^{n+} \rightleftharpoons R_nM_2 + nM_1^+$$

where R stands for the resin anion and M_1 a monovalent cation and M_2 an n-valent cation, can be shown to be:

$$K'_{ex.} = \frac{(\gamma_{m_1^+})^n_{resin}}{(\gamma_{m_2^{++}})_{resin}} = \frac{(M_2^{n+})\gamma_{m_2}^n \times X^n_{M_1}}{[m_1^+]^n \gamma^n_{M_1} \times X_{M_2}}$$

From this expression it is obvious that the highly charged ions will have higher $K'_{ex.}$ values.

These considerations apply to the operation of an ion exchange column since the rate of flow of eluent in most cases is slow enough to permit the establishment of exchange equilibria to a reasonable approximation.

If for a series of metal ions values of $K'_{ex.}$ are not sufficiently different to provide for a good separation on a column of reasonable length, complexing agents can be used to improve the separability of the metals. Such a case might well arise with two metal ions of the same charge and similar ionic radii. The addition of sufficient complexing agent to the eluent would result in an increase in the total concentration of the metals in the aqueous phase, C_{M_2} and C_{M_3}. It would be possible to evalute the new position of equilibrium by substituting $\beta_o \cdot C_M$ for $[M^{n+}]$ for each of the metal ions. The ratio of their distributions can be shown to be: X_M/C_M

$$\frac{X_{M_2}/C_{M_2}}{X_{M_3}/C_{M_3}} = \frac{K'_{ex\,2}}{K'_{ex\,3}} \cdot \frac{\beta_{o_2}}{\beta_{o_3}} \qquad \text{(XII-19)}$$

From equation (XII-19) it can be seen that the difference in degree of complexation (i.e. the β_o ratio) can serve to amplify the difference in the exchange constants to bring about a greater degree of separation of the two metal ions.

If the pair of metals involved each forms a single complex having the same stoichiometry with the ligand and this is in much greater concentration than the free metal ion in solution, then equation XII-19 simplifies to:

$$\frac{X_{M_2}/C_{M_2}}{X_{M_3}/C_{M_3}} = \frac{K'_{ex.\,2}}{K'_{ex.\,3}} \; \frac{K'_{f\,3}}{K'_{f\,2}} \qquad \text{(XII-20)}$$

where K'_f values are the conditional formation constants of the metal complexes.

Adjacent lanthanide cations have been separated by cation exchange methods employing citrate complex formation to improve their separability.

All of the foregoing considerations apply to the single equilibration in a one-step contacting of the resin and aqueous phases. Under the conditions of operation of an ion exchange column, a very large number of such equilibration steps take place which results in magnifying small differences in ion exchange behavior to the point where good separations are feasible. Calculations concerning multistage separation processes are outside the scope of this text.

SUGGESTIONS FOR FURTHER READING

1. G. H. Morrison and H. Freiser, Solvent Extraction in Analytical Chemistry, John Wiley and Sons, Inc., New York (1957).
2. H. A. Laitinen, Chemial Analysis, McGraw Hill Book Co., Inc., (1960), Chapter 25.

XII-4. PROBLEMS

1. Describe the distribution of 8-hydroxyquinoline between $CHCl_3$, and water as a function of pH with the aid of the following data: $K_{D_R} = 720$, $K_{a_1} = 8 \times 10^{-6}$, $K_{a_2} = 1.4 \times 10^{-10}$

   Construct a diagram of log D vs pH.
2. Derive an expression for D, the distribution ratio, for the distribution of OsO_4 between CCl_4 and water taking into account the following equilibria:
   (a) aqueous phase

$$OsO_4 + H_2O \overset{K}{\rightleftharpoons} H_2OsO_5$$

$$H_2OsO_5 \overset{K_{a1}}{\rightleftharpoons} H^+ + HOsO_5^-$$

$$HOsO_5^- \overset{K_{a2}}{\rightleftharpoons} H^+ + OsO_5^{=}$$

(b)organic phase: $4OsO_4 \overset{K_p}{\rightleftharpoons} (OsO_4)_4$

Assume that OsO_4 is the only species that distributes.

3. Suppose in a given system D = 1 for a solute of interest. Starting with 100 ml of aqueous phase containing 0.1 M of the solute, what is the final concentration after (a) extracting with 100 ml of organic phase (b) extracting with a total of 1000 ml organic phase added at one time (c) extracting with 10 consecutive 100 ml portions of organic phase?
4. What is the minimum difference in $pH_{1/2}$ values for the extraction of two metal ions of the same charge as metal chelates which is required for good separation of the pair ($<$ 1% of one and $>$ 99% of the other metal)? How can this difference be described in terms of the various constants that comprise the extraction constants for the two metals?
5. The extraction of Zn^{2+} with dithizone into $CHCl_3$ at 25° is characterized by the following data: $K_a/K_{D_R} = 1.3 \times 10^{-10}$ $K_f = 1.1 \times 10^{15}$, $K_{D_C} = 6 \times 10^4$.
   a) If equal volumes of organic and aqueous phases are used, what per cent extraction will be obtained at pH 6.0, pH 8.0, pH 10.0 if the dithizone concentration in the organic phase is 10^{-3} M?
   b) Calculate the % extraction at pH 8.0 and 10.0 if the aqueous phase had an equilibrium NH_3 concentration of 0.05 M.
6. Acetylacetone, a liquid chelating agent of limited solubility in water, is useful as a metal extractant. The extraction behavior may be described by an expression similar to equation (XII-7). With the help of the following data calculate the overall formation constants of the metal chelates involved.
   pK_a (acetylacetone) = 8.95; aqueous solubility of acetylacetone = 1.7 M

Metal Ion	$pH_{1/2}$	K_D
Al^{3+}	1.75	9.5
Be^{2+}	0.67	41.0
Cu^{2+}	1.10	6.5
UO_2^{++}	1.66	65.5

7. The extraction behavior of metals by cupferron (N-nitrosophenylhydroxylamine) is described by equation (XII-7). From the extraction constant, K, defined as

$$K = \frac{K_{D_C} K_f K_a^n}{K_{DR}^n}$$

calculate the $pH_{1/2}$ values for the extractions in which a 0.002 M solution cupferron in $CHCl_3$ is employed as extractant.

Metal Ion	Fe(III),	Pu(IV),	Th(IV)	Cu(II)	Al(III)	Pb(II)	Ni(II)
log K	8.7	7.0	4.4	2.7	2.5	-1.0	-4.0

8. For the cation exchange resin Dowex 50 the value of $K'_{ex.}$ given by
$K' = \frac{[Na^+]\,X_{Cs^+}}{[Ca^+]\,X_{Na^+}}$ (like eqn XII-16) is 1.75. If a solution (500 ml) that is 0.1 M in Na^+ and 0.05 M in Cs^+ is equilibrated with 1 g. of Dowex 50 (4.3 milliequivalents per g.) what are the final concentrations of Na^+ and Cs^+?

XIII

Acid-Base Titrations

A titration curve is a graphical representation of the manner in which the logarithm of a critical concentration variable changes with the amount of titrant added. In every useful titration curve, this critical concentration variable undergoes a great change in the immediate vicinity of equivalence points (the equivalence point is the point at which a stoichiometric amount of titrant has been added). This change is invaluable in locating the equivalence point. Sometimes it is convenient to plot titration data so as to give the slope, (Δ Log C / Δ Volume of titrant), as ordinate against the volume of titrant as abscissa. In this manner it is possible to locate accurately the more difficultly accessible end-points (the end-point is the point at which an indicator system, visual or instrumental, signals the end of the titration; ideally the end-point and the equivalence point coincide).

In acid-base titrations, the appropriate concentration variable is the pH. This is plotted against either the volume of titrant added or the percent of total titrant stoichiometrically required. One of the best ways of studying factors affecting the reliability of acid-base titrations is to consider titration curves. In addition we can enlarge our understanding of buffer action by an examination of certain features of titration curves. Therefore in this and successive chapters we will discuss methods for the derivation or construction of titration curves and we will also describe their interesting properties. Although in effect we have developed all of the necessary equations in previous chapters for calculating the various points on titration curves, it is worthwhile to briefly review these calculations in the present context.

XIII-1. TITRATIONS OF STRONG ACIDS WITH STRONG BASES

Let us consider the titration of a solution of HCl with a solution of KOH. Any mixture of HCl and KOH can be described by the following charge balance:

$$[Cl^-] + [OH^-] = [H^+] + [K^+] \quad \text{(XIII-1)}$$

This equation may be used to obtain the hydrogen ion concentration at any point on the titration curve. Thus when the acid is present in excess,

$$[H^+] = ([Cl^-] - [K^+]) + [OH^-]$$

or

$$[H^+] = ([Cl^-] - [K^+]) + \frac{K_w}{[H^+]} \quad \text{(XIII-2)}$$

In most cases of practical importance, essentially the entire portion of the titration curve in which the acid is present in excess, the last term in equation (XIII-2) may be omitted.

On the other hand beyond the equivalence point, the form of equation (XIII-1) that may be used is:

$$[OH^-] = ([K^+] - [Cl^-]) + [H^+[\quad \text{(XIII-3)}$$

This equation can be rearranged to:

$$[H^+] = \frac{K_w - [H^+]^2}{[K^+] - [Cl^-]} \quad \text{(XIII-4)}$$

and in most cases the term $[H^+]^2$ is sufficiently smaller than K_w and can be ignored.

Of course at the equivalence point, i.e., when $[K^+] = [Cl^-]$, $[H^+] = [OH^-]$ as may be seen from equation (XIII-1) and therefore

$$[H^+] = K_w^{\frac{1}{2}} .$$

Now let us consider a typical titration in which 50 ml of 0.10 M HCl is titrated with 0.2 M KOH. Table XIII-1 summarizes the

Table XIII-1

Calculation of pH Values in HCl vs. KOH Titration

Volume of KOH added (V ml.)	Millimoles of KOH added (V×0.2)	Millimoles acid in excess	Total Volume (ml.)	$[Cl^-]-[K^+]$	$[H^+]$	pH
0	0	50×0.10	50.0	$\frac{5.00}{50}$	0.10	1.00
10.0	2.00	5.00-2.00	60.0	$\frac{3.00}{60}$	0.05	1.30
24.9	4.98	5.00-4.98	74.9	$\frac{0.02}{74.9}$	$10^{-3.58}$	3.58
25.0	5.00	5.00-5.00	75.0	0	$K_w^{\frac{1}{2}}$	7.00
		Millimoles base in excess		$[K^+]-[Cl^-]$		
25.1	5.02	5.02-5.00	75.1	$\frac{.02}{75.1}$	$\frac{K_w}{.02/75.1}$	10.62
30.0	6.00	6.00-5.00	80.0	$\frac{1.0}{80.0}$	$\frac{K_w}{1.0/80}$	12.10

calculations involved in finding pH values at a number of typical points.*

From these points the titration curve for the strong acid vs. strong base can be drawn, as shown in Fig. XIII-1. It is important to note that the pH of the solution does not change very much during most of the neutralization. The pH does not rise more than one unit until well over 80% neutralization. In this sense the strong acid exhibits some degree of buffer action in this region. Another feature of significance is the region of very rapid pH change from 98% to 102% neutralization where the pH change is approximately six units from pH 4 to 10. Acid-base indicators which show a color change anywhere between pH 4 and 10, are suitable for this titration. Naturally, the best indi-

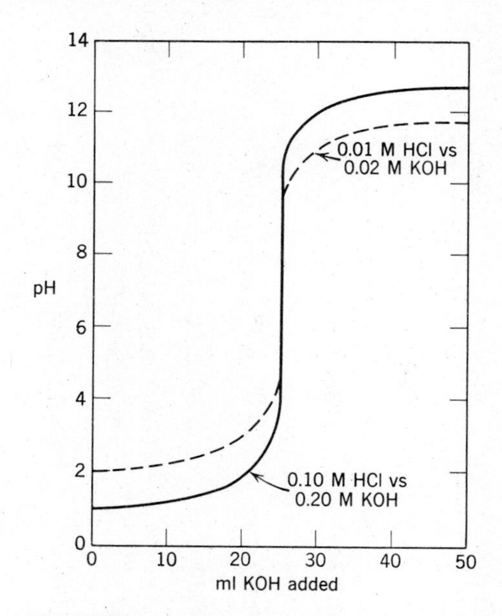

Fig. XIII-1. Titration of a Strong Acid with a Strong Base

*In these and subsequent titration curve calculations a single value of each equilibrium constant will be assumed to apply throughout the titration. Ionic strength corrections for K values may be applied at individual points at the discretion of the reader.

cator would be one whose color change is as close as possible to the pH of the equivalence point (i.e., 7.0). After the region of rapidly changing pH in the immediate vicinity of the equivalence point, the titration curve again assumes a relatively slow rate of pH rise.

If the acid and base involved in the titration are at another concentration, then the resulting titration curve will be somewhat altered. For example, as shown by the dotted curve in Fig. XIII-1 which represents the titration curve of 0.01 M HCl vs. 0.01 M KOH the flat portions of the curve are displaced parallel to the original curve to give a smaller vertical region. Similarly if more concentrated reagents were used, the vertical region would be larger.

XIII-2. TITRATION OF WEAK MONOPROTIC ACIDS WITH STRONG BASES

Let us consider the titration of any weak monoprotic acid HA with a solution of NaOH. Any mixture of these two substances can be described by the following two equations:

Charge Balance: $[Na^+] + [H^+] = [OH^-] + [A^-]$ (XIII-5)

Mass Balance: $C_a = [HA] + [A^-]$ (XIII-6)

Where C_a is the analytical concentration of the acid as defined by equation (XIII-6). In order to calculate the hydrogen ion concentration at any point on the titration curve, we need only to solve for $[HA]$ and $[A^-]$ using these equations and substitute them in the expression for the acid dissociation constant. This leads to the following expression:

$$K_a = \frac{[H^+] \times ([Na^+] + [H^+] - [OH^-])}{C_a - ([Na^+] + [H^+] - [OH^-])} \qquad \text{(XIII-7)}$$

Equation (XIII-7) simplifies to equation (V-9) for the weak monoprotic acid, before any NaOH is added, i.e. $[Na^+] = 0$.

Once NaOH has been added, we can distinguish three convenient forms of equation (XIII-7), (a) for the region before the equivalence point, (b) the equivalence point and (c) the region after the equivalence point.

In region (a) equation (XIII-7) will simplify in most cases to:

$$K_a = \frac{[H^+] \times [Na^+]}{C_a - [Na^+]} \qquad \text{(XIII-8)}$$

or $$[H^+] = K_a \times \frac{C_a - [Na^+]}{[Na^+]}$$

Since the hydrogen ion concentration in this expression is given by a ratio of concentrations it is obvious that the value of $[H^+]$ will be independent of the total volume of the solution, as long as this simplified equation is valid. Hence

$$[H^+] = K_a \times \frac{\text{(Millimoles acid - millimoles NaOH added)}}{\text{Millimoles NaOH added}} \quad \text{(XIII-9)}$$

At the equivalence point where $[Na^+] = C_a$, equation (XIII-7) usually simplifies to that of the weak monoacidic base (V-13)

$$K_a = \frac{[H^+] \times [Na^+]}{[OH^-]}$$

or $$\frac{K_a K_w}{[Na^+]} = [H^+]^2 \quad \text{(XIII-10)}$$

In the region beyond the equivalence point, where $[A^-]$ is substantially equal to C_a equation (XIII-5) may be used to show that:

$$[OH^-] = [Na^+] - C_a$$

or $$[H^+] = \frac{K_w}{[Na^+] - C_a} \quad \text{(XIII-11)}$$

The same result could be obtained from equation (XIII-7).

Let us illustrate the above discussion by calculating some typical pH values in the titration of 50 ml of 0.2 M HOAc with 0.1 M NaOH. These are summarized in Table XIII-2, and graphically shown in Fig. XIII-2.

It is interesting to compare the shape of the titration curve in Fig. XIII-2 with that of the strong acid-strong base curve in Fig. XIII-1. Several points of contrast are immediately evident:

(a) Whereas the vertical portion in the vicinity of the equivalence point in the case of the strong acid-strong base extended approximately 6 pH units from pH 4 to 10, in the case of the weak acid-strong base, this range has narrowed to approximately 3 pH units, from pH 7 to 10. Hence an indicator that has a color transition in this pH range should be used.

(b) The equivalence point is obtained at a pH value significantly higher than 7.0, in this case about 8.8.

Table XIII-2

Calculation of pH Values in HOAc vs. NaOH Titration

Millimoles of HOAc = 50 × 0.20 = 10.0; pK_a = 4.70; pK_w = 14.00

Volume of NaOH added	Millimoles of NaOH added	(Millimoles acid-Millimoles NaOH)	Total Volume	$[H^+]$ from Equation		pH
0	0	10.0	50	$(0.20\,K_a)^{\frac{1}{2}}$	(V-9)	2.70
20	2.0	8.0	70	$\frac{K_a \times 8.0}{2.0}$	(XIII-9)	4.10
100	10.0	0	150	$\left(\frac{K_a K_w}{10/150}\right)^{\frac{1}{2}}$	(XIII-10)	8.76
150	15.0	−5.0	200	$\frac{K_w}{5.0/200}$	(XIII-11)	12.40

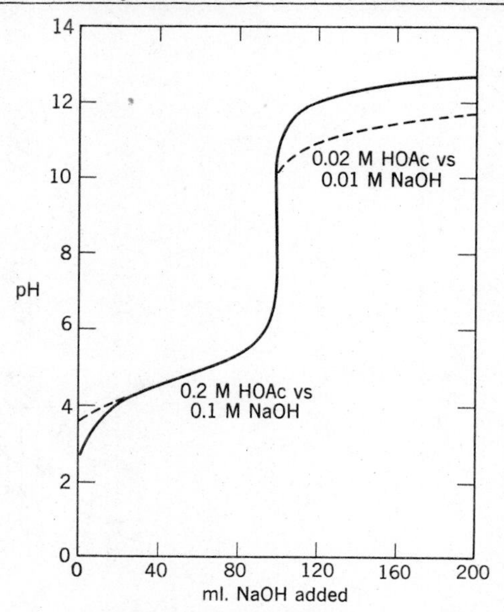

Fig. XIII-2 — Titration of Weak Acid with a Strong Base

(c) The region between 0 and 100% neutralization is S-shaped in contrast with the flatter appearing corresponding portion of the strong acid-strong base curve. The main difference in these two curves is the initial portion of the titration. This is due to the fact that the strong acid in this region acts as a much better buffer than the mixture of the weak acid and a relatively small amount of its conjugate base.

(d) The 0 to 100% neutralization range of the titration curve of the strong acid-strong base was shown to be quite sensitive to the effects of dilution whereas, except for a small initial portion, this region, for a weak acid-strong base titration curve, is essentially insensitive to dilution. This shows as did equation (XIII-9), the buffer action of a mixture of a weak acid and its conjugate base. It is of interest to note that this portion of the titration curve, when experimentally determined, serves as a basis for the calculation of pK_a values of weak acids. Typically, the pH at 50% neutralization might be used, since at this point, $pH = pK_a$.

(e) A point of similarity in this titration curve, with that of the strong acid-strong base curve may be seen in the region following the equivalence point, where the two curves become identical. (Why?)

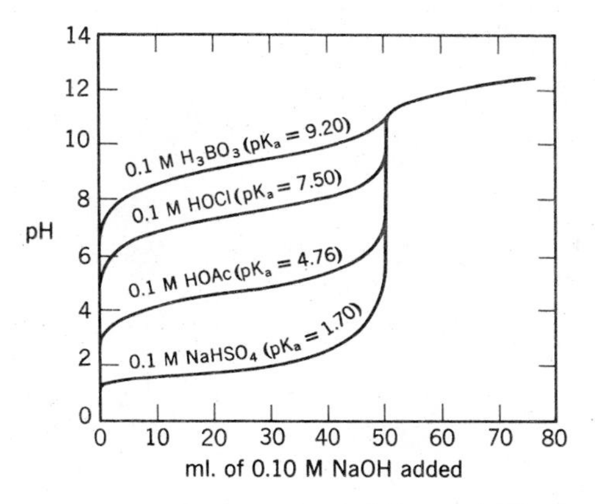

Fig. XIII-3 — Titration of Several Weak Acids with a Strong Base

An interesting general view of titrations of weak acids with a strong base may be seen from Fig. XIII-3 in which the titrations of several weak acids, of increasing pK_a values, with a strong base, are shown. If 0.10 M solutions are employed, then it would appear that for all practical purposes, acids whose pK_a values are greater than about 8 may not be titrated with any degree of accuracy, using a visual indicator, since the vertical region in the vicinity of the equivalence point is too small in such cases.

An examination of Fig. XIII-3 also serves to demonstrate the possibility of titrating a mixture of acids whose pK_a values are widely separated. Thus at a pH slightly greater than 4, the titration of HSO_4^- is essentially complete whereas that of the HOCl is just beginning. Using a pH meter as an indicator we could very easily titrate separately HSO_4^- and HOCl in a mixture of the two.

XIII-3. TITRATION OF WEAK MONOACIDIC BASES WITH STRONG ACIDS

The derivation of titration curves for weak monoacid bases vs. strong acids is quite similar to that of the weak acid given in Sec. XIII-2. A mixture of a base B and HCl for example may be described by the following charge and mass balance equations:

$$[BH^+] + [H^+] = [OH^-] + [Cl^-] \qquad \text{(XIII-12)}$$

$$C_b = [B] + [BH^+] \qquad \text{(XIII-13)}$$

which may be combined with the basic dissociation constant to give the following expression:

$$K_b = \frac{K_w}{K_a} = \frac{[OH^-] \times ([OH^-] + [Cl^-] - [H^+])}{C_b - ([OH^-] + [Cl^-] - [H^+])} \qquad \text{(XIII-14)}$$

In the absence of HCl this equation simplifies to equation (V-20) for the weak monoacid base, since $[Cl^-] = 0$.

In the region before the equivalence point, equation XIII-14 will simplify in most cases to:

$$[OH^-] = \frac{K_b \times (C_b - [Cl^-])}{[Cl^-]} \qquad \text{(XIII-15)}$$

$$[H^+] = \frac{K_a \cdot [Cl^-]}{C_b - [Cl^-]}$$

which further simplifies in the manner of equation (XIII-9) to:

$$[H^+] = \frac{K_a \times \text{(millimoles acid added)}}{\text{(millimoles base - millimoles acid added)}} \qquad \text{(XIII-16)}$$

At the equivalence point where $C_b = [Cl^-]$, equation (XIII-14) usually simplifies to that of the weak monoprotic acid (Equation V-13)

$$[H^+] = (K_a \times [Cl^-])^{\frac{1}{2}} \qquad \text{(XIII-17)}$$

In the region beyond the equivalence point where $[BH^+]$ is substantially equal to C_b, equation (XIII-12) may be used to show that

$$[H^+] = [Cl^-] - C_b \qquad \text{(XIII-18)}$$

The reader will find it useful to construct a Table similar to Table XIII-2 for the titration of 100 ml of 0.1 M NH_3 with 0.1 M HCl.

A detailed discussion of the titration curve shown in Fig. XIII-4 is unnecessary in the light of its similarity to the titration curve of the weak acid with a strong base shown in Fig. XIII-2 and described in Sec. XIII-2.

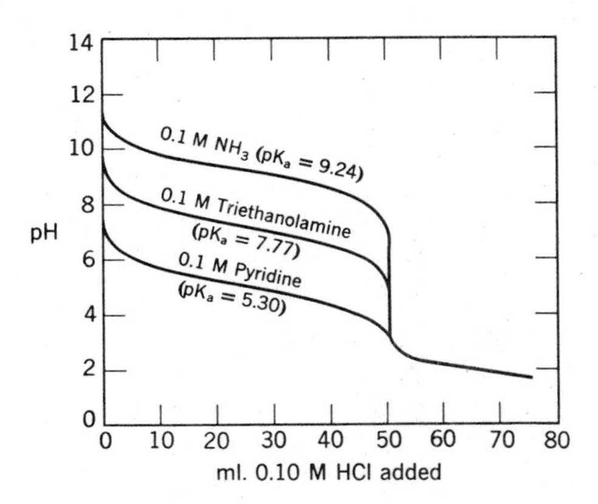

Fig. XIII-4 — Titration of Several Weak Bases with a Strong Acid

For example, in an NH_3 - HCl titration using approximately 0.1 M solutions, the S-shaped curve shows a buffer region in the range of pH from 8.2 to 10.2, a region of rapid change from pH 4 to 7, and an equivalence point at a pH of approximately 5.2.

Fig. XIII-4 also illustrates the effect of changing base strength on the shape of the titration curve. It is interesting to note at this point that when we encounter a base that is so weak (pK_a below 6) that we cannot obtain a sharp end-point with a visual indicator, such as is the case with pyridine, we can always turn to the conjugate acid which will necessarily be strong enough to titrate against a base. Similar considerations apply to extremely weak acids. Although boric acid cannot be usefully titrated with NaOH, its conjugate base in borax, sometimes used as a primary standard for strong acids, is readily titrated with HCl.

XIII-4. TITRATIONS OF WEAK ACIDS WITH WEAK BASES

The titration of a weak acid with a weak base or vice versa is never encountered in practice because analysts always use strong acids or bases as titrants. The titration of a weak acid and a weak base is characterized by a relatively small pH change in the vicinity of the equivalence point which renders end-point detection extremely difficult. For this reason no quantitative examination of the titration curve calculations for this case will be made here. However, in dealing with titrations of mixtures of acids and bases such a case might have to be considered (see XIII-7).

XIII-5. ACID-BASE INDICATORS

Acid-base indicators are usually organic acids or bases in which either the conjugate acid or base is highly colored. For example, phenolphthalein is a colorless weak acid (HIn) whose conjugate base (In^-) is red. Such indicators are referred to as one-color indicators. Indicators in which both conjugate acid and base are colored are called two-color indicators. For example, methyl orange is a base which is yellow and whose conjugate acid is red. Two-color indicators are far more frequently encountered as may be seen from Table VI in the Appendix.

As we have already seen, titration curves exhibit a sharp pH change in the vicinity of the equivalence point. By the addition of a small quantity of properly selected indicator we can observe a color change close to the equivalence point and thus obtain the endpoint. It is of primary interest to determine the pH range in which any particular indicator can be used. Since the appearance or change of color is related to dissociation the pK_a value of the indicator is of paramount importance. Thus, if HIn is a typical indicator acid and In^- its conjugate base, then

$$HIn \rightleftharpoons H^+ + In^-$$

and

$$K_{In} = \frac{[H^+][In^-]}{[HIn]} \qquad \text{(XIII-19)}$$

If we are dealing with a one-color indicator such as phenolphthalein, i.e. only In^- is colored, then the appearance or disappearance of color depends solely on the concentration of In^- which the eye can detect. For phenolphthalein this concentration is approximately 5×10^{-6} molar. The pH at which this occurs depends on the total concentration of indicator used. Thus if C_{HIn} is the total indicator concentration, then

$$[HIn] = C_{HIn} - [In^-] \qquad \text{(XIII-20)}$$

and the $[H^+]$ at which color appears may be calculated from equations (XIII-19) and (XIII-20).

$$[H^+] = K_{In} \left\{ \frac{C_{HIn}}{[In^-]} - 1 \right\} \qquad \text{(XIII-21)}$$

Example 1.

At what pH will the pink color of phenolphthalein appear in 100 ml of a solution to which 3 drops of 0.03 M phenolphthalein have been added? pK_a (phenolphthalein) = 9.70.

If we assume 30 drops to the milliliter then 0.10 ml have been used.

$$C_{HIn} = \frac{0.10 \times 0.03}{100} = 3 \times 10^{-5}$$

$$[H^+] = 10^{-9.70} \left[\frac{3 \times 10^{-5}}{5 \times 10^{-6}} - 1 \right] = 10^{-9.00}$$

Hence the pH at which color appears is 9.00. If 30 drops rather than 3 drops of the indicator solution were added, the pH at which color appears, would be 7.92.

The behavior of two-color indicators contrasts sharply with that of one-color indicators in that the pH of the color transition is essentially independent of total indicator concentration. Let us consider the case of methyl orange, for example, in which the acid form is red colored and the basic form has a yellow color of almost equal intensity. The solution containing an equal concentration of each of these forms will obviously be orange in color. Hence with the help of equation (XIII-19) we can see that the pH at which the orange color appears is equal to the pK of the indicator. In solutions in which the ratio of the concentration of the red and yellow forms is close to unity, the color will also appear orange. Of course, if one of these forms predominates let us say in a ratio of 3 : 1, then the color of this solution will appear to be about the same as that of the predominant form alone. If $[HIn] : [In^-] = 3$ or greater, the solution will be decidedly red. The pH of this solution is given by:

$$pH = pK_{HIn} + \log \tfrac{1}{3}\,.$$

If the ratio of $[In^-] : [HIn]$ is 3 or greater, the solution will be decidedly yellow. The pH of this solution will be given by:

$$pH = pK_{HIn} + \log 3$$

At all intermediate pH values

$$pH = pK_{HIn} \pm 0.5$$

the solution is orange.

Although the ratio 3:1 is fairly reasonable for many two-color indicators, the specific value that applies which may vary from one indicator to another depends on the color intensities of the two conjugate forms.

From this discussion we can see that the pH of the indicator transition occurs over a range of approximately one unit and is centered on a value determined by the pK of the indicator. Obviously, too, the pH of the color change is determined by the concentration ratio of the two indicator forms rather than by the total concentration.

It is interesting that although in both one and two-color indicators the transition range is closely related to the pK of the indicator, in the former the color change can occur at a pH removed by as much as one unit from the pK_{HIn} whereas in the latter the pH of the center of the color transition coincides with the pK_{HIn}.

Another observation of importance is that the indicator, itself an acid or base, consumes titrant. Hence even though only a small amount of indicator is used, a lower limit of the concentration of the substance being titrated is imposed. If extremely dilute solutions must be titrated, then indicators which are especially sensitive must be used, or replaced entirely by a pH meter. The pH meter can be considered to be the universal indicator.

XIII-6. TITRATIONS OF POLYPROTIC ACIDS

The titration of a polyprotic acid with a strong base is quite similar to the titration of several monobasic acids of different strengths, because the differences in the successive dissociation constants of the acid are usually large enough so that neutralization occurs a step at a time. For example, it is obvious that if we add 3 millimoles of NaOH to 5 millimoles of H_3PO_4, the solution will contain 3 millimoles of NaH_2PO_4 and 2 millimoles of H_3PO_4 rather than 1 millimole of Na_3PO_4 and 4 millimoles of H_3PO_4. Hence the treatment of titration calculations for polyprotic acids that follows will be quite similar to that developed above for monobasic acids. The differences that do exist will be seen in the vicinity of the equivalence points.

Let us consider the titration of any weak diprotic acid H_2A with a solution of NaOH. Proceeding as we did in Sec. XIII-2 with monoprotic acids, we may write the charge and mass balance equations as follows:

$$[Na^+] + [H^+] = [OH^-] + [HA^-] + 2[A^=] \qquad \text{(XIII-22)}$$

and

$$C_a = [H_2A] + [HA^-] + [A^=] \qquad \text{(XIII-23)}$$

where C_a is the analytical concentration of the acid as defined by equation (XIII-23). Combining these equations with the stepwise dissociation constant equilbria we have:

$$[H^+] - [OH^-] = \frac{C_a (K_1 [H^+] + 2K_1 K_2)}{[H^+]^2 + K_1 [H^+] + K_1 K_2} - [Na^+] \qquad \text{(XIII-24)}$$

Before the first equivalence point it will be seen that $[H^+]$ as well as K_1 is greater than K_2. Hence $K_1 K_2$ will be negligible in comparison to the other terms in equation (XIII-24), resulting in the simplification of this equation to:

$$[H^+] - [OH^-] = \frac{C_a \times K_1 [H^+]}{[H^+]^2 + K_1 [H^+]} - [Na^+]$$

$$= \frac{C_a \times K_1}{[H^+] + K_1} - [Na^+] \qquad \text{(XIII-25)}$$

Equation (XIII-25) is a rearranged form of equation (XIII-7) for the monoprotic acid titration.

Similarly, beyond the first equivalence point when $[H^+]$ as well as K_2 is less than K_1, $[H^+]^2$ in the denominator of equation (XIII-24) becomes negligible and we obtain:

$$[H^+] - [OH^-] = \frac{C_a K_2}{[H^+] + K_2} - ([Na^+] - C_a) \qquad \text{(XIII-26)}$$

Equation (XIII-26) is also seen to represent the equation for the titration of a monoprotic acid, in which in place of $[Na^+]$, the total base added, there is $([Na^+] - C_a)$ the base in excess of that required in the first neutralization step. This demonstrates that the various neutralization steps of the titration curve of a diprotic or polyprotic acid will duplicate the titration curves of monoprotic acids.

The titration curve calculation at the intermediate equivalence points of di- or polyprotic acids will of course be the same as that developed in Chapter VI. In most titration curve calculations the simplified equation will apply. For instance for a diprotic acid, $[H^+] = (K_1 K_2)^{\frac{1}{2}}$ at the first equivalence point.

In calculating pH values in the vicinity of the equivalence point, most cases can be treated by a simplified version of equation (XIII-24) obtained by neglecting $([H^+] - [OH^-])$. Thus for the diprotic acid:

$$[H^+]^2 \times [Na^+] - [H^+] \ \{K_1 (C_a - [Na^+])\} - K_1 K_2 (2C_a - [Na^+]) = 0 \qquad \text{(XIII-27)}$$

As an illustration of the above treatment the calculation of some typical pH values in the titration of 50 ml of 0.05 M H_2CO_3 with 0.1 M NaOH are summarized in Table XIII-3 on following page. Inasmuch as pK_2 is greater than 8, the titration curve of H_2CO_3 with NaOH would show only one distinct endpoint corresponding to the neutralization of the first proton. In Table XIII-3, the pH value at the second equivalence point, 11.35, rises less than one pH unit in the next 25 ml addition of NaOH. The titration of Na_2CO_3 with HCl, however, shows both endpoints distinctly since both pK_a values are above 6. (See Fig. XIII-5.)

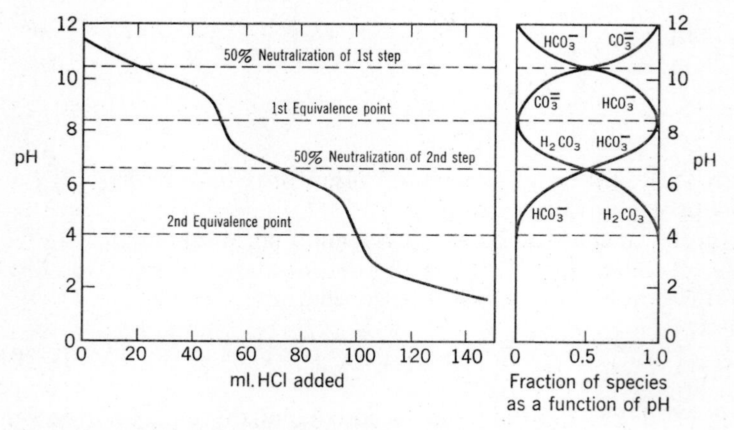

Fig. XIII-5 — Titration of Na_2CO_3 with HCl

XIII-7. TITRATION OF MIXTURES OF ACIDS

It is possible to distinguish two kinds of acid mixtures. The first type consists of acids having essentially indistinguishable pK values. In such cases calculations can be carried out as if only one acid having a pK equal to the average pK, and a concentration equal to the sum of the concentrations, were present. Also in such cases it is practically impossible to distinguish, by titration with a base, the component acids. For example, the following pairs of acids fall into this category: HCl and HNO_3; HCl and sulfamic acid; CH_3COOH and HCOOH; NH_4Cl and phenol.

Table XIII-3

Calculation of pH Values in H_2CO_3 vs. NaOH Titration

Millimoles $H_2CO_3 = 50 \times 0.05 = 2.50$ $pK_{a_1} = 6.30$, $pK_{a_2} = 10.30$, $pK_w = 14.00$

Volume of NaOH added	Millimoles of NaOH added	(Millimoles acid-Millimoles NaOH)	Total Volume	$[H^+]$ from	Equation	pH
0	0	2.50	50	$(0.05\ K_{a_1})^{\frac{1}{2}}$	V-13	3.80
20	2.0	0.50	70	$\frac{K_{a_1} \times 2.0}{0.5}$	VIII-9	5.70
25	2.5	0	75	$(K_{a_1} K_{a_2})^{\frac{1}{2}}$	VI-30	8.30
30	3.0	*2.0	80	$\frac{K_{a_2} \times 0.5}{2.0}$	VIII-9	10.90
50	5.0	*0.0	100	$\left(\frac{K_w K_{a_2}}{0.025}\right)^{\frac{1}{2}}$	VI-22	11.35
75	7.5	*-2.5	125	$\frac{K_w}{2.5/125}$	XIII-11	12.30

* 2 (mmoles acid) - mmoles NaOH gives the remaining HCO_3^-

The second type of mixture includes acids having significantly different pK values. In such instances, calculations similar to those described for the polyprotic acid may be carried out. That is to say the addition of base will be considered to result in neutralization of only the strongest acid in the mixture with the others being neutralized serially in accord with their pK values. A typical titration curve obtained for a mixture of this type is shown in Fig. XIII-6.

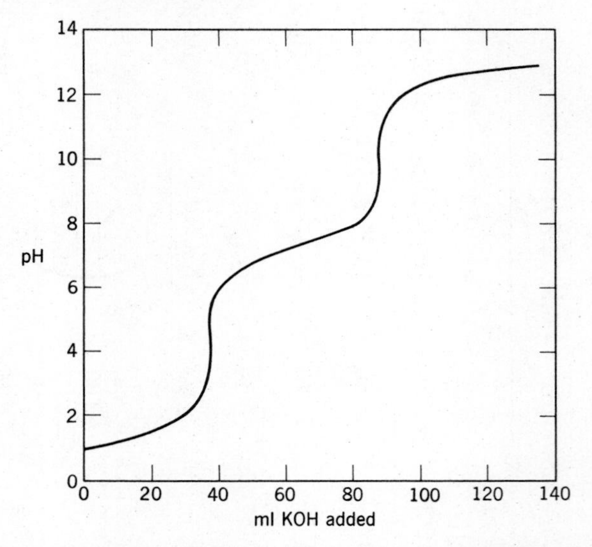

Fig. XIII-6 — Titration of 100 ml of a mixture of 0.075 M HCl and 0.10 M HOCl with 0.2 M KOH.

A further point of comparison with the titration curves of polyprotic acids lies in the fact that the pH values of the intermediate equivalence points in titrations of equimolar mixtures are given by expressions such as: $[H^+] = (K_1K_2)^{1/2}$, provided that both acids are weak. (where the subscripts $_1$ and $_2$ refer to different acids rather than the successive steps of the same acid) as has been previously shown in Equation VII-17.

XIII-8. TITRATION ERRORS

The term "titration error" will be considered to mean the error which occurs when the color change of the visual indicator takes place either before or after the equivalence point. We will omit from consideration the errors arising from uncertainties in reading the burette, and so forth. If the equivalence point is obtained by locating the inflection point in a titration curve (as for example via a differential titration curve), then another sort of titration error will be incurred. This results from the fact that not in all titrations do the equivalence and inflection points coincide. Since the majority of titrations involve the use of visual indicators, we will concentrate our attention on factors affecting the first type of titration error.

In the case of a titration of a strong acid with a strong base, the percentage titration error can be obtained simply by calculating the number of milliequivalents of acid remaining (or excess base added), dividing this quantity by the total number of milliequivalents of substance being titrated and multiplying by 100.

Example 2.

50 ml of 0.1 M HCl is titrated with 0.2 M NaOH. If methyl orange is used as the indicator, the end-point will be observed at a pH of 4.0. Calculate the titration error.

In this case the end-point is reached before the equivalence point and therefore the titration error is negative. The number of milliequivalents of $[H^+]$ left unneutralized at pH 4.0 = $10^{-4} \times 75$ (the total volume of the solution is 75 ml at the end-point).

$$\text{The titration error} = \frac{10^{-4} \times 75 \times 100}{0.1 \times 50} = -0.15\%$$

In this calculation we have intentionally ignored $[H^+]$ from the dissociation of H_2O, since this is negligible in comparison to the total $[H^+]$ at pH 4.0. If the end-point were much closer to the equivalence point it would be necessary for accurate calculation of the titration error to subtract the $[H^+]$ contribution from the dissociation of water.

If in the above titration the phenolphthalein end-point were used, i.e. the color change occurs at a pH of 9.0, then the excess NaOH added would be $10^{-5} \times 75$ milliequivalents, (using a value of $pK_w = 14.0$, pOH = 5.0).

$$\text{Therefore the titration error} = +\frac{75 \times 10^{-5} \times 100}{0.1 \times 50}$$

$$= +0.02\%$$

With the help of the above examples we may generalize to

give the formula that describes the titration error for strong acid-strong base titrations.

$$\text{Titration Error} = \frac{x \cdot V_t}{C_a \cdot V_a} \cdot 100\% \qquad \text{(XIII-28)}$$

Where x = -(concentration of unneutralized acid remaining at the end-point)

or x = +(concentration of excess NaOH added)

C_a = concentration of the acid being titrated

V_a = volume of the acid being titrated

V_t = total volume of the solution at the end-point

Equation (XIII-28) shows that as the solutions being titrated become more dilute (C_a decreases, or the total volume V_t, increases) the titration error increases.

In the case of the titration of a weak acid and a strong base, the titration error can be described in terms of the percentage of unneutralized acid that remains at the end-point.

$$\text{i.e., Titration Error} = -\frac{100 \times [HA]}{[HA] + [A^-]}\%$$

Substituting the acid dissociation constant expression for HA in this definition of the titration error we obtain:

$$\text{Titration Error} = \frac{-100\,[H^+]}{[H^+] + K_a}\% \qquad \text{(XIII-29)}$$

Equation (XIII-29) which applies to titrations in which the end-point comes ahead of the equivalence point, neglects the contribution of [HA] from the hydrolysis of the salt formed. For most purposes, however, equation (XIII-29) is sufficiently accurate.

It is interesting to note that the titration error in the case of weak acids, unlike that of strong acids, is independent of concentrations used. It must be realized that upon increasing dilution, hydrolysis becomes relatively more important and must be considered in the calculation. For example with acetic acid titrations the titration error is independent of concentration in solutions greater than 0.001 M. Below this concentration, hydrolysis effects must be considered.

With the help of equation (XIII-29) we can learn whether an acid of given concentration can be titrated to a predetermined pH without incurring a given error, provided the dissociation constant of the acid is known.

For example if we were to use the the phenolphthalein end-point i.e. pH = 9.0, what is the weakest acid that we could titrate with a maximum error of 0.1% ?

$$0.1 = \frac{100 \times 10^{-9}}{10^{-9} + K_a}$$

Therefore $K_a = 10^{-6}$
For a 1% error, $K_a = 10^{-7}$

It is important to realize that if the end-point is one that follows the equivalence point in a weak acid-strong base titration, that equation (XIII-29) does not hold, since the titration curve after the equivalence point becomes identical with a strong acid-strong base titration. In such situations equation (XIII-28) must be used.

XIII-9. BUFFER INDEX

The effectiveness of buffers can be described in terms of a quantity called the buffer index, β, which is the reciprocal of the slope of a neutralization curve. This reciprocal slope gives the instantaneous value of the amount of strong base that must be added to obtain a unit increase in pH.

i.e. $$\beta = \frac{dC_B}{dpH} \qquad \text{(XIII-30)}$$

where C_B is the molar concentration of strong base titrant added. It is obvious that this definition could be restated in terms of the concentration of added acid, C_A, to give the equivalent definition:

$$\beta = -\frac{dC_A}{dpH}$$

The negative sign occurs since the change in pH on the addition of acid is negative.

It may be asked why entire titration curves are being considered in a discussion of buffer effectiveness. It should be realized that testing the ability of buffer solutions to absorb quantities of base over wide ranges amounts to a titration.

Let us consider the titration of a strong acid, HCl, with a strong base, NaOH. The proton balance is given by:

$$[Na^+] + [H^+] = [OH^-] + [Cl^-]$$

i.e. $C_B + [H^+] = [OH^-] + C_A$ where C_A and C_B are the analytical concentrations of HCl and NaOH.

$$C_B = \frac{Kw}{[H^+]} - [H^+] + C_A$$

Differentiating the above equation with respect to $[H^+]$ we obtain:

$$\frac{dC_B}{d[H^+]} = \frac{-Kw}{[H^+]^2} - 1$$

However, $dpH = -\frac{1}{2.3} d \ln [H^+] = -\frac{d[H^+]}{2.3[H^+]}$ (XIII-31)

$$\text{Hence, } \beta = \frac{dC_B}{dpH} = -2.3[H^+] \frac{dC_B}{d[H^+]} = 2.3[H^+]\left\{\frac{Kw}{[H^+]^2} + 1\right\}$$

$$= \frac{2.3\,(Kw + [H^+]^2)}{[H^+]}$$

$$= 2.3\,([OH^-] + [H^+]) \qquad \text{(XIII-32)}$$

As may be seen from equation (XIII-32) the buffer effectiveness of a solution of a strong acid or a strong base is proportional to the concentration of the strong acid or strong base. The minimum buffer index is obtained at $[H^+] = [OH^-]$.

In a similar fashion we can obtain an expression for the buffer index of solutions of weak acids. In a mixture of a weak acid HA and a strong base, NaOH, the following proton balance may be written:

$$[Na^+] + [H^+] = [OH^-] + [A^-]$$

or

$$C_B = \frac{Kw}{[H^+]} - [H^+] + \frac{C_A K_a}{K_a + [H^+]}$$

where C_A and K_a are the analytical concentration and acid dissociation constant respectively, of HA. Differentiating then,

$$\frac{dC_B}{d[H^+]} = \frac{-Kw}{[H^+]^2} - 1 - \frac{C_A K_a}{(K_a + [H^+])^2}$$

$$\text{and } \beta = \frac{dC_B}{dpH} = 2.3\left\{\frac{Kw}{[H^+]} + [H^+] + \frac{C_a K_a [H^+]}{(K_a + [H^+])^2}\right\} \qquad \text{(XIII-33)}$$

Equation (XIII-33) is represented graphically in Fig. XIII-7. A maximum value of β is obtained at a pH = pK_a. At this point $\beta \approx \frac{2.3}{4} \cdot C_A$. It will also be observed that at very high pH values corresponding to an excess of NaOH, that the buffer index rises

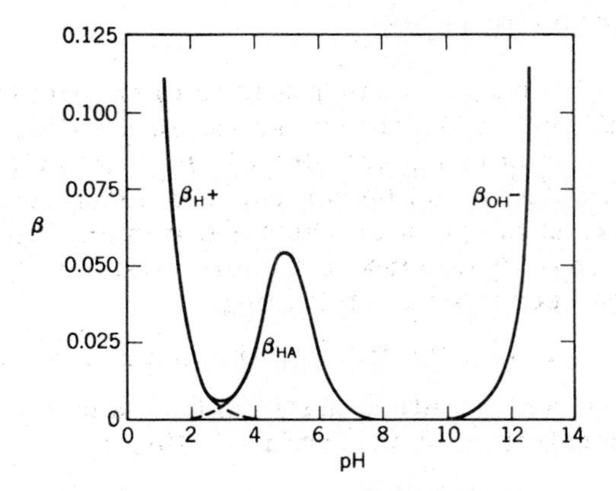

Fig. XIII-7 — Variation of Buffer Index during the titration of a mixture of a strong and a weak acid with a strong base.

proportionally to the $[OH^-]$. The branch of the curve at low pH values has been dotted since these pH values cannot be attained in HA and NaOH mixtures. This part of the curve can be realized by the addition of a strong acid to the solution. Thus we can observe a most interesting property of the buffer index, namely, the buffer index of a mixture is the sum of the buffer indexes of each of the acid-base couples present.

i.e. $\beta = \beta_{H+} + \beta_{OH^-} + \beta_{HA}$

represents equation (XIII-33).

An appreciation of this property of the buffer index enables us to design suitable buffers or mixtures of buffers having desirable β values over any pH range that is required.

Table XIII-4 presents a summary of several typical β values.

Table XIII-4

Solution	pH	β
0.10 M HCl	1.00	0.23
*0.10 M HO Ac	2.85	0.006
*0.05 M HO Ac + 0.05 M NaOAc	4.70	0.058
H_2O	7.00	4.6×10^{-7}
0.10 M NaOH	13.0	0.23

* $C_A = 0.10$

XIII-10. SHARPNESS INDEX

The sharpness index, η, which is used to characterize the sharpness of the end-point of any neutralization titration, is defined as the rate of change of pH with the fraction of acid titrated. The end-point can be defined, if a differential titration curve is drawn, as the point at which a maximum value of η is obtained. The sharpness index is closely related to the buffer index as can be seen from the following:

$\eta = \frac{dpH}{dT}$ where T is the fraction titrated and is equal to $\frac{C_B}{C_A}$; C_B is the molar concentration of strong base and C_A is the analytical concentration of acid titrated. Hence,

$$\eta = \frac{dpH}{dC_B} \times C_A = \frac{C_A}{\beta} \qquad \text{(XIII-34)}$$

Maximum values of η for various types of titrations can be found from the corresponding minimum β values, found either by differentiation of equation XIII-33 or by inspection of the β-pH curve (Fig. XIII-7). For example, in a weak acid-strong base titration, $\beta_{minimum}$ is obtained at the intersection of the β_{OH^-} and β_{HA} curves.

$$\beta_{min} = \beta_{HA} + \beta_{OH^-} = 2\beta_{HA} \text{ or } 2\beta_{OH^-}$$

$$= 2 \times 2.3\,[OH^-]$$

Since the $[OH^-]$ at the equivalence point of the weak acid-strong base titration is given by:

$$[H^+] = \left(\frac{K_w K_a}{C_A}\right)^{\frac{1}{2}} = \frac{K_w}{[OH^-]}$$

Therefore $\beta_{min} = 4.6\left(\frac{C_A K_w}{K_a}\right)^{\frac{1}{2}}$

Hence $\eta_{max.} = \frac{C_A}{\beta_{min.}} = 0.22\left(\frac{C_A \cdot K_a}{K_w}\right)^{\frac{1}{2}}$

In a similar manner we can obtain values of $\eta_{max.}$ for other analytically important titrations.

SUGGESTIONS FOR FURTHER READING

1. S. Bruckenstein and I. M. Kolthoff in Treatise on Analytical Chemistry, Editors, I. M. Kolthoff and P. J. Elving, Interscience Publishers (1959), Part I, Vol. I, Chapter 12.
2. T. B. Smith, Analytical Processes, Edward Arnold Publishers Ltd., (1940), Chapter 10.
3. H. A. Laitinen, Chemical Analysis, McGraw Hill Book Co. Inc. (1960), Chapter 5.

XIII-11. PROBLEMS

1. Draw the titration curves for the following titrations with the aid of calculated pH values of the mixture after the addition of 0, 20, 50, 99, 100, 101, 120 and 200% of the stoichiometric quantity of titrant required. Assume that the titrants are each 0.10 M.
   (a) 50 ml of 0.10 M $HClO_4$ with KOH
   (b) 100 ml of 0.02 M NH_3 with HCl
   (c) 50 ml of 0.03 M HOAc with NaOH
   (d) 25 ml of 0.20 M NaCN with HCl
   (e) 75 ml of 0.05 M $Na_2B_4O_7$ with HCl
   (f) 50 ml of 0.01 M H_2SO_4 with 0.02 M NaOH.
2. 50.0 ml of a solution of 0.25 M ethylenediamine is titrated with 0.50 M HCl. Calculate the pH values of the mixture after the addition of 0, 20, 24, 25, 26, 49, 50, 51, and 60 ml of HCl. Is there a suitable indicator that can be used at each of the equivalence points?
3. Draw the titration curve obtained by titrating 50.0 ml of 0.03 M H_3PO_4 with 0.01 M KOH, with the help of pH values calculated for mixtures near and at each of the three equivalence points. How many inflection points does this curve show?
4. 50.0 ml of a mixture that is 0.01 M in HOAc and 0.02 M in HCl is titrated with 0.04 M NaOH. Calculate the pH values after the addition of the following titrant volumes: 0, 20, 25, 30, 30, 37.5 and 40 ml. Is there a suitable indicator for detection of the first equivalence point?
5. Calculate the titration errors in the following titrations:
   (a) 0.10 M HCl vs. 0.05 M NaOH
   (b) 0.03 M HOAc vs. 0.02 M KOH
   (c) 0.06 M HOAc vs. 0.04 M KOH
   (d) 0.50 M NaOH vs. 0.10 M $HClO_4$
   (e) 0.25 M NH_3 vs. 0.20 M HCl

if in each case the pH at the end-point is (i) 3.5, (ii) 7.0 and (iii) 9.0.

6. (a) The following set of data was obtained in the titration of 0.1058 g of a monoprotic acid in 100 ml of water with 0.0998 M NaOH. Plot the titration curve and calculate the equivalent weight of the acid as well as its pK_a value.
   (b) Plot the derivative curve for this titration and compare the end-point with that obtained in (a).

ml NaOH	0.	0.50	1.00	1.50	2.00	2.50	3.00	3.50	4.00	
pH	3.28	4.07	6.26	6.60	6.83	7.02	7.17	7.28	7.42	
ml NaOH	4.50	5.00	5.50	6.00	6.50	7.00	7.55	8.00	8.50	9.05
pH	7.52	7.67	7.78	7.89	8.10	8.25	8.58	9.04	10.02	10.62
ml NaOH	9.50	10.00								
pH	10.80	11.00								

7. The titration of an unknown acid with 0.10 M NaOH exhibits only one detectable equivalence point. This occurs when 60 ml of NaOH have been added. The solution has a pH of 4.48 when only 15 ml of NaOH have been added and a pH of 6.12 when 40 ml of NaOH have been added. Calculate the dissociation constant or constants of the acid.
8. Calculate the buffer index values for the following solutions:
   (a) 0.50 M HCl
   (b) 0.05 M HCl
   (c) 0.05 M H_3PO_4
   (d) 0.05 M NaOH
   (e) 0.05 M NH_3
   (f) 0.05 M NH_3 and 0.05 M NH_4Cl
   (g) 0.05 M NH_3 and 0.20 M NH_4Cl
9. Calculate the sharpness index at the equivalence points of the titration curves in problem 1.

XIV

Precipitation Titrations

Unlike acid-base reactions, reactions involving formation of precipitates are not always sufficiently rapid to permit their use in titrimetry. Furthermore the occurrence of coprecipitation during such titrations might seriously limit the accuracy of the stoichiometry. For these reasons relatively few precipitation reactions have been adapted to titrimetric determinations. By and large precipitation titrations are employed for the determination of various anions since more convenient methods, notably complexometric titrations (Chapter XV), are available for most metal ions. The formation of the silver halide precipitates is widely used for the determination of halides, particularly chloride, in the well known Mohr or Volhard titrations. The titration of sulfate with barium ion and the titration of fluoride with thorium ion are also used to some extent.

Titration curves for precipitation reactions are based on the use of the pM ($-\log [M^{n+}]$) in place of the pH, used in the acid base titrations, as the critical concentration variable.

Since titrations involving silver halides are so widely used the development of titration curve calculations will be illustrated with the silver-chloride titration.

XIV-1 TITRATION OF CHLORIDE ION WITH SILVER ION.

Let us considera typical titration in which 50.00 ml. of 0.10 M NaCl is titrated with 0.05 M $AgNO_3$. The calculations involved in the construction of the titration curve fall into the following categories.

(1) The region between 0 and 99% reagent added.
(2) The region very close to the equivalence point, (i.e. > 99% or < 101% reagent added.)
(3) The equivalence point. (100% reagent added.)
(4) The region containing a significant excess of Ag^+ (i.e. > 101% reagent added.)

To illustrate these calculations the following points on the titration curve will be considered in detail.

(a) 20 ml of 0.05 M $AgNO_3$ added.

Amount of Cl^- originally present = 50 × 0.10
= 5.0 m moles

Amount of Ag^+ added = 20 × 0.05 = 1.0 m moles

Amount of Cl^- left in solution = 4.0 m moles

Total volume of solution = 50 + 20 = 70 ml

$$\text{Therefore, } [Cl^-] = \frac{4.0}{70} = 5.7 \times 10^{-2}\ M$$

This concentration of Cl^- is taken as the total concentration because the contribution from the dissolved AgCl is negligible in comparison.

From the equation: $K_{sp} = [Ag^+][Cl^-]$

$$pAg = pK_{sp} - pCl$$

if $pK_{sp} = 9.55$*, then $pAg = 9.55 - 1.25 = 8.30$

(b) 99.95 ml of 0.05 M $AgNO_3$ added.

$$[Cl^-] \text{ in excess} = \frac{50. \times 0.10 - 99.95 \times .05}{50 + 99.95}$$

$$= \frac{0.0025}{150} = 1.7 \times 10^{-5}\ M$$

*In these and subsequent titration curve calculations, a single value of the solubility product will be assumed to apply throughout the titration. Ionic strength corrections for K values may be applied at individual points at the discretion of the reader.

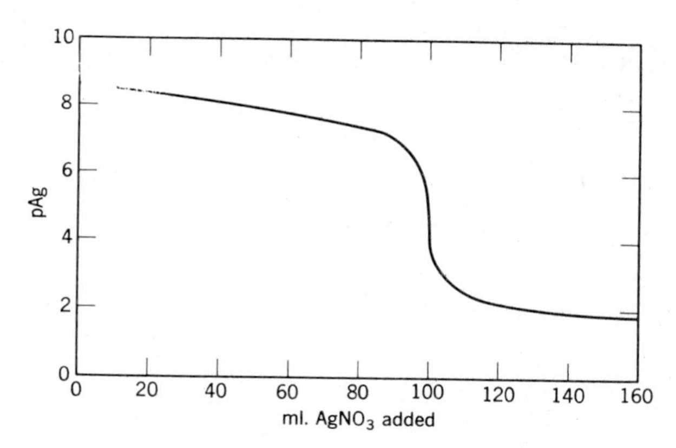

Fig. XIV-1. Titration of NaCl with $AgNO_3$

Since this concentration is so small we cannot neglect the concentration of Cl^-, X, from the dissolved AgCl.

Hence,

$$K_{sp} = [Ag^+][Cl^-] = X.(1.7 \times 10^{-5} + X)$$

The resulting quadratic equation may be solved to give:

$$X = 1.1 \times 10^{-5} \text{ and } pAg = 4.95$$

Notice that the contribution of the dissolved AgCl to the chloride ion concentration, 1.1×10^{-5}, is a considerable fraction of the total chloride ion concentration, 2.8×10^{-5}M.

(c) 100 ml of 0.05 M $AgNO_3$ added.

With the addition of an exactly equivalent amount of $AgNO_3$ there is no excess of Cl^- or Ag^+. Both the $[Ag^+]$ and $[Cl^-]$ in solution are derived solely from the dissolved AgCl.

$$K_{sp} = [Ag^+][Cl^-]$$

Therefore $[Ag^+] = [Cl^-] = (K_{sp})^{\frac{1}{2}}$

and $pAg = \frac{9.55}{2} = 4.78$

(d) 100.05 ml of 0.05 M $AgNO_3$ added.

The amount of Ag^+ added in excess is equal to 0.05×0.05 or 2.5×10^{-3} m moles. This is equivalent to $[Ag^+] = 1.7 \times 10^{-5}$ M since the total volume is 150 ml. Analogous to (b), this concentration is small enough to require consideration of the Ag^+ from the dissolved AgCl.

$$K_{sp} = [Ag^+][Cl^-] = (1.7 \times 10^{-5} + X)X$$

where X is the solubility of AgCl in this solution. This is seen to be the same equation as in (b), so that

$$X = 1.1 \times 10^{-5}.$$

Therefore $[Ag^+] = 2.8 \times 10^{-5}$ M

and $\qquad pAg = 4.55$

Note the formal similarity of this calculation with that of (b) at a point which is equally distant from the equivalence point. This illustrates the fact that the titration curve of AgCl, as with every electrolyte containing one cation and one anion, (AB), is symmetrical about the equivalence point in its immediate vicinity. This would not be the case for electrolytes of the type A_2B or AB_2, for example Ag_2CrO_4 or $PbCl_2$.

(e) 150 ml of 0.05 M $AgNO_3$ added.

$$\text{The excess } [Ag^+] = \frac{150 \times 0.05 - 50 \times 0.10}{200}$$

$$= 1.25 \times 10^{-2} \text{ M}$$

and pAg = 1.90

Once again the Ag^+ from the dissolved AgCl is negligible in comparison with the excess Ag^+.

The titration curve of Cl^- vs. Ag^+ which is shown in Fig. XIV-1 is seen to resemble the strong acid strong base titration curve in Fig. XIII-1. This is to be expected if one realizes that the expression for the solubility product of AgCl is exactly analogous to the ion product of water. Just as the OH^- is titrated with H^+, the Cl^- here is titrated with Ag^+. The height of the vertical portion of the titration curve of a strong acid-strong base is determined by the ion product of water; similarly the height of the vertical portion of the titration curve of Cl^- vs Ag^+ depends on the K_{sp} of AgCl. The smaller the K_{sp} the larger the break. Hence as might be expected, the vertical portions in the titration curves of Br^- or I^- vs Ag^+ are larger and these titrations are theoretically capable of greater precision. Of course the concentrations of reagents used will also affect the height of the vertical portion of the titration curve as well as the precision of the titration.

XIV-2. TITRATION OF CHROMATE ION WITH SILVER ION

This titration has been chosen primarily to illustrate the effect of electrolyte type on the slope of the titration curve. The calculations are similar to those that have been carried out in the AgCl case, just described. The only notable difference occurs in the immediate vicinity of the equivalence point where there is need to solve a cubic equation.

Just before the equivalence point in the presence of an excess of chromate of concentration C,

$$K_{sp} = [Ag^+]^2[CrO_4^=] = (2X)^2 \cdot (C + X)$$

where X is the molar solubility of Ag_2CrO_4. Just beyond the equivalence point, the appropriate equation is:

$$K_{sp} = (C + 2X)^2 \cdot X$$

where C is the concentration of excess Ag^+.

Because of the difference in the forms of these equations, the titration curve is unsymmetrical in the vicinity of the equivalence point. The point of inflection in the titration curve is reached somewhat beyond the equivalence point.

XIV-3. INDICATORS FOR PRECIPITATION TITRATIONS

Indicators for precipitation titrations fall into one of two general categories.

(a) Those which react with an excess of titrant and

(b) Those which interact with the precipitate itself.

In the latter case advantage is taken of the fact that the charge on the surface of a precipitate changes radically at the equivalence point, making it capable of adsorbing an oppositely charged dye. In the former case the indicator might be such as to form with the excess titrant, a distinctively colored precipitate or a colored, soluble compound. Naturally the indicator must not be capable of reaction with the titrant until the ion being titrated has reacted completely. Hence a consideration of the solubility products and other equilibrium constants are of importance in the selection of such indicators. Most of these types of indicators may be illustrated with the various titrations used in the determination of halides. For example, eosin, which is strongly adsorbed on the surface of AgBr or AgI immediately after the equivalence point (in the presence of a slight excess of Ag^+) may be used for the titration of Br^-, I^- and SCN^- with Ag^+. Since eosin is too strongly adsorbed on AgCl to be useful in Ag^+-Cl^- titrations dichlorofluorescein is frequently employed. Eosin remains anionic down to a pH of 1, but dichlorofluorescein may not be used below a pH of 4.

In the Mohr titration of chloride, the indicator $CrO_4^=$ forms a brick-red precipitate of Ag_2CrO_4. If the solution is sufficiently acidic ($pH < 6.5$) to transform $CrO_4^=$ to $Cr_2O_7^=$, the Mohr method cannot be used.

The Fe^{+3} indicator in the Volhard titration forms a soluble red-colored complex with excess CNS^-. Since HCNS is a strong acid, CNS^- exists even at very low pH permitting the colored complex to form. Hence the Volhard titration may be used in solutions that are more acidic than would be permissible with adsorption indicators.

In all of the above titrations with Ag^+, the pH must be kept below 10.5, because at this point Ag_2O begins to precipitate.

Other examples of indicators that react with excess titrant include phenolphthalein which turns pink upon the addition of a slight excess of potassium palmitate in the determination of water hardness, and diphenylamine, a redox indicator (see Chapter XV) which turns from blue to colorless on the addition of a slight excess of ferrocyanide initially containing a small amount of ferricyanide in the titration of zinc.

XIV-4. TITRATION ERRORS

In this section we will consider the quantitative aspects of the Mohr and Volhard titrations and discuss the factors that will affect the magnitude of the titration errors involved therein.

XIV-4 (a). THE MOHR TITRATION.

In a typical Mohr titration 4 m moles of NaCl will have been titrated using 0.1 M $AgNO_3$ in a solution whose total volume is 100 ml, and to which 0.3 ml of 1.0 M K_2CrO_4 has been added as indicator. In this solution pK_{sp} of AgCl is 9.56

and pK_{sp} of Ag_2CrO_4 is 10.85. When Ag^+ is added to this solution, AgCl precipitates first (see Chapter VIII). The major question here is how close to the equivalence point in the Ag^+-Cl^- titration will the appearance of the Ag_2CrO_4 precipitate (i.e. the end point), occur.

At the end-point, i.e. when Ag_2CrO_4 just begins to precipitate,

$$10^{-10.85} = [Ag^+]^2[CrO_4^{=}]$$

Since $[CrO_4^{=}] = 3.0 \times 10^{-3}$ M at the end-point, the $[Ag^+]$ at the end point is

$$\left(\frac{10^{-10.85}}{3.0 \times 10^{-3}}\right)^{\frac{1}{2}} = 6.9 \times 10^{-5} \text{ M.}$$

It might now be asked, how this concentration of $[Ag^+]$ required to obtain the first trace of Ag_2CrO_4, was obtained. Not all of this $[Ag^+]$ represents an excess of titrant added from the buret inasmuch as the dissolution of AgCl contributes to the total $[Ag^+]$. Hence it will be necessary to subtract the $[Ag^+]$ from the AgCl, from 6.9×10^{-5}. This amount will be equal to the $[Cl^-]$.

$$[Cl^-] = \frac{10^{-9.56}}{6.9 \times 10^{-5}} = 4.1 \times 10^{-6}$$

Hence the concentration of silver ion added in excess at the end-point is $(6.9 \times 10^{-5} - 4.1 \times 10^{-6}) = 6.5 \times 10^{-5}$ M. Since the volume of the solution is 100 ml, this represents an amount of Ag^+ of $6.5 \times 10^{-5} \times 100 = 6.5 \times 10^{-3}$ m moles. The volume of titrant containing this quantity of Ag^+, is

$$\frac{6.5 \times 10^{-3}}{0.10} = 0.07 \text{ ml.}$$

and is called the titration error. The percentage titration error in this case is

$$\frac{0.065 \times 100}{40} = 0.16\%$$

From the foregoing calculations it will be noticed that the magnitude of the titration error will depend on the quantity of $CrO_4^{=}$ added, the concentration of titrant, the volume of the titration mixture at the end-point and factors which affect the solubility product constants, namely temperature and ionic strength. The titration error may be seen to increase with decreasing quantities of K_2CrO_4 added and with increasing solution volumes at the end-point. Although it is theoretically possible to increase the $[CrO_4^=]$ and decrease the volume, until the titration error is eliminated, this is not practical because the solution becomes so deeply yellow that the end-point is obscured.

In actual practice the end-point errors are found to be somewhat greater than the calculated errors because a minimum amount of Ag_2CrO_4 must form for the color change to be visible. Of course it is good laboratory practice to titrate unknown samples under conditions that closely approximate those used in the standardization titrations so as to compensate almost entirely for titration errors.

An obvious requirement for anions that can be titrated with Ag^+ using $CrO_4^=$ as an indicator is that the solubilities of the silver salts must be sufficiently smaller than that of Ag_2CrO_4 to permit their quantitative precipitation prior to the onset of the Ag_2CrO_4 precipitation. It will be instructive to calculate a critical K_{sp} value for a hypothetical anion that would give rise to 0.1% titration error under conditions paralleling those in the example discussed at the beginning of Section XIV-3-(b).

Example 1.

In a titration of 4 m moles of NaX using 0.1 M $AgNO_3$ in a solution whose total volume is 100 ml. and to which 0.3 ml of 1.0 M K_2CrO_4 has been added, what is the maximum possible value of the $K_{sp}AgX$, if the titration error is to be limited to 0.1% ?

Since the $[Ag^+]$ at the end point is fixed at 6.9×10^{-5} M (see above), then the magnitude of the end-point error will depend on how far from this concentration will be that which obtains at the equivalence point. Hence two answers to this question are possible, one in which the concentration of the anion at the equivalence point is greater than 6.9×10^{-5} M (i.e. a negative error) and one in which the concentration of the anion at the equivalence point is less than 6.9×10^{-5} M (i.e. a positive error).

For a solution in which the final anion concentration, $[X^-]$, is greater than 6.9×10^{-5}, we recognize that only part of the total will arise from excess anion and part from the solubility of AgX. The X^- concentration from the solubility of AgX must be equal to the $[Ag^+]$ from the AgX. In the presence of excess X^-, this is 6.9×10^{-5} M which is the total $[Ag^+]$. Since we are postulating 0.1% of the original X^- is untitrated at the end-point, this corresponds to 0.1% of 4 m moles which in 100 ml amounts to 4.0×10^{-5} M.

Hence, Total $[X^-] = 4.0 \times 10^{-5} + 6.9 \times 10^{-5}$

$$= 1.09 \times 10^{-4}\ M$$

and, K_{sp} of AgX $= [Ag^+][X^-]$

$$= 6.9 \times 10^{-5} \times 1.09 \times 10^{-4}$$

$$= 7.5 \times 10^{-9}$$

For a solution in which the final anion concentration $[X^-]$, is less than 6.9×10^{-5} M, we recognize that the Ag^+ is present in excess. Postulating a positive titration error of 0.1%, this amounts to a concentration of untitrated Ag^+ of 4×10^{-5} M. The difference between this value and the total, i.e. 2.9×10^{-5} M arises from the solubility of AgX and is also equal to the $[X^-]$ from the same source.

Hence K_{sp} of AgX $= 2.9 \times 10^{-5} \times 6.9 \times 10^{-5} = 2.0 \times 10^{-9}$

From this example it will be seen that for univalent anions whose silver salts have a solubility product between the values 7.5×10^{-9} and 2.0×10^{-9}, the titration error in the Mohr titration would be 0.1%or less. If the K_{sp} is larger than the upper limit then the titration will become increasingly negative without limit. In contrast, if the K_{sp} is below 2.0×10^{-9} the error becomes increasingly positive up to a limit imposed by the maximum $[Ag^+]$ required to initiate the precipitation of Ag_2CrO_4, i.e. 6.9×10^{-5} M which under the conditions of the titration outlined above corresponds to a titration error of 0.17%. Therefore in comparing Cl^-, Br^- and I^- titration with Ag^+, the titration errors vary by a negligible amount (from 0.16% with Cl^- up to the limiting value of 0.17% for Br^- and I^-).

Since $CrO_4^=$ is converted into $HCrO_4^-$ and then to $Cr_2O_7^=$ with increasing acidity, the effectiveness of the CrO_4 indicator will depend on the pH of the solution being titrated. From the following equilibrium expressions the $[CrO_4^=]$ can be calculated at any given pH.

$$\frac{[H^+][CrO_4^=]}{[HCrO_4^-]} = 3.2 \times 10^{-7}$$

$$\frac{[Cr_2O_7^=]}{[HCrO_4^-]^2} = 43$$

At the end point, $[CrO_4^=] + [HCrO_4^-] + 2[Cr_2O_7^=] = 3.0 \times 10^{-3}$

Assuming that the titration is carried out at a pH of 5.0,

$$[CrO_4^{=}]\left(1+\frac{10^{-5}}{3.2\times10^{-7}}\right)+[CrO_4^{=}]^2\times\frac{2\times43\times10^{-10}}{(3.2\times10^{-7})^2}=3.0\times10^{-3}$$

and it follows that, $[CrO_4^{=}] = 7.5 \times 10^{-5}$ M.

Hence the $[Ag^+]$ at the end point $= \left(\frac{10^{-10.85}}{7.5\times10^{-5}}\right)^{\frac{1}{2}}$

At this point, $[Cl^-] = \frac{10^{-9.56}}{4.3\times10^{-4}} = 6.4\times10^{-7}$ M.

The concentration of Ag^+ from the dissolved AgCl, i.e. 6.4×10^{-7} is negligible compared to 4.3×10^{-4}.

The titration error is therefore given by,

$$\frac{4.3\times10^{-4}\times100\times100}{0.1\times40}=1.1\%$$

This calculation indicates the need for keeping the pH close to neutrality in the Mohr titration.

XIV-4-(b). THE VOLHARD TITRATION

Let us first consider the titration error that would arise in the titration of 4 m moles of $AgNO_3$ with 0.10 M KCNS using enough Fe^{+++} to give a concentration of 0.02 M in a solution whose final volume is 100 ml. In this solution, K_{sp} of AgCNS is 2.0×10^{-12} and K for the formation of $FeCNS^{++}$ is 30. The end-point of the titration corresponds to the appearance of a pink color due to the formation of $FeCNS^{++}$. This color is visible when $[FeCNS^{++}]$ reaches 3.0×10^{-6} M. From the equilibrium expression for the formation of this complex, the $[CNS^-]$ at the end-point can be calculated

$$K=30=\frac{[FeCNS^{++}]}{[Fe^{+++}][CNS^-]}=\frac{3.0\times10^{-6}}{0.02\times[CNS^-]}$$

and $[CNS^-] = 5.0\times10^{-6}$ M. From this total concentration of CNS^-, that which arises from the dissolved AgCNS, must be subtracted in order to obtain the excess of CNS^- added. This will be equal to $[Ag^+]$ from the dissolved AgCNS.

$$K_{sp}=[Ag^+][CNS^-]$$

$$2\times10^{-12}=[Ag^+][5.0\times10^{-6}]$$

Therefore $[Ag^+] = .40 \times 10^{-6}$ M

and the excess $[CNS^-] = 5.0 \times 10^{-6} - 0.4 \times 10^{-6}$

$= 4.6 \times 10^{-6}$ M.

Hence the titration error is $\frac{4.6 \times 10^{-6} \times 100}{0.1} = +0.005$ ml

corresponding to a titration which is +0.01%.

In the Volhard titration, which is the principal application of this titration, Cl^- is determined by the addition of a known excess of Ag^+, which is then back-titrated with standard CNS^-. Here, the presence of the AgCl precipitate which is more soluble than AgCNS greatly increased the titration error.

At the end-point of the Volhard titration, the solution contains both precipitates, AgCl and AgCNS. Hence,

$$[Ag^+][Cl^-] = 4.0 \times 10^{-10}$$

and

$$[Ag^+][CNS^-] = 2.0 \times 10^{-12}$$

so that, $\frac{[Cl^-]}{[CNS^-]} = 200$

Since the total $[CNS^-]$ at the end-point was found to be 5.0×10^{-6} M, the $[Cl^-]$ at the end point must be $200 \times 5.0 \times 10^{-6} = 1.0 \times 10^{-3}$ M, if equilibrium is attained. This means that the excess CNS^- reacts with the AgCl:

$$AgCl + CNS^- \rightarrow AgCNS + Cl^-$$

In order to prevent this reaction from causing a large titration error, the AgCl must be either filtered off or its rate of reaction with CNS^- drastically reduced by the addition of nitrobenzene, prior to the back-titration. If this were not done, as much as 0.10 m moles ($1.0 \times 10^{-3} \times 100$) of excess CNS^- would have to be added before a permanent end-point would be reached.

XIV-5. TITRATIONS OF HALIDE MIXTURES

If the absence of coprecipitation is assumed the titration of a mixture of halides containing iodide, bromide and chloride with silver ion will proceed at first (from points A to B, Fig. XIV-2) in exactly the same manner as if only iodide ion were present. From B to C the titration curve is calculated in exactly the same manner as if Br^- were present alone and beyond point C as if only Cl^- were present. It should be noted that the points, i.e. the pAg values, at which AgBr and AgCl start to precipitate are also precisely the same as they would be if each of these ions were present alone. For points A, B and C in Figure XIV-2 the pAg values are obtained from:

$$[Ag^+] = \frac{K_{sp}}{[X^-]}$$

where $[X^-]$ stands for the halide ion in question. Thus it may be seen that none of the halide ions will interfere with each other, i.e. the extent of the precipitation of each halide depends solely on the equilibrium Ag^+ concentration.

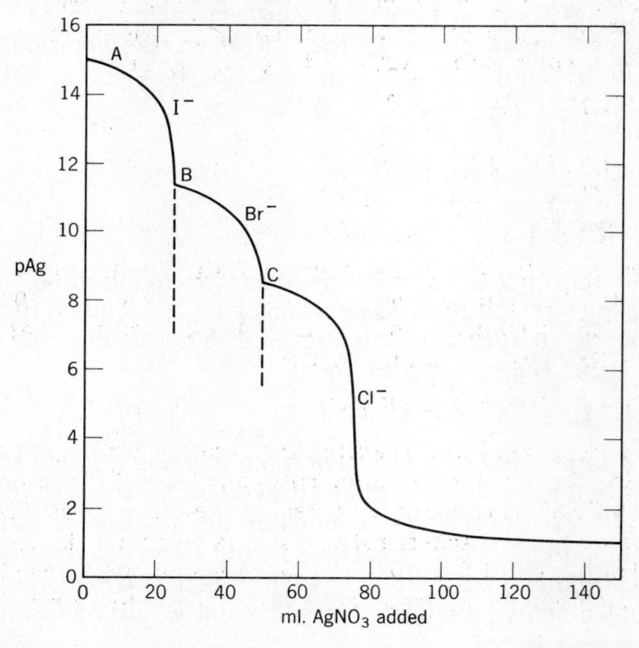

Fig. XIV-2. Idealized Titration Curve for 0.10 M Halides titrated with 0.20 M $AgNO_3$.

The major difference in the titration of halide mixtures is that the end-points or break-points (points B & C) in the curve occur before the equivalence point. This may be seen from the following calculation. Let the analytical concentrations of I^-, Br^-, and Cl^- be represented by C_{I^-}, C_{Br^-}, and C_{Cl^-}. AgBr will begin to precipitate, at point B, when

$$[Ag^+] = \frac{K_{spAgBr}}{C_{Br^-}}$$

At this point, the

$$[I^-] = \frac{K_{spAgI}}{[Ag^+]} = \frac{K_{spAgI}}{K_{spAgBr}} \cdot C_{Br^-}$$

in most titrations of analytical interest, C_{Br^-} is greater than 10^{-5} M so that $[I^-]$ is greater than $(K_{spAgI})^{1/2}$ which is the value that $[I^-]$ would have at the equivalence point of the Ag^+ vs I^- titration.

For example if C_{Br^-} is 0.05 M and K_{spAgI} is 10^{-16}, K_{spAgBr} is 10^{-13},

At point B, $[I^-] = \frac{10^{-16}}{10^{-13}} \times 0.05 = 5 \times 10^{-5}$ M.

This is much greater than $(10^{-16})^{1/2} = 10^{-8}$ M. This therefore represents a negative error in the iodide titration. The relative error in the iodide titration expressed as a percentage error is,

$$\text{Percentage Error} = \frac{-100 \times [I^-]}{C_{I^-}} = -100 \frac{K_{spAgI}}{K_{spAgBr}} \frac{C_{Br^-}}{C_{I^-}}$$

In the preceding illustration if C_{I^-} at point B (assuming no precipitation) is 10^{-2} M, the percentage error is -0.5%.

From the general equation for the titration error it can be seen that the error increases with increasing ratios of initial concentrations of Br^- to I^- in the mixture. It may be noted that the dilution factor cancels out because the ratio of C_{Br^-} to C_{I^-}, rather than the individual concentrations, is involved. If the error expression is generalized to include other suitable pairs of ions, then the error will depend on the ratio of the K_{sp} values.

The difference between the end-point B and the equivalence point of the iodide titration, which represents a negative error for iodide, obviously results in a positive error for the bromide titration, i.e. the excess iodide at point B will be titrated together with the bromide ion. However the bromide end-point C will also occur before the equivalence point giving rise to a compensating negative error. The relative error in the bromide titration therefore is:

$$100 \cdot \frac{[I^-] \text{ at B} - [Br^-] \text{ at C}}{C_{Br^-}} \cdot \%$$

As before,

$$[I^-] \text{ at B} = C_{Br^-} \cdot \frac{K_{spAgI}}{K_{spAgBr}}, \text{ Similarly}$$

$$[Br^-] \text{ at } C = C_{Cl^-} \cdot \frac{K_{spAgBr}}{K_{spAgCl}}$$

Substituting for [I] at B and $[Br^-]$ at C, the percentage error in the bromide titration is:

$$100 \cdot \frac{K_{spAgI}}{K_{spAgBr}} - \frac{K_{spAgBr}}{K_{spAgCl}} \cdot \frac{C_{Cl^-}}{C_{Br^-}}$$

It is interesting to note that the error arising from point B, (shown as the first term in the expression above), is independent of the initial concentrations of any of the halides. For example if C_{Cl^-} is 0.05 M and K_{spAgCl} is 10^{-10}, and the other conditions remain as described in the previous illustration, the relative error in the bromide portion of the titration is

$$100 \left(10^{-3} - 10^{-3} \times \frac{0.05}{0.05}\right) = 0\%$$

Titration errors that are observed in practice are generally much larger than those described in this section because of coprecipitation phenomena. The effect of coprecipitation will be described in the next section.

XIV-6. EFFECT OF COPRECIPITATION ON THE TITRATION OF HALIDES.

Precipitation titration curves obtained in practice deviate from the ideal curves described above because of coprecipitation phenomena such as adsorption and solid solution formation. (Such deviations are more pronounced in dilute solutions.) The titration of I^- vs Ag^+ represents an illustration of the effects of adsorption. In Fig. XIV-3 the solid line represents the idealized curve. The slope of the experimental curve is less than that of the ideal curve and the inflection point of the experimental curve occurs before a stoichiometric amount of Ag^+ has been added. In the region to the left of point X, the adsorption of excess I^- ions results in the presence of a higher silver ion concentration in solution, than predicted from the ideal curve. This explains why the experimental curve to the left of X is lower than the ideal curve. At point X, the extents of adsorption of Ag^+ and I^- are equal. At this point, X, called the isoelectric point of this precipitate, the real and ideal curves coincide. Since AgI crystals have a greater adsorption affinity for I^- than for Ag^+, the isoelectric point X occurs at a pAg value below that at the equivalence point. The difference between the pM at the isoelectric point and equivalence point depends on the nature of the precipitate. The relatively large difference observed for AgI makes AgI a good example for the illustration of adsorption effects in precipitation titrations. The portion of the experimental curve beyond point X, is higher than the ideal curve because the adsorption of excess Ag^+ on the precipitate is reflected in a higher pAg value for the solution.

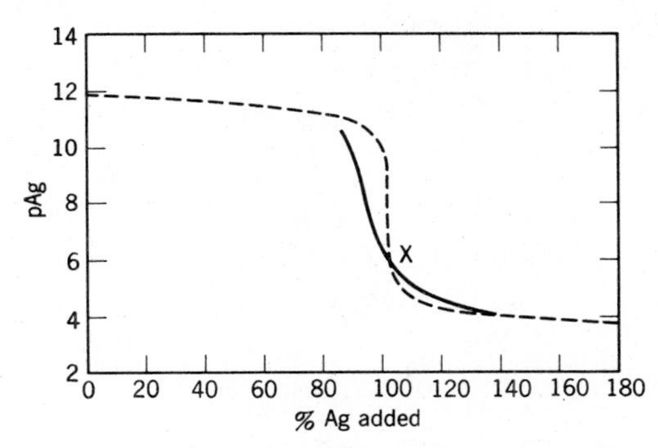

Fig. XIV-3. Effect of Adsorption in Titration of 10^{-4} M I^- with 10^{-4} M $AgNO_3$. Solid line, actual titration. Dashed line, calculated. Based on Fig. XII-2 in H. A. Laitinen Chemical Analysis, McGraw Hill, 1960, p. 207.

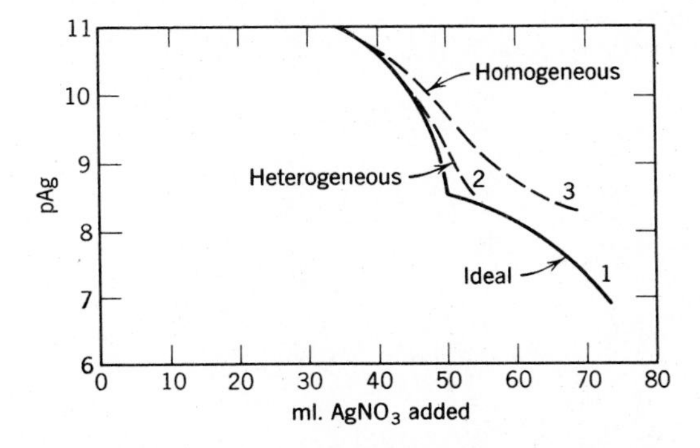

Fig. XIV-4. Effect of Solid Solution Formation on the Titration of a Mixture of Halides.

In the titration of a mixture of Br^- and Cl^- with Ag^+, significant differences between the theoretical and experimental titration curves occur because of the formation of solid solutions between AgBr and AgCl. (See Chap. VIII.)

The magnitude of these differences will depend not only on the ratio of the chloride and bromide ion concentrations, but also on the nature of the solid solution formation process.

If the precipitate can equilibrate at any instant with the solution, then the extent of chloride incorporation with the AgBr lattice at the end-point will depend on the ratio of $[Cl^-]$ to $[Br^-]$ at the end-point also. Since essentially all of the Br^- has been removed from the solution at the end-point, the resultant high $[Cl^-]/[Br^-]$ value will cause a relatively high degree of contamination and a large titration error. (See Curve 3, Fig. XIV-4.)

If the precipitate formation is of the Doerner-Hoskins type, i.e. the equilibration processes in the solid are very slow, then the extent of contamination is relatively small since the bulk of the precipitate is formed at values of $[Cl^-]/[Br^-]$ that are far smaller than obtain at the end-point. (Curve 2, Fig. XIV-4)

Under conditions normally encountered in the titration of bromide and chloride mixtures, the precipitate formation is governed by homogeneous solid solution formation.

SUGGESTIONS FOR FURTHER READING

H. A. Laitinen, Chemical Analysis, McGraw Hill Book Co. Inc. (1960)† Chapter 12.

J. F. Coetzee, in Treatise on Analytical Chemistry, Editors I. M. Kolthoff and P. J. Elving, Interscience Publishers (1959) Part I, Vol. I, Chapter 19.

R. C. Bowers, L. Hsu and J. A. Goldman, Anal. Chem., 33, 190, (1961).

XIV-7. PROBLEMS

1. Calculate the concentration of dissolved I^-, IO_3^-, SCN^- and $C_2O_4^=$ when 25.00 ml of 0.10 M $AgNO_3$ has been added to the following solutions.
   (a) 25.00 ml of 0.10 M KI
   (b) 24.49 ml of 0.10 M KIO_3
   (c) 25.52 ml of 0.10 M KSCN
   (d) 30.00 ml of 0.10 M KI

(e) 23.00 ml of 0.10 M KSCN
(f) 25.00 ml of 0.05 M $Na_2C_2O_4$
(g) 50.00 ml of 0.10 M $Na_2C_2O_4$

2. The following solutions are titrated with 0.05 M $AgNO_3$:
(a) 50 ml of 0.03 M KBr
(b) 25 ml of 0.03 M KIO_3
(c) 50 ml of 0.03 M $K_2C_2O_4$
Calculate and plot pAg vs. V, the number of ml of $AgNO_3$ added, at the following percentages of the stoichiometric amount of titrant: 0, 20, 50, 99.0, 99.2, 99.6, 99.8, 100.0, 100.2, 100.4, 100.6, 100.8, 101, and 150%.
Plot $\frac{\Delta pAg}{\Delta V}$ vs V and determine the maximum value of $\frac{\Delta pAg}{\Delta V}$ in each curve.
Do these maxima coincide with the equivalence points of the titrations? Explain.

3. (a) Fifty ml of a 0.01 M solution of KIO_3 is titrated with a 0.10 M solution of $AgNO_3$ by the Mohr method. What is the theoretical concentration of chromate ions required in order that Ag_2CrO_4 will just begin to precipitate at the equivalence point?
(b) If the chromate ion concentration at the end-point is 2.0×10^{-3} M calculate the iodate ion concentration at the end-point.
(c) Calculate the theoretical titration error in (b).

4. Calculate the chromate ion concentration that should be present at the end-point in the titration of 50 ml of 0.02 M KBr with 0.01 M $AgNO_3$ (by the Mohr method), if the maximum permissible theoretical titration error is 1.0%.

5. In a titration of 10 millimoles of NaX/100 ml with 0.20 M $AgNO_3$ in the presence of 5.0×10^{-3} M K_2CrO_4, what is the value of the solubility product of AgX if the theoretical titration error is a) + 0.02%? b) −0.20%?

6. Calculate the theoretical titration error in problem 3(b) if the pH of the solution at the end-point is 4.5.

7. Fifty ml of a 0.15 M solution of $AgNO_3$ is titrated with 0.10 M KCNS using Fe^{+++} as indicator. Calculate the theoretical titration error if the initial Fe^{+++} concentration present is 0.01 M and the concentration of $FeSCN^{++}$ at the end-point is 2.5×10^{-6} M.

8. In a Volhard titration, 50 ml of 0.10 M $AgNO_3$ is added to 25 ml of 0.05 M NaCl and the excess Ag^+ titrated with 0.10 M KCNS in

the presence of 0.03 M Fe^{+++}. Calculate the theoretical titration error if the concentration of $FeSCN^{++}$ at the end-point is 3.0×10^{-6} M.

9. A solution is 0.05 M in KBr and 0.10 M in KI. Calculate the theoretical titration curve if 50 ml of this solution is titrated with 0.10 M $AgNO_3$, and plot pAg vs ml $AgNO_3$ added.
10. Calculate the iodide ion concentration that remains in solution when AgBr starts to precipitate and hence calculate the theoretical titration error in problem (9). What would this titration error be if KCl were substituted for KBr in problem (9)?

XV

Complexometric Titrations

XV-1. SUITABILITY OF COMPLEXING AGENTS AS METAL TITRANTS

The development in the last fifteen years of complexing agents which can be effectively employed as titrants has extended the scope of titrimetric techniques to include determinations of most metal ions. The understanding of such titrations may be best approached in terms of their analogy to the Bronsted acid-base titrations.

For example the base NH_3 can be neutralized with the hydronium ion or the hydrated zinc ion.

$$H^+ + NH_3 \rightleftharpoons NH_4^+ \ ; \ K = 10^{9.3} \tag{1}$$

$$Zn^+ + 4NH_3 \rightleftharpoons Zn(NH_3)_4^{++} \ ; \ K_f = k_1k_2k_3k_4 = 10^{9.1} \tag{2}$$

Although these two reactions have approximately the same overall formation constant, the titration curves shown in Fig. XV-1 clearly indicate that only the hydrogen ion may be successfully titrated with NH_3. The variation of pZn with added NH_3 is much more gradual than the corresponding pH variation, mainly because of the formation of intermediate complexes whose stepwise k values are both small and close together.

Another point of contrast between the reaction of NH_3 with the hydronium ion and the hydrated zinc ion is the much greater effect that dilution has in the latter case since the overall equilibrium expression contains the factor $[NH_3]$ raised to the

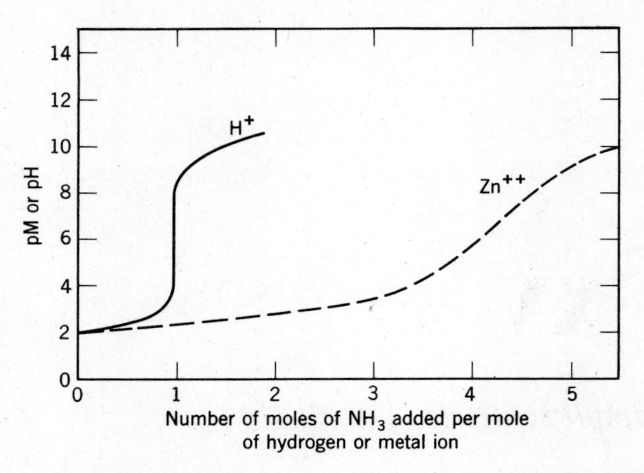

Fig. XV-1 – Titration of 10^{-2} M H^+ and 10^{-2} M Zn^{++} with NH_3.

power of four. Thus in the region well beyond the equivalence point, a tenfold decrease in $[NH_3]$, which would result in only a tenfold increase in the ratio $[H^+]/[NH_4^+]$, and therefore the $[H^+]$, but would raise the ratio $[Zn^{++}]/[Zn(NH_3)_4^{++}]$ by a factor of 10,000.

Chelating agents form complexes with metals of somewhat greater stability than simpler monodentate ligands, but more important, do so with fewer intermediates. Hence each stepwise constant has a larger value of k, since it is equivalent to the product of two or more such steps with a monodentate ligand. (Table 1)

Table 1

Formation Constants of Zinc Complexes with Nitrogen Containing Ligands

Ligand					
NH_3	$\log k_1 = 2.3$	$\log k_2 = 2.3$	$\log k_3 = 2.4$	$\log k_4 = 2.1$	$\log K_f = 9.1$
$NH_2CH_2CH_2NH_2$ (en)	$\log k_1 = 5.9$		$\log k_2 = 5.2$		$\log K_f = 11.1$
$(NH_2CH_2CH_2)_3N\cdot$(tren)		$\log k_1 = 14.7$			

Fig. XV-2 shows the titration curves of Zn^{++} with NH_3, en, and tren. The most suitable titrant, namely the one which gives

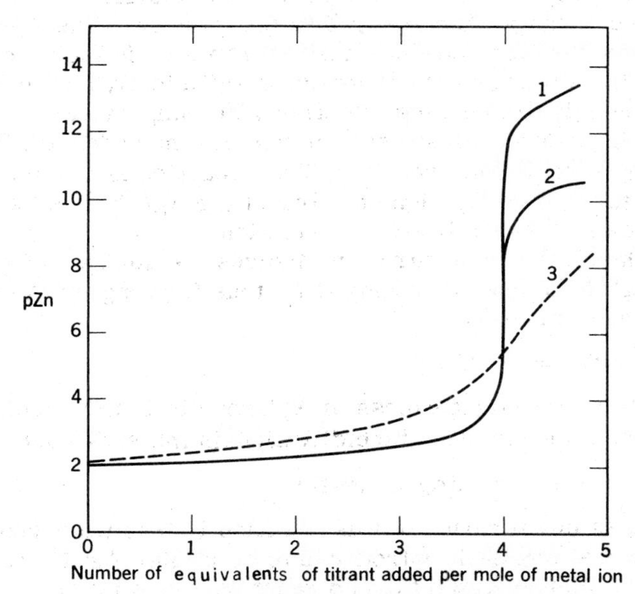

Fig. XV-2 — Titration of Zn^{++} with Ammonia, Ethylenediamine and Triaminotriethylamine.

the largest vertical break, is tren. Although en represents a far more suitable metal titrant than does NH_3 (Fig. XV-2) the formation of an intermediate $Zn(en)^{++}$ results in the rise of pZn before the equivalence point and a smaller vertical region.

Another consideration of primary importance in the suitability of a complexing agent as a metal titrant is the rate of its reaction with a metal ion. In contrast to proton transfer reactions which are all fairly rapid, certain metal complex formations proceed relatively slowly. The hydrated chromium (III) ion is notoriously slow in its reactions with almost any ligand which effectively prevents it being determined by direct titration with metal complexing agents. Metal ions whose complex formations and dissociations are relatively slow (such complexes are called "inert" in contrast with those formed rapidly, called "labile"), might be determined by a back-titration procedure.

In summary, therefore, a metal complex formation is a suitable titration reaction if (a) the complex is sufficiently stable to bring about a large change in pM in the vicinity of the equivalence point reaction and (b) is formed rapidly. It should be kept in mind that as long as the complex is sufficiently stable, monodentate ligands can be used effectively in complexation titrations. The Liebig titration illustrates the use of a monodentate ligand in a complex formation reaction. This reaction is suitable because of the extremely high stability of the $Ag(CN)_2^-$ ($K_f \approx 10^{20}$) and its relatively rapid rate of formation.

The classical Liebig titration involves the addition of a solution of $AgNO_3$ to one containing CN^-, thus forming the $Ag(CN)_2^-$ complex quantitatively.

$$2CN^- + Ag^+ \rightleftharpoons Ag(CN)_2^-$$

The addition of a slight excess of Ag^+ results in the precipitation of AgCN and the resulting turbidity signals the end-point.

$$Ag(CN)_2^- + Ag^+ \rightleftharpoons 2AgCN \text{ (solid)}$$

The Deniges modification of this titration is less time-consuming because ammonia is introduced to solubilize AgCN. KI is used as the indicator and the end-point occurs when the first precipitate of AgI causes a turbidity.

The calculations involved in deriving metal complexation titration curves will now be illustrated with reference to EDTA and other chelating titrants.

XV-2. METAL-EDTA TITRATION CURVES

As an example let us consider the titration of a 50.0 ml solution of 0.01 M Ca^{++} (strongly buffered at pH 10.00), with a 0.01 M solution of the disodium salt of EDTA. The titration curve (pCa vs ml EDTA added) can be derived by calculating pCa values at a number of points.

(a) Before the equivalence point, e.g., 25.0 ml EDTA added.

At this point the Ca^{++} which is in excess has a concentration equal to

$$\frac{50 \times 0.01 - 25 \times 0.01}{75} = 10^{-2.52}\text{ M}$$

(b) At the equivalence point. 50.0 ml EDTA added.

At this point the concentration of the calcium-EDTA complex is equal to

$$\frac{50 \times 0.01}{100} = 5.0 \times 10^{-3} \text{ M}$$

The calcium ion in solution which is formed on the dissociation of the complex is given by the expression:

$$K_f' = K_f \alpha_4 = \frac{[CaY^{=}]}{[Ca^{++}]\, C_{H_4Y}}$$

(See Chapter IX Example 4.)

In this problem $\log K_f = 10.70$

From Fig. IX-6 at pH 10.0, $\alpha_4 = 10^{-0.46}$

Hence $K_f' = 10^{10.24}$

Let $x = [Ca^{++}] = C_{H_4Y}$

Then, $10^{10.24} = \dfrac{5 \times 10^{-3} - x}{x^2}$

and $x = 10^{-6.27} = [Ca^{++}]$

(c) After the equivalence point, e.g. 75 ml of EDTA added.

The concentration of excess EDTA $= \dfrac{75 \times 0.01 - 50 \times 0.01}{125}$

$= 2.0 \times 10^{-3}$ M

The concentration of the Ca^{++} - EDTA complex is

$$\frac{50.0 \times .01}{125} = 4.0 \times 10^{-3} \text{M}$$

Therefore $\dfrac{4.0 \times 10^{-3} - X}{x\,(2.0 \times 10^{-3} + x)} = 10^{10.24}$

and $x = 10^{-9.94} = [Ca^{++}]$

It is important to note that the volume of this solution, 125 ml, has no effect on the pCa since it was used both in the numerator and denominator of the above expression in the calculation of x. This situation is exactly analogous to that with the Bronsted acid and its conjugate base and provides the basis of so-called metal-ion buffers. (See Sec. XV-4.)

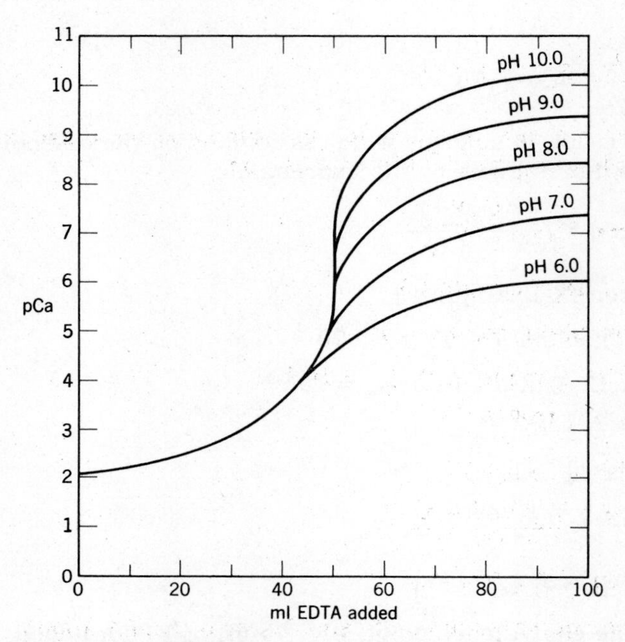

Fig. XV-3 — Effect of pH on the Titration of Ca^{++} with EDTA.

By calculations similar to those in (a), (b), and (c), values for pCa at the following points can be determined and used to draw the titration curve (Fig. XV-3).

Volume of EDTA added	0	25.0	45.0	49.9	50.0	50.1	75.0	100
pCa	2.00	2.52	4.30	5.00	6.27	7.54	9.94	10.24

Fig. XV-3 also shows the titration curves that would result if the solutions were buffered at lower pH values. It will be noted that the breaks in the titration curve get increasingly smaller as the pH is lowered. The reason for this is the effect of pH on α_4 and hence on K'_f. A study of this figure shows the importance of having a sufficiently high pH for the titration. Below a pH of 8.0 the Ca^{++}-EDTA titration does not have a sufficient change in pCa in the region of the equivalence point to give a sharp end point.

It is possible to calculate the minimum permissible pH for a metal-EDTA titration by adopting the criterion of 99.9% reaction when 0.1% excess titrant has been added. If 0.01 M metal ion solution is used this excess corresponds to

$$C_{H_4Y} = 10^{-5}\,M. \quad \text{Since } \frac{[MY^{(4-n)-}]}{C_M} = 1000,$$

$$K'_f = \frac{[MY^{(4-n)-}]}{C_M \cdot C_{H_4Y}} = \frac{10^3}{10^{-5}} = 10^8$$

The pH value at which this K'_f value of 10^8 is achieved can be obtained for various metal ions from values of K_f and α_4 of EDTA. A useful representation of such data is shown in Fig. XV-4.

In quite a number of instances metal-EDTA titrations are carried out in the presence of auxiliary complexing agents which

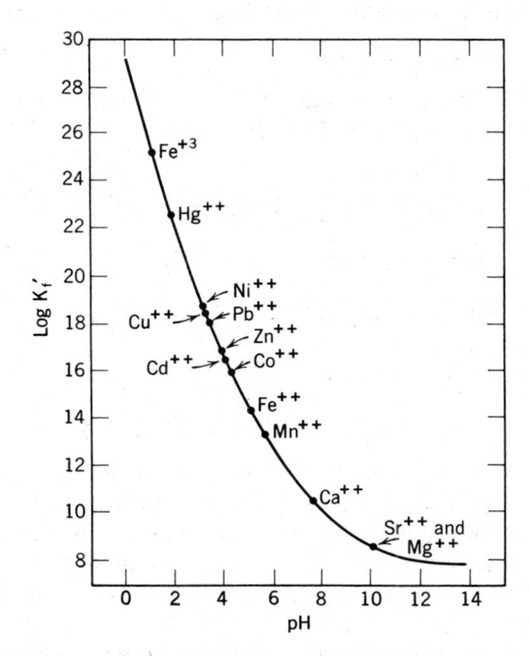

Fig. XV-4 — The Minimum pH Permissible for a Metal-EDTA Titration Curve. [With permission from Reilley, C. N. and Schmid, R. W., Anal. Chem. 30, 947 (1958).]

might have been added to act as buffers, to prevent the precipitation of the metal hydroxides, or both. With only a slight modification, calculations involved in deriving the titration curves in such instances are essentially the same as those outlined in the previous example.

The titration of 50.0 ml of a solution containing 1.0×10^{-3} M Cu^{++}, a mixture of NH_4^+ and NH_3 whose total concentration is 0.1 M and whose pH is 8.00 with 0.1 M EDTA can be used to illustrate this type of titration.

(a) Before the equivalence point, e.g. 0.10 ml of EDTA added. At this point the copper which has not reacted with EDTA is given by

$$C_{Cu} = \frac{50 \times .001 - 0.10 \times .10}{50.1}$$

$$= 8.0 \times 10^{-4} = 10^{-3.10} \text{ M}$$

In order to find pCu it is of course necessary to take into account the copper-ammonia complexation. From Equation IX-2,

$$[Cu^{++}] = \beta_0 \cdot C_{Cu}$$

More specifically, since we have to use the K_a or α to find $[NH_3]$, the expression developed in Sec. IX-3 applies here.

$$\beta_0 = \frac{[Cu^{++}]}{C_{Cu}} = \frac{1}{1 + k_1 \alpha_1 C_{NH_3} + k_1 k_2 (\alpha_1 C_{NH_3})^2 + (k_1 k_2 k_3 (\alpha_1 \cdot C_{NH_3})^3 + k_1 k_2 k_3 k_4 (\alpha_1 C_{NH_3})^4 + k_1 k_2 k_3 k_4 k_5 \cdot (\alpha_1 C_{NH_3})^5}$$

Taking pKa of NH_4^+ as 9.27, at pH 8.00

$$\alpha_1 = \frac{10^{-9.27}}{10^{-9.27} + 10^{-8.00}} = 10^{-1.27}$$

Hence $[NH_3] = \alpha_1 \cdot C_{NH_3} = 10^{-1.27} \times 0.10 = 10^{-2.27}$

Incorporating this value of $[NH_3]$ into the expression for β_0, using logarithms of the stepwise formation constants of Cu^{++}-NH_3 complexes, we obtain $\beta_0 = 10^{-4.56}$, (the values of the stepwise formation constants for the Cu^{++}-NH_3 complexes are the same as those used in Example 1, Chapter IX). From this $[Cu^{++}] = 10^{-4.56} \times 10^{-3.10} = 10^{-7.66}$ and pCu = 7.66.

As long as C_{NH_3} and the pH remain constant, the value of β_0 remains the same. Hence if volume changes during the titration are negligible the entire portion of the titration curve before the equivalence point will have exactly the same shape as that in which no ammonia is present (i.e., $\beta_0 = 1$) except for a displacement of this entire portion of the curve to higher pCu values by an amount equal to log β_0.

This suggests an alternative method of drawing this part of the titration curve. This method is of particular value in describing titrations under varying conditions of pH and concentration of auxiliary complexing agent. First, pCu values at a few selected points are calculated as if no auxiliary complexing agent were present just as was done above for the Ca^{++}-EDTA titration. Second, the β_0 value is calculated from the appropriate values of C_{NH_3} and pH. The titration curve that applies to these conditions may now be readily drawn, parallel to the first but higher by an amount equal to $-\log \beta_0$. The segments of the various titration curves before the equivalence point shown in Fig. XV-5 at various pH values were drawn in this fashion.

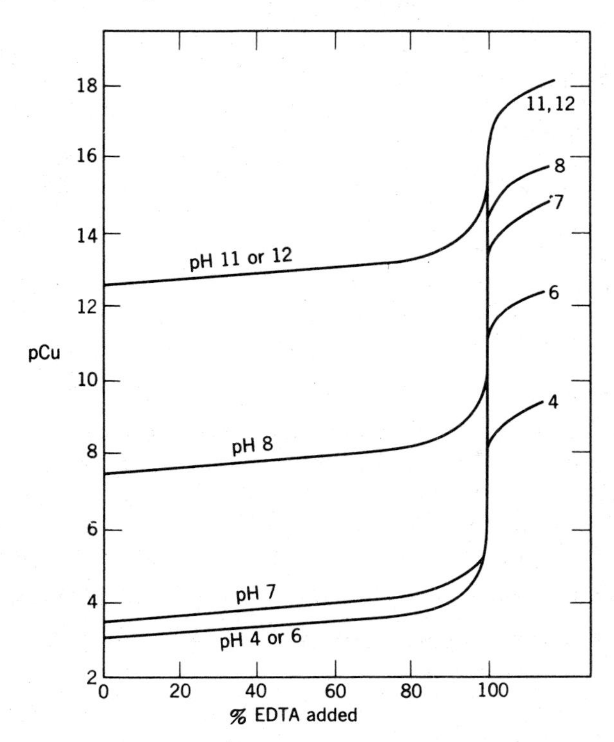

Fig. XV-5 — Titration Curves of 10^{-3} M Copper with EDTA in the Presence of 0.1 M NH_3 at Various pH Values.

(b) At the equivalence point, i.e., 0.50 ml of EDTA added.

At this point the concentration of the copper-EDTA complex is essentially equal to the total copper concentration = $\frac{50 \times 0.001}{50.5} = 1.0 \times 10^{-3}$ M. Assuming that the pH as well as the total volume has remained constant, then it follows that the values of α and β calculated in point (a) still apply. The concentration of the copper not associated with EDTA, C_{Cu}, is given by the expression:

$$K'_f = K_f \alpha_4 \beta_0 = \frac{[CuY^{=}]}{C_{Cu} \cdot C_{H_4Y}}$$

where $K_f = 10^{18.80}$

$\alpha_4 = 10^{-2.29}$ (from Fig. IX-5)

and $\beta_o = 10^{-4.56}$ as before.

At the equivalence point $C_{Cu} = C_{H_4Y}$

Substituting we obtain:

$$K'_f = 10^{11.95} = \frac{1.0 \times 10^{-3}}{(C_{Cu})^2}$$

i.e., $C_{Cu} = 10^{-7.48}$ M

and $[Cu^{++}] = \beta_0 C_{Cu}$

$$= 10^{-4.56} \times 10^{-7.48} = 10^{-12.04}$$

Hence pCu = 12.04

Notice that the value of pCu at the equivalence point is still dependent on the value of β_0 although to a lesser extent (i.e. to the one-half power) than it did in a region before the equivalence point, (i.e. to the first power). As will be shown again below, pCu becomes independent of β_0 after the equivalence point.

(c) After the equivalence point, e.g. 2.0 ml of EDTA added.

The concentration of excess EDTA $= \frac{2.0 \times 0.1 - 50 \times .001}{52}$

$$= 2.9 \times 10^{-3} \text{ M}$$

The concentration of the Cu-EDTA complex is

$$\frac{50 \times 0.001}{52} = 9.6 \times 10^{-4} \text{ M}$$

Therefore, $C_{Cu} = \frac{[CuY^{=}]}{K_f' \cdot C_{H_4Y}} = 10^{-12.43}$

$$[Cu^{++}] = \beta_0 \, C_{Cu} = 10^{-4.56} \times 10^{-12.43} = 10^{-16.99}$$

and pCu = 16.99

It should be again noticed that pCu here has the same value as if no ammonia were present (i.e., $\beta_0 = 1$). This follows from the proportionality of $[Cu^{++}]$ to β_0 along with its inverse proportionality to K_f' which in turn is directly proportional to β_0. (See section on metal buffers below.)

A family of titration curves for Cu-EDTA at varying pH values which have been derived by calculations that we have just described, is shown in Fig. XV-5. It will be noticed that the portion of the curves beyond the end-point depend only on K_f and α_4 and are independent of β_0. Regardless of the nature or concentration of the auxiliary complexing agent, these portions of the titration curves will remain the same, shifting only with pH and the concentration ratio $[CuY^{=}]/C_{H_4Y}$. The sharpness of the titration which is dependent on the height of the vertical portion at the equivalence point is seen to improve when the pH is raised from 4 to 6. This improvement is caused by an increase in α_4. An increase in pH to 8 does not give a corresponding improvement because the decrease in β_0 counteracts the increase in α_4. The optimum pH for this particular titration is in the vicinity of pH 7.

XV-3. METAL BUFFERS

We have seen how a mixture of a Bronsted acid and its conjugate base acts as a hydrogen ion buffer. (See V-5.) In an analogous manner a metal ion buffer consists of a mixture of a Lewis acid i.e. a metal complex and its conjugate base, i.e., excess ligand. Such a mixture is useful in those situations where it is desirable to maintain the concentration of the hydrated metal ion at low, constant values without resorting to the use of a second phase. (In a manner of speaking a solution of KCl saturated with AgCl will act as a silver ion buffer as long as it is in contact with solid AgCl.)

In cases where the desired level of metal ion concentration is 10^{-2} or greater the use of a salt containing the simple hydrated metal ion is suitable. This recalls the suitability of the use of strong acids as buffers at pH values below 2.

Metal buffer calculations are carried out in exactly the same manner as hydrogen ion buffers. (See V-5.) It is important to recall that the log K_f of the metal complex system chosen be as close as possible to the desired pM value.

Example 1.

Compare the effect of adding 1.0 millimole of EDTA to a liter of each of the following two solutions whose total calcium concentration is 10^{-2} M and whose pH remains constant at 12.0 and whose initial pCa is 7.0.

(a) pCa adjusted with EDTA (log K_f = 10.7)
(b) pCa adjusted with nitrilotriacetic acid (log K_f = 6.4)

(a) $$K_f = \frac{[CaY^{=}]}{[Ca^{++}][Y^{-4}]}$$

Since $[Ca^{++}] = 10^{-7.0}$, $[CaY^{=}] = C_{Ca} = 10^{-2}$

Hence $$[Y^{-4}] = \frac{10^{-2}}{10^{-7.0} \times 10^{10.7}} = 10^{-5.7} \text{ M}$$

Upon the addition of 1.0 millimole of EDTA per liter, $[Y^{-4}]$ increases to 10^{-3} M. (The initial $[Y^{-4}]$ may be ignored.) Solving for $[Ca^{++}]$ in the expression above $[Ca^{++}]$ =

$$\frac{10^{-2}}{10^{-3} \times 10^{10.7}} = 10^{-9.7}$$

Hence pCa has risen almost three units to 9.7.

(b) $$K_f = \frac{[CaZ^{-}]}{[Ca^{++}][Z^{-3}]}$$

where Z^{-3} represents the nitrilotriacetate anion. In this case the initial $[Z^{-3}]$ is given by:

$$[Z^{-3}] = \frac{10^{-2}}{10^{-7.0} \times 10^{6.4}} = 10^{-1.4}$$

The addition of 1 millimole of EDTA per liter will result in the transformation:

$$CaZ^{-} + Y^{-4} \rightarrow CaY^{=} + Z^{-3}$$

Hence $$[CaZ^{-}] = 10^{-2} - 10^{-3} = 10^{-2.05}$$

$$[Z^{-3}] = 10^{-1.4} + 10^{-3} = 10^{-1.4}$$

Therefore the new pCa is 7.05. This much smaller change demonstrates the advantage of using a metal buffer system whose $\log K_f$ (more generally $\log K'_f$ the conditional formation constant) is close to the desired pM.

XV-4. METALLOCHROMIC INDICATORS

The development of visual indicators for use in complexometric titrations even in very dilute solutions has provided a stimulus for the increasing popularity of titrimetric determinations of metal ions. Such indicators behave in an exactly analogous manner as the acid-base indicators (See XIII-5.)

A metallochromic indicator is a substance capable of giving a highly colored, water-soluble complex whose stability is such that its $\log K'_f$ is close to the value of pM at the equivalence point of the titration. Most metallochromic indicators are colored, weak Bronsted acids having acid-base indicator properties as well as metal ion indicator properties so that color changes will depend on pH. This is one of the reasons for maintaining pH control through the use of buffers during a complexometric titration. In some instances, the indicator is colorless, e.g., sulfosalicylic acid, in analogy with one-color acid-base indicators.

One of the widely used indicators for calcium is Eriochrome Black T which is a sulfonated, o,o'-dihydroxy azo dye that can be represented by the formula H_2In^-. The dissociation of the two protons can be represented by the equations:

$$\underset{\text{wine-red}}{H_2In^-} \rightleftharpoons \underset{\text{blue}}{HIn^=} + H^+ \quad (pK_1 = 6.3)$$

$$\underset{\text{blue}}{HIn^=} \rightleftharpoons \underset{\text{orange}}{In^{\equiv}} + H^+ \quad (pK_2 = 11.5)$$

Most metal ions form intensely colored, red complexes with Eriochrome Black T. However, a number of these are so stable that the metal is not readily displaced from the dye by the addition of a small amount of chelating titrant beyond the end-point. ($\log K_{f_{MIn}} \gg$ pM at the equivalence point.) In other instances, although the stabilities are appropriate the rates of transformation of the metal complexes are too slow.

The choice of conditions for the use of Eriochrome Black T as an indicator for the titration, for example, of magnesium with EDTA may be made on the basis of the following considerations.

The K'_f values of both the Mg-EDTA and Mg-Indicator vary with pH. From Fig. XV-4 the minimum pH at which magnesium may be titrated with EDTA is 10. We must also keep in mind that in order to avoid having the uncomplexed indicator in the orange form the pH should be kept sufficiently below 11.5. (The color change from red to blue is more easily seen than that from red to orange.) Hence for all practical purposes, the pH must be adjusted between 10 and 11. Within this range the optimum pH would be one in which the $\log K'_{f_{MIn}}$ is closest to pM at the equivalence point. In the titration of 10^{-2} M magnesium, pMg at the equivalence point is given by:

$$pMg = \frac{1}{2} (\log K'_{MgY} - \log C_{Mg})$$

This expression is formally analogous to the expression

$$pH = \frac{1}{2} (pK_a - \log C_{HA})$$

for a weak Bronsted acid, and has been used above for the calculation of pM at the equivalence points of the titration curves of Ca^{++} and Cu^{++} vs. EDTA. At pH 10, pMg at the equivalence point is equal to 5.12, since $\log K_{MgY} = 8.69$, $\log \alpha_4 = -0.46$ and $\log C_{Mg} = -2.0$.

At the minimum pH = 10.00, $\log K'_{f_{MgIn}} = \log K_{f_{MgIn}} + \log \alpha_2$

Since $K_2 = 10^{-11.50}$, $\alpha_2 = \dfrac{K_2}{K_2 + H^+} = 10^{-1.51}$

Since $K_{f_{MgIn}} = 10^{6.95}$, $K'_{f_{MgIn}} = 10^{5.44}$

In order to reduce the value of $\log K'_{f_{MgIn}}$ to where it would equal pMg at the equivalence point (i.e., 5.12) it would be necessary to reduce the pH below the minimum value of 10. Hence in this titration the optimum pH is 10. Since the difference between pMg at the equivalence point and $\log K'_{f_{MgIn}}$ is small, the theoretical titration error is small.

In the titration of calcium with EDTA, Eriochrome Black T is not a good indicator unless some magnesium is present. (Usually a small amount of a magnesium salt is added to the EDTA titrant prior to standardization). In the titration of 0.01 M Ca^{++} with EDTA, in the absence of Mg^{++}, the value of pCa at the equivalence point varies between 5.2 at pH 8 and 6.3 at pH 11. In this pH range the value of the logarithm of the apparent indicator constant ($\log K'_{f_{CaIn}}$) varies between 2.9 and 4.7. Since log $K'_{f_{CaIn}}$ is much smaller than pCa at the equivalence point, the color change occurs much too soon and is too gradual to be of use. Raising the pH above 11 will increase $K'_{f_{CaIn}}$ and thus bring it closer to pCa; however, this is not practical because the free indicator color changes from blue to orange at pH values higher than 11. The presence of Mg^{++} (of Mg-EDTA which is transformed by Ca^{++} to Mg^{++}) renders Eriochrome Black T a useful indicator for the titration of Ca^{++} at pH 10, since the indicator complex which it forms is not dissociated until after all of the Ca^{++} is titrated. At pH 10, pCa at the equivalence point is 6.1. The value of $\log K'_{f_{MgIn}} = 5.4$ is much closer to this pCa than the corresponding value of $\log K'_{f_{CaIn}} = 3.8$.

The use of Eriochrome Black T as an indicator in the Zn^{++}-EDTA titration illustrates a case in which the indicator metal-complex is so stable that the color change occurs after the equivalence point, ($\log K'_{f_{ZnIn}} >$ pZn at equivalence point). Since $\log K'_{f_{ZnIn}}$ changes more rapidly with pH than does pZn (Why?), it is possible to adjust the pH to reduce the difference between them to a reasonably small value. (Table XV-1)

Table XV-1

Titration Characteristics of 0.01 M Zn^{++} with EDTA Using Eriochrome Black T as Indicator

pH	pZn at Equivalence Point	$\log K'_{f_{ZnIn}}$
9	8.6	10.4
8	8.1	9.4
7	7.6	8.4

A practical lower limit is reached at pH 7 since at values lower than this pH, the free indicator changes color to red.

It has been tacitly assumed in this section that the maximum color change will occur at the point when free and metal complexed indicator concentrations are equal, i.e., $pM = \log K'_{f_{MIn}}$.
This is only an approximation since the color intensities of these two forms are not necessarily the same.

XV-5. TITRATION ERRORS

Titration errors which occur when the color change of the metallochromic indicator does not coincide with the equivalence point, may be calculated in a similar manner to that employed in Chapter XIII, Sec. 8 and Chapter XIV, Sec. 4. As an illustration let us calculate the error arising in the titration of Mg^{++} with EDTA described in the previous section.

At pH 10.0 we have seen that $\log K'_{f_{MgIn}} = 5.44$. Assuming that the color change takes place at 50% conversion of the metal-indicator complex, pMg = 5.44. Hence the total concentration of EDTA, C_{H_4Y} is obtained from

$$K'_{f_{MgY}} = \frac{[MgY^{=}]}{[Mg^{++}] \cdot C_{H_4Y}}$$

If the titrant used was 0.01 M EDTA, then

$$[MgY^{=}] = \tfrac{1}{2} \times 10^{-2} \text{ because of dilution.}$$

$$\text{Hence,} \quad 10^{8.23} = \frac{5 \times 10^{-3}}{10^{-5.44} \times C_{H_4Y}}$$

$$\text{i.e.} \quad C_{H_4Y} = 10^{-5.09}$$

There are two contributing terms to C_{H_4Y}; (a) the EDTA from the excess titrant added and (b) the EDTA from the dissociation of the Mg-EDTA complex. This latter term will be equal to the $[Mg^{++}]$ which in this case is $10^{-5.44}$. Therefore the concentration of excess EDTA added = $C_{H_4Y} - [Mg^{++}] = 10^{-5.09} - 10^{-5.44} = 4.51 \times 10^{-6}$ M. Assuming that the volume of the solution at the end-point is 150 ml, the titration error is

$$\frac{4.51 \times 10^{-6} \times 100}{10^{-2}} = +0.05 \text{ ml.}$$

This may also be expressed as a percentage titration error of

$$\frac{0.05}{50} \times 100 = +0.1\%$$

If the color change had occurred before the equivalence point (pMg = 5.12), for example, at pMg = 4.5, the titration error calculation may be obtained simply as:

$$[Mg^{++}] \text{ untitrated} = 10^{-4.5} - 10^{-5.12} = 3.0 \times 10^{-5}$$

The volume of EDTA required to titrate 100 ml containing this concentration of Mg^{++} is $\frac{3.0 \times 10^{-5} \times 100}{10^{-2}} = 0.3$ ml. Hence the the titration error = −0.3 ml.

SUGGESTIONS FOR FURTHER READING

1. G. Schwarzenbach, Complexometric Titrations, Interscience Publishers, Inc., New York, Translated by H. Irving, 1957.
2. A. Ringbom, Chapter 14, Part I, Vol. I, Treatise on Analytical Chemistry, Interscience, New York (1959), I. M. Kolthoff and P. J. Elving, Editors.
3. H. A. Laitinen, Chemical Analysis, McGraw Hill Book Co., Inc. (1960), Chapter 13.

XV-6. PROBLEMS

1. A solution initially containing 2.0×10^{-3} M Mg^{++} is buffered at pH 10 and titrated with 0.10 M EDTA.
   (a) Calculate conditional formation constant of the Mg-EDTA complex.
   (b) Calculate the value of pMg at the following percentages of the stoichiometric amount of EDTA added: 0, 25, 50, 75, 90, 99, 100, 101, 125.
   (c) Construct the titration curve.
   (d) Calculate the theoretical titration error in this titration using Eriochrome Black T as the indicator. Assume that the end-point is taken at 10% conversion of the indicator from the complexed form $MgIn^-$ to the form $HIn^=$.
   (e) What would the titration error be if the end-point is taken at 90% conversion of $MgIn^-$ to $HIn^=$?
2. Repeat the calculations in problem 1. (a) and (b) if the titration is carried out at pH 7.0, 8.0, and 9.0.
3. A solution is 1.0×10^{-3} M in Cu^{++}, and is buffered by having 0.03 M NH_3 and 0.09 M NH_4Cl.
   (a) Calculate the conditional formation constant of the Cu-EDTA complex in this solution.
   (b) Calculate the value of pCu at the following percentages of the stoichiometric amount of EDTA added (neglect dilution): 0, 25, 50, 75, 90, 99, 100, 101, 125.

(c) Construct the titration curve.

4. Calculate the titration error for the Deniges modification of the Liebig titration of 1.0 M KCN with 0.10 M $AgNO_3$ assuming that the final concentration of I^- is 0.01 M and NH_3 is 0.2 M.
5. 4.0 millimoles of NaCl in 100 ml of solution is titrated with 0.10 M Hg^{++} using an indicator which changes color as soon as $[Hg^{++}]$ reaches 1.0×10^{-5} M. Calculate the titration error
6. (a) A solution is 0.02 M in Cd^{++} and contains a mixture of NH_3 and NH_4Cl at a total concentration of 0.20 M. The proportions of NH_3 and NH_4Cl can be varied to give pH values of 8.0, 9.0 and 10.0. Calculate the titration curves of Cd^{++} with 0.10 M EDTA at each of the pH values using points corresponding to 0, 25, 50, 75, 90, 99, 99.9, 100, 100.1, and 125% of the stoichiometric amount of EDTA added.

   (b) What would the theoretical titration errors be at each of these pH values using Eriochrome Black T assuming that the end-point is at 50% conversion of the metal indicator?
7. (a) A 1.0×10^{-2} M Pb^{++} solution is mixed with EDTA at pH 9.0 to give a pPb of 9.0. What is the change in pPb effected by the addition of 1.0 millimoles of EDTA per liter of solution?

   (b) Repeat the calculation in (a) if the initial pPb is 15.0. Account for the difference in the answers in (a) and (b).
8. Design metal buffers to function at the following pM values: 6, 8, 10, and 12 at a pH of 7.0 and total metal ion concentration of 10^{-2} M for the following metal ions: Cu^{++}, Ni^{++}, Co^{++}, Zn^{++}.

XVI

Redox Titrations

In redox titrations the change in the potential, rather than a change in the logarithm of a critical concentration variable, is plotted against the amount of titrant added. Redox titration curves nevertheless have the same general shape as other titration curves. This is what we would expect from the manner in which the electrode potential varies with the logarithm of a concentration term as shown in the Nernst equation.

XVI-1. TITRATION CURVES

Whenever two redox couples are mixed in any proportion the electrode potentials of the two couples become identical when equilibrium is attained. This potential may be calculated from the Nernst equation for the couple present in excess (i.e. having measurable quantities of both oxidized and reduced forms). This principle is employed in redox titration curve calculations.

There will be two types of calculations involved in the derivation of redox titration curves. One type applies on either side of the equivalence point and the second type applies at the equivalence point.

Let us illustrate the method of calculation by considering the titration curve of 50 ml of 0.10 M $FeCl_3$ with 0.10 M $SnCl_2$. It will be assumed that the reaction mixture is always 0.1 M in HCl.

Even before any $SnCl_2$ is added the solution containing $FeCl_3$ has a small but indefinite amount of Fe^{+2} present, otherwise the potential of this redox couple would be infinite. For this reason the potential of a $FeCl_3$ solution is indeterminate. Let us instead calculate the electrode potential of a $FeCl_3$ system when $[Fe(III)]/[Fe(II)] = 10^3$. This requires so small a volume of titrant that it can be used to represent the starting point of the titration. For this point the Nernst equation written in terms of the formal potentials and concentrations is:

$$E = E^{\circ\prime}_{Fe^{+3}, Fe^{+2}} + 0.059 \log \frac{[Fe(III)]}{[Fe(II)]}$$

$$= 0.73 + 0.059 \log 10^3$$

$$= 0.91 \text{ v.}$$

A typical point before the equivalence point is obtained when say 10 ml of $SnCl_2$ have been added.

Millimoles Fe(III) initially = 5.0

Millimoles Sn(II) added = 1.0

From the reaction, $2\,Fe^{+3} + Sn^{+2} \rightleftharpoons 2\,Fe^{+2} + Sn^{+4}$

$$[Fe(III)] = \frac{5.0 - 2.0}{60} \text{ M}$$

and $$[Fe(II)] = \frac{2.0}{60} \text{ M}$$

$$[Sn(IV)] = \frac{1.0}{60} \text{ M}$$

but $$[Sn(II)] \sim 0$$

Since the potentials of the two couples are the same it is equally valid to calculate the potential from the Nernst equation for either. Inasmuch as the concentration ratio for the iron couple is known and that for the tin is not, we must use the expression:

$$E = E^{\circ\prime}_{Fe^{+3}, Fe^{+2}} + 0.059 \log \frac{3.0}{2.0}$$

$$= 0.74 \text{ v.}$$

Although not needed for the titration curve, it is now possible to calculate [Sn(II)] if desired.

$$E = E^{\circ\prime}_{Sn^{+4}, Sn^{+2}} + \frac{0.059}{2} \log \frac{1.0/60}{[Sn(II)]}$$

$$0.74 = 0.15 + \frac{0.059}{2} \log \frac{1.0/60}{[Sn(II)]}$$

Therefore $[Sn(II)] = 10^{-25.2}$ M.

This situation is reversed when enough titrant has been added to pass the equivalence point, e.g. 50 ml of $SnCl_2$ have been added.

$$[Sn(II)] = \frac{2.5}{100} \text{ M}$$

$$[Sn(IV)] = \frac{2.5}{100} \text{ M}$$

$$[Fe(II)] = \frac{5.0}{100} \text{ M}$$

$$[Fe(III)] \sim 0$$

In this case the Nernst equation for the tin couple must be used.

$$E = E^{o\prime}_{Sn^{+4},Sn^{+2}} + \frac{0.059}{2} \log \frac{[Sn(IV)]}{[Sn(II)]}$$

$$= 0.15 \text{ v.}$$

Now let us calculate the potential at the equivalence point. A certain fraction, X, of each of the reactants will remain unreacted here.

$$2\,Fe^{+3} + Sn^{+2} \rightleftharpoons 2\,Fe^{+2} + Sn^{+4}$$

$$X \qquad X \qquad I\text{-}X \qquad I\text{-}X$$

Note that the use of the fraction X eliminates the need to consider the different coefficients in the equation.

$$\frac{[Fe(III)]}{[Fe(II)]} = \frac{X}{I\text{-}X} = \frac{[Sn(II)]}{[Sn(IV)]}$$

The Nernst equations are therefore

$$E_{eq.} = E^{o\prime}_{Fe^{+3},Fe^{+2}} + 0.059 \log \frac{X}{I - X}$$

$$E_{eq.} = E^{o\prime}_{Sn^{+4},Sn^{+2}} + \frac{0.059}{2} \log \frac{I - X}{X}$$

The logarithmic terms may be eliminated from these equations by multiplying the second equation by 2 and adding.

$$E_{eq.} + 2\,E_{eq.} = E^{o\prime}_{Fe^{+3},Fe^{+2}} + 2\,E^{o\prime}_{Sn^{+4},Sn^{+2}}$$

$$\therefore\ E_{eq.} = \frac{E^{o\prime}_{Fe^{+3},Fe^{+2}} + 2\,E^{o\prime}_{Sn^{+4},Sn^{+2}}}{3}$$

$$= 0.34 \text{ v.}$$

This value is seen to be independent of concentrations. In general the potential of a system at the equivalence point is readily seen to be:

$$E_{eq} = \frac{n_1 E_1^{\circ\prime} + n_2 E_2^{\circ\prime}}{n_1 + n_2} \qquad \text{(XVI-1)}$$

(provided that both E_1 and E_2 depend on concentration <u>ratios</u> only), where n and $E^{\circ\prime}$ are the number of electrons and formal potential of the half-cell reactions.

The analytical usefulness of a redox reaction will depend in part on the extent of completion of the reaction at the equivalence point. This may be calculated as X, the fraction unreacted, from the equations used above. Thus for the Fe(III)-Sn(II) titration just considered:

$$0.34 = 0.15 + \frac{0.059}{2} \log \frac{I - X}{X}$$

Therefore $X = 10^{-6.43}$

XVI-2. PROPERTIES OF REDOX TITRATION CURVES

By means of such calculations as have been shown above the titration curve (Fig. XVI-1) can be obtained for the Fe(III)-Sn(II) titration. Redox titration curves are formally analogous to weak base-weak acid titration curves in the sense that S-shaped 'buffer' regions are found both before and after the equivalence

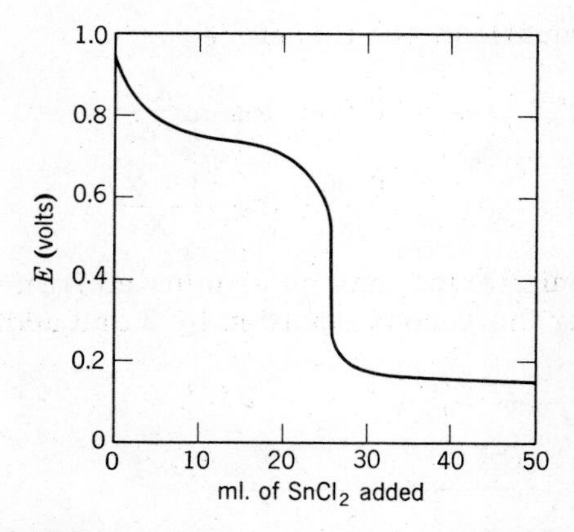

Fig. XVI-1 — Titration of $FeCl_3$ with $SnCl_2$

point. Redox titrations are practical, i.e. the change in E in the vicinity of the equivalence point is sufficiently large whenever the difference in the formal potentials of the reactant and titrant are sufficiently widely separated.

Solutions containing both oxidized and reduced forms of a given redox couple in reasonable ratios (between 10:1 and 1:10) will have potentials within 59/n millivolts of the $E^{\circ\prime}$ of the couple and can act therefore in a manner analogous to acid-base buffers as redox poising systems. Such systems are useful whenever it is desirable to control the oxidizing or reducing environment of a reaction under study. From the nature of the Nernst equation, it is obvious that the larger $\underline{n}$ is, the more effective a poising system can be at a given concentration, i.e. the flatter this region of the titration curve will be. For this same reason a reactant whose half-cell reaction involves a larger number of electrons will give rise to a sharper break in the titration curve.

Although the effect of pH and complexing agents on formal potentials have been discussed previously, it might be well to point out that the height of the vertical region of a titration curve of a given reactant and titrant which depends on the difference in $E^{\circ\prime}$ values will therefore depend on the total composition of the reaction medium in which the titration is carried out. In the case of a redox couple involving two cations, complexation generally lowers $E^{\circ\prime}$ since the higher oxidation state is likely to form more stable complexes. In a case where the oxidized form is an oxyanion and the reduced form a cation, the effect of complexation will generally raise $E^{\circ\prime}$ since the anion is not likely to be complexed. Thus in the titration of Fe^{++} vs. MnO_4^-, the presence of H_3PO_4 improves the titration by lowering $E^{\circ\prime}_{Fe^{+3}, Fe^{+2}}$ and raising $E^{\circ\prime}_{MnO_4^-, Mn^{++}}$. It is interesting to note that in the presence of EDTA which forms very stable complexes, $E^{\circ\prime}$ values are sometimes altered quite dramatically.

Fig. XVI-2 which shows the $E^{\circ\prime}$ values for the As(V)-As(III) and I_3^- - I^- systems demonstrates an analytically important reaction in which pH control is necessary for successful operation. In solutions of very low pH, arsenic (V) is a stronger oxidant than I_2 or I_3^-. In order to titrate As(III) with I_3^- as an oxidant, pH values from about 4 to 9 must be employed in order to obtain a sufficiently large difference in the $E^{\circ\prime}$ values of the two couples to ensure completeness of reaction at the equivalence point.

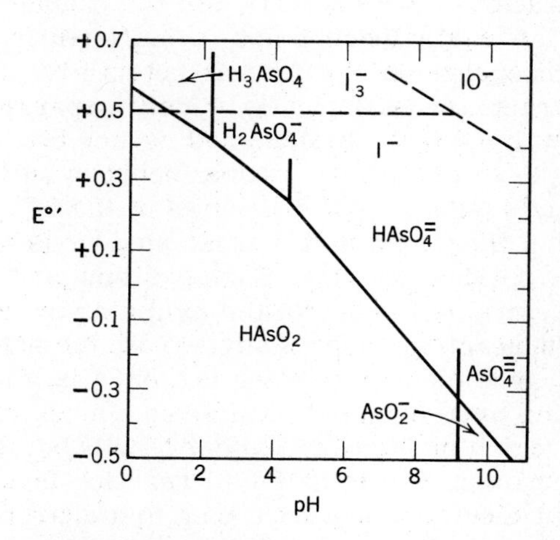

Fig. XVI-2 — Formal Potentials of the Iodine and Arsenic Systems.

XVI-3. REDOX INDICATORS

Aside from those substances which react specifically with either the oxidized or reduced form of the redox couple to give a distinctive color (e.g. starch-iodine color and Fe^{+++}-CNS^- color), redox indicators are themselves redox couples in which one or both forms are distinctively and differently colored. Thus for the half-reaction:

$$In_{Ox} + ne \rightleftharpoons In_{Red}$$

we can write the potential

$$E = E^{\circ\prime}_{In} + \frac{0.059}{n} \log \frac{[In_{Ox}]}{[In_{Red}]}$$

The color of the solution which depends on the ratio of the concentrations of the oxidized and reduced forms of the indicator, thus varies with the potential of the solution. Just as in two-color acid base indicators in which the maximum color change occurs at $pH = pK_{HIn}$, here the maximum color change occurs at $E = E^{\circ\prime}_{In}$. Hence a suitable redox indicator is one whose $E^{\circ\prime}_{In}$ is as close as possible to the equivalence point potential of the titration in which it is used. A wide variety of such indicators is now available and is shown in Table VII in the Appendix.

If we accept the criterion that an effective indicator must change color within 0.1% of the equivalence point, then it is possible to derive a set of limiting potentials within which the color change must occur.

At 99.9% oxidation of a reactant the potential of the titration mixture is given by

$$E = E_1^{\circ\prime} + \frac{0.059}{n_1} \log \frac{99.9}{0.1} = E_1^{\circ\prime} + \frac{3 \times 0.059}{n_1}$$

At 0.1% beyond the equivalence point, the potential of the system is controlled by the titrant (oxidant)

$$E = E_2^{\circ\prime} + \frac{0.059}{n_2} \log \frac{0.10}{100} = E_2^{\circ\prime} - 3 \times \frac{0.059}{n_2}$$

Hence the color change must occur between the two potential values $\left(E_1^{\circ\prime} + \frac{0.177}{n_1}\right)$ and $\left(E_2^{\circ\prime} - \frac{0.177}{n_2}\right)$, i.e. the $E_{In}^{\circ\prime}$ must lie between these two values.

It should be noted that if the $E^{\circ\prime}$ values are so close that the two limits overlap, then the reaction does not proceed to within 0.1% of completion at the equivalence point; (obviously this fact is not dependent on the presence of an indicator.)

XVI-4. TITRATION ERRORS

In the previous section the criteria for selection of a redox indicator for a maximum titration error of 0.1%, was outlined. We will now describe a method of calculating the titration error for indicators that obey these criteria.

As an illustration let us consider the use of m-bromophenol-indophenol ($E_{In}^{\circ\prime}$ = 0.25 v.) in the Fe(III)-Sn(II) titration described in Sec. XVI-1. In using this indicator an excess of Sn(II) will be added since the end-point is below the equivalence point potential (0.34 v.). If the titration error is X%, the concentration ratio of Sn(IV)/Sn(II) $= \frac{100}{X}$, ignoring the amount of Sn(II) left unreacted; (this amount, equal to the [Fe(III)] left in solution is small and may be calculated from the Nernst equation for Fe(III)/Fe(II) if verification is needed).

$$\text{Hence} \quad 0.25 = 0.15 + \frac{0.059}{2} \log \frac{100}{X}$$

and $X = 10^{-1.39}$

$= 0.041\%$

This theoretical titration error is far less than errors encountered from other sources of experimental error.

SUGGESTIONS FOR FURTHER READING

1. H. A. Laitinen, Chemical Analysis, McGraw Hill Book Co., Inc., New York (1960), Chapter 17.
2. G. Charlot, Qualitative Inorganic Analysis, John Wiley and Sons, New York (1954).

XVI-5. PROBLEMS

1. Fifty ml of a 0.1 M solution of $FeSO_4$ is titrated with 0.10 M $Ce(SO_4)_2$. (Assume that the solution is 1.0 M in H_2SO_4 throughout the titration.) Calculate the potential of a platinum electrode in the solution vs. the standard hydrogen electrode, when the following volumes of titrant have been added. 10.0, 25.0, 30.0, 40.0, 49.4, 49.6, 49.8, 50.0, 50.2, 50.4, 50.8, 51.0, 60.0, 70.0 and 80.0 ml. Draw the titration curve and select a suitable redox indicator for this titration.
2. Calculate the solution composition and the potential at the equivalence points of the following titrations
   (a) 25.0 ml of 0.02 M Fe^{+3} vs. 0.01 M Sn^{++}
   (b) 50.0 ml of 0.10 M H_3AsO_3 vs. 0.10 M Ce^{+4}
   (c) 30.0 ml of 0.05 M $Cr_2O_7^{=}$ vs. 0.10 M Sn^{++}
   (d) 25.0 ml of 0.10 M Fe^{++} vs. 0.10 M Ce^{+4}
   (e) 50.0 ml of 0.01 M $Fe(CN)_6^{-4}$ vs. 0.01 M Ce^{+4}
   Assume that the solution is 1.0 M in HCl throughout the titration.
3. Calculate the potential of the titration mixtures in question 2 when (a) 25%, (b) 50%, (c) 75%and (d) 150% of the stoichiometric quantity of the titrant has been added.
4. Calculate the percentage of the substances left untitrated at the equivalence points in each of the titrations in question 2.

5. (a) Construct a titration curve for the titration described in question 2(e).
   (b) If the solution was 1 M in $HClO_4$ throughout the titration (rather than 1 M in HCl), what would be the equivalence point potential and the solution composition at the equivalence point?
   (c) Select a suitable redox indicator for the titration in 1 M HCl.
   (d) Calculate the titration error if the same indicator was used for the titration in 1 M $HClO_4$.
6. In the titration of a standard $Na_2S_2O_3$ solution, 50 ml of 0.10 M I_2 solution in 0.20 M KI was reduced to I^- and then 0.03 ml of the I_2 solution was added before the starch-iodine color was observed. Assuming that the final volume was 150 ml calculate the final potential set up in the solution.
7. One liter of a 1.0 M solution of HCl containing 6.3 millimoles of cerium and 2.4 millimoles of tin was found to have a potential of 600 mv. Calculate the concentrations of Ce^{+4}, Ce^{+3}, Sn^{+2}, Sn^{+4} in the solution.
8. (a) Calculate the theoretical titration errors if diphenylamine sulfonate were used in the titrations in question 2.
   (b) Select where necessary more suitable redox indicators for these titrations.

Appendix

A liberal selection of available equilibrium constants is included here for use in problem solving. Comprehensive tabulations of analytically useful equilibrium constants can be found in:

Handbook of Analytical Chemistry, by Louis Meites, McGraw Hill Book Company, New York (1963).

Stability Constants of Metal Ion Complexes with Solubility Products of Inorganic Substances, Parts I and II, by J. Bjerrum, G. Schwarzenbach and L. G. Sillen, The Chemical Society, London (1957).

All of the constants tabulated are the thermodynamic constants for the reactions in water unless otherwise specified.

Table I

ACID DISSOCIATION CONSTANTS AT 25° C

Acid	Equilibrium	pK_a
Acetic acid	$CH_3COOH \rightleftarrows CH_3COO^- + H^+$	4.76
α-Aminoacetic acid(glycine)	$H_3{}^+NCH_2COO^- \rightleftarrows H_2NCH_2COO^- + H^+$	9.78
Ammonium ion	$NH_4{}^+ \rightleftarrows NH_3 + H^+$	9.24
Anilinium ion	$C_6H_5NH_3{}^+ \rightleftarrows C_6H_5NH_2 + H^+$	4.60
Arsenic acid	$H_3AsO_4 \rightleftarrows H_2AsO_4{}^- + H^+$	2.22
	$H_2AsO_4{}^- \rightleftarrows HAsO_4{}^= + H^+$	6.98
	$HAsO_4{}^= \rightleftarrows AsO_4^{3-} + H^+$	11.4
Arsenious acid	$HAsO_2 \rightleftarrows AsO_2{}^- + H^+$	9.22
Benzoic acid	$C_6H_5COOH \rightleftarrows C_6H_5COO^- + H^+$	4.21
Boric acid	$B(OH)_3 + H_2O \rightleftarrows B(OH)_4{}^- + H^+$	9.23
Carbonic acid	$CO_2 + H_2O \rightleftarrows HCO_3{}^- + H^+$	6.35
	$HCO_3{}^- \rightleftarrows CO_3{}^= + H^+$	10.33
Chloroacetic acid	$ClCH_2CO_2H \rightleftarrows ClCH_2CO_2{}^- + H^+$	2.87
Chromic acid	$H_2CrO_4 \rightleftarrows HCrO_4{}^- + H^+$	0.75
	$HCrO_4{}^- \rightleftarrows CrO_4{}^= + H^+$	6.52
Citric acid	$H_3C_6H_5O_7 \rightleftarrows H_2C_6H_5O_7{}^- + H^+$	3.13
	$H_2C_6H_5O_7{}^- \rightleftarrows HC_6H_5O_7{}^= + H^+$	4.76
	$HC_6H_5O_7{}^= \rightleftarrows C_6H_5O_7^{3-} + H^+$	6.40
Cyanic acid	$HOCN \rightleftarrows OCN^- + H^+$	3.66
Dichloroacetic acid	$Cl_2CHCO_2H \rightleftarrows Cl_2CHCO_2{}^- + H^+$	1.26

ACID DISSOCIATION CONSTANTS - Page 2

Acid	Equilibrium	pK_a
Dichromic acid	$H_2Cr_2O_7 \rightleftarrows HCr_2O_7^- + H^+$	- 1.4
	$HCr_2O_7^- \rightleftarrows Cr_2O_7^= + H^+$	1.64
Dimethylammonium ion	$(CH_3)_2NH_2^+ \rightleftarrows (CH_3)_2NH + H^+$	10.77
Ethylammonium ion	$C_2H_5NH_3^+ \rightleftarrows C_2H_5NH_2 + H^+$	10.63
Ethylenediamine	$^+H_3N(CH_2)_2NH_3^+ \rightleftarrows {}^+H_3N(CH_2)_2NH_2 + H^+$	7.52
	$^+H_3N(CH_2)_2NH_2 \rightleftarrows NH_2(CH_2)_2NH_2 + H^+$	10.65
Ethylenediaminetetra acetic acid(EDTA)	$H_4Y \rightleftarrows H_3Y^- + H^+$	2.0
	$H_3Y^- \rightleftarrows H_2Y^= + H^+$	2.67
	$H_2Y^= \rightleftarrows HY^{-3} + H^+$	6.16
	$HY^{-3} \rightleftarrows Y^{-4} + H^+$	10.26
Formic acid	$HCOOH \rightleftarrows HCOO^- + H^+$	3.75
Glycinium ion	$^+H_3NCH_2CO_2H \rightleftarrows H_3{}^+NCH_2CO_2^- + H^+$	2.35
	$^+H_3NCH_2CO_2^- \rightleftarrows H_2NCH_2CO_2^- + H^+$	9.78
Hydrazoic acid	$HN_3 \rightleftarrows H^+ + N_3^-$	4.72
Hydrocyanic acid	$HCN \rightleftarrows CN^- + H^+$	9.4
Hydrofluoric acid	$HF \rightleftarrows F^- + H^+$	3.17
Hydrogen sulfide	$H_2S \rightleftarrows SH^- + H^+$	7.0
	$SH^- \rightleftarrows S^= + H^+$	12.9
Hydrogen sulfate ion	$HSO_4^- \rightleftarrows SO_4^= + H^+$	1.99
Hypobromous acid	$HBrO \rightleftarrows BrO^- + H^+$	8.60
Hypochlorous acid	$HClO \rightleftarrows ClO^- + H^+$	7.53
Hypoiodous acid	$HIO \rightleftarrows IO^- + H^+$	10.4
Methylammonium ion	$CH_3NH_3^+ \rightleftarrows CH_3NH_2 + H^+$	10.62
Monochloroacetic acid	$ClCH_2CO_2H \rightleftarrows ClCH_2CO_2^- + H^+$	2.87
Nitrous Acid	$HNO_2 \rightleftarrows NO_2^- + H^+$	3.29

ACID DISSOCIATION CONSTANTS - Page 3

		pK_a
Oxalic acid	$H_2C_2O_4 \rightleftharpoons HC_2O_4^- + H^+$	1.27
	$HC_2O_4^- \rightleftharpoons C_2O_4^= + H^+$	4.27
Phenol	$C_6H_5OH \rightleftharpoons C_6H_5O^- + H^+$	10.0
Phosphoric acid	$H_3PO_4 \rightleftharpoons H_2PO_4^- + H^+$	2.15
	$H_2PO_4^- \rightleftharpoons HPO_4^= + H^+$	7.20
	$HPO_4^= \rightleftharpoons PO_4^{-3} + H^+$	12.4
Propionic acid	$CH_3CH_2COOH \rightleftharpoons CH_3CH_2COO^- + H^+$	4.87
Pyridinium ion	$C_5H_5NH^+ \rightleftharpoons C_5H_5N + H^+$	5.22
Salicylic acid	$HOC_6H_4CO_2H \rightleftharpoons HOC_6H_4CO_2^- + H^+$	2.96
Succinic acid	$H_2C_4H_4O_4 \rightleftharpoons HC_4H_4O_4^- + H^+$	4.21
	$HC_4H_4O_4^- \rightleftharpoons C_4H_4O_4^= + H^+$	5.64
Sulfamic acid	$HNH_2SO_3 \rightleftharpoons NH_2SO_3^- + H^+$	1.0
Sulfurous acid	$SO_2 + H_2O \rightleftharpoons HSO_3^- + H^+$	1.76
	$HSO_3^- \rightleftharpoons SO_3^= + H^+$	7.19
Tartaric acid	$H_2C_4H_4O_6 \rightleftharpoons HC_4H_4O_6^- + H^+$	3.04
	$HC_4H_4O_6^- \rightleftharpoons C_4H_4O_6^= + H^+$	4.37
Trichloroacetic acid	$Cl_3CCO_2H \rightleftharpoons Cl_3CCO_2^- + H^+$	0.2
Triethanolammonium ion	$(C_2H_4OH)_3NH^+ \rightleftharpoons (C_2H_4OH)_3N + H^+$	7.76
Trimethylammonium ion	$(CH_3)_3NH^+ \rightleftharpoons (CH_3)_3N + H^+$	9.80

Table II

Solubility Product Constants at 25° C

Substance	Formula	K_{sp}
Aluminum arsenate	$AlAsO_4$	1.6×10^{-16}
Aluminum hydroxide	$Al(OH)_3$	2.0×10^{-32}
Aluminum phosphate	$AlPO_4$	5.8×10^{-19}
Aluminum sulfide	Al_2S_3	2.0×10^{-7}
Barium arsenate	$Ba_3(AsO_4)_2$	1.1×10^{-13}
Barium bromate	$Ba(BrO_3)_2$	5.5×10^{-6}
Barium carbonate	$BaCO_3$	5.1×10^{-9}
Barium chromate	$BaCrO_4$	1.2×10^{-10}
Barium fluoride	BaF_2	1.0×10^{-6}
Barium iodate	$Ba(IO_3)_2$	1.5×10^{-9}
Barium oxalate	BaC_2O_4	1.5×10^{-8}
Barium phosphate	$Ba_3(PO_4)_2$	6.0×10^{-39}
Barium sulfate	$BaSO_4$	1.0×10^{-10}
Cadmium arsenate	$Cd_3(AsO_4)_2$	2.2×10^{-33}
Cadmium carbonate	$CdCO_3$	5.2×10^{-12}
Cadmium hydroxide	$Cd(OH)_2$	2.8×10^{-14}
Cadmium oxalate	CdC_2O_4	2.8×10^{-8}
Cadmium sulfide	CdS	7.0×10^{-27}
Calcium arsenate	$Ca_3(AsO_4)_2$	6.8×10^{-19}
Calcium carbonate	$CaCO_3$	4.7×10^{-9}
Calcium chromate	$CaCrO_4$	7.1×10^{-4}
Calcium fluoride	CaF_2	4.9×10^{-11}

Table II — Continued

Solubility Product Constants at 25° C

Substance	Formula	K_{sp}
Calcium hydroxide	$Ca(OH)_2$	5.5×10^{-6}
Calcium iodate	$Ca(IO_3)_2$	7.0×10^{-7}
Calcium oxalate	CaC_2O_4	2.1×10^{-9}
Calcium phosphate	$Ca_3(PO_4)_2$	2.0×10^{-29}
Calcium sulfate	$CaSO_4$	2.4×10^{-5}
Cerium(III) oxalate	$Ce_2(C_2O_4)_3$	2.5×10^{-29}
Cerium(III) sulfide	Ce_2S_3	6.0×10^{-11}
Chromium arsenate	$CrAsO_4$	7.8×10^{-21}
Chromium hydroxide	$Cr(OH)_3$	7.0×10^{-31}
Cobalt carbonate	$CoCO_3$	8.0×10^{-13}
Cobalt hydroxide	$Co(OH)_2$	2.0×10^{-16}
Cobalt oxalate	CoC_2O_4	4.0×10^{-8}
Cobalt sulfide	CoS	8.0×10^{-23}
Copper(I) bromide	$CuBr$	5.3×10^{-9}
Copper(I) chloride	$CuCl$	3.2×10^{-7}
Copper(I) cyanide	$CuCN$	1.0×10^{-11}
Copper(I) iodide	CuI	1.1×10^{-12}
Copper(I) sulfide	Cu_2S	1.0×10^{-48}
Copper(I) thiocyanate	$CuNCS$	1.6×10^{-11}
Copper(II) carbonate	$CuCO_3$	2.5×10^{-10}
Copper(II) chromate	$CuCrO_4$	3.6×10^{-6}
Copper(II) hydroxide	$Cu(OH)_2$	2.2×10^{-20}
Copper(II) sulfide	CuS	8.0×10^{-36}
Iron(II) carbonate	$FeCO_3$	3.5×10^{-11}
Iron(II) hydroxide	$Fe(OH)_2$	8.0×10^{-16}
Iron(II) oxalate	FeC_2O_4	2.0×10^{-7}

Table II — Continued

Solubility Product Constants at 25° C

Substance	Formula	K_{sp}
Iron(II) sulfide	FeS	5.0×10^{-18}
Iron(III) arsenate	$FeAsO_4$	5.8×10^{-21}
Iron(III) hydroxide	$Fe(OH)_3$	6.0×10^{-38}
Iron(III) phosphate	$FePO_4$	1.3×10^{-22}
Iron(III) sulfide	Fe_2S_3	1.0×10^{-88}
Lead arsenate	$Pb_3(AsO_4)_2$	4.1×10^{-36}
Lead bromide	$PbBr_2$	4.6×10^{-6}
Lead carbonate	$PbCO_3$	1.5×10^{-13}
Lead chloride	$PbCl_2$	1.6×10^{-5}
Lead chromate	$PbCrO_4$	2.0×10^{-16}
Lead fluoride	PbF_2	2.7×10^{-8}
Lead formate	$Pb(CHO_2)_2$	2.0×10^{-7}
Lead hydroxide	$Pb(OH)_2$	4.0×10^{-15}
Lead iodate	$Pb(IO_3)_2$	2.6×10^{-13}
Lead iodide	PbI_2	7.1×10^{-9}
Lead molybdate	$PbMoO_4$	4.0×10^{-6}
Lead oxalate	PbC_2O_4	8.0×10^{-12}
Lead sulfate	$PbSO_4$	1.7×10^{-8}
Lead sulfide	PbS	8.0×10^{-28}
Lead thiocyanate	$Pb(NCS)_2$	3.0×10^{-8}
Magnesium ammonium phosphate	$MgNH_4PO_4\cdot$	2.5×10^{-13}
Magnesium arsenate	$Mg_3(AsO_4)_2$	2.1×10^{-20}
Magnesium carbonate	$MgCO_3$	1.0×10^{-5}

Table II — Continued

Solubility Product Constants at 25° C

Substance	Formula	K_{sp}
Magnesium fluoride	MgF_2	6.4×10^{-9}
Magnesium hydroxide	$Mg(OH)_2$	1.1×10^{-11}
Magnesium oxalate	MgC_2O_4	8.6×10^{-5}
Manganese(II) arsenate	$Mn_3(AsO_4)_2$	1.9×10^{-29}
Manganese(II) carbonate	$MnCO_3$	8.8×10^{-11}
Manganese(II) hydroxide	$Mn(OH)_2$	1.6×10^{-13}
Manganese(II) oxalate	MnC_2O_4	1.0×10^{-15}
Manganese(II) sulfide	MnS	1.0×10^{-11}
Mercury(I) acetate	$Hg_2(C_2H_3O_2)_2$	3.5×10^{-10}
Mercury(I) bromide	Hg_2Br_2	1.3×10^{-22}
Mercury(I) carbonate	Hg_2CO_3	8.9×10^{-17}
Mercury(I) chloride	Hg_2Cl_2	1.3×10^{-18}
Mercury(I) oxide	Hg_2O	1.6×10^{-23}
Mercury(I) iodate	$Hg_2(IO_3)_2$	2.5×10^{-14}
Mercury(I) chromate	Hg_2CrO_4	2.0×10^{-9}
Mercury(I) iodide	Hg_2I_2	4.0×10^{-29}
Mercury(I) sulfate	Hg_2SO_4	6.8×10^{-7}
Mercury(II) bromide	$HgBr_2$	1.1×10^{-19}
Mercury(II) chloride	$HgCl_2$	6.1×10^{-15}
Mercury(II) iodate	$Hg(IO_3)_2$	3.0×10^{-13}
Mercury(II) iodide	HgI_2	4.0×10^{-29}
Mercury(II) oxide	HgO	3.0×10^{-26}
Mercury(II) sulfide	HgS (black)	3.0×10^{-52}

Table II — Continued

Solubility Product Constants at 25° C

Substance	Formula	K_{sp}
Nickel arsenate	$Ni_3(AsO_4)_2$	3.1×10^{-26}
Nickel hydroxide	$Ni(OH)_2$	2.0×10^{-15}
Nickel iodate	$Ni(IO_3)_2$	1.4×10^{-8}
Nickel oxalate	NiC_2O_4	4.0×10^{-10}
Nickel sulfide	NiS	2.0×10^{-21}
Silver acetate	$AgC_2H_3O_2$	2.3×10^{-3}
Silver arsenate	Ag_3AsO_4	1.0×10^{-22}
Silver bromide	$AgBr$	5.2×10^{-13}
Silver carbonate	Ag_2CO_3	8.2×10^{-12}
Silver chloride	$AgCl$	1.0×10^{-10}
Silver chromate	Ag_2CrO_4	2.4×10^{-12}
Silver cyanide	$AgCN$	2.0×10^{-16}
Silver iodate	$AgIO_3$	3.1×10^{-8}
Silver iodide	AgI	8.3×10^{-17}
Silver molybdate	Ag_2MoO_4	2.8×10^{-12}
Silver oxalate	$Ag_2C_2O_4$	1.0×10^{-11}
Silver oxide	Ag_2O	2.6×10^{-8}
Silver phosphate	Ag_3PO_4	1.0×10^{-21}
Silver sulfate	Ag_2SO_4	1.7×10^{-5}
Silver sulfide	Ag_2S	7.0×10^{-50}
Silver thiocyanate	$AgNCS$	1.0×10^{-12}
Strontium carbonate	$SrCO_3$	7.0×10^{-10}
Strontium chromate	$SrCrO_4$	5.0×10^{-6}
Strontium fluoride	SrF_2	7.9×10^{-10}

Table II — Continued

Solubility Product Constants at 25° C

Substance	Formula	K_{sp}
Strontium iodate	$Sr(IO_3)_2$	3.3×10^{-7}
Strontium oxalate	SrC_2O_4	5.6×10^{-8}
Strontium phosphate	$Sr_3(PO_4)_2$	1.0×10^{-31}
Strontium sulfate	$SrSO_4$	7.6×10^{-7}
Thallium(I) bromide	$TlBr$	3.9×10^{-6}
Thallium(I) chloride	$TlCl$	1.8×10^{-4}
Thallium(I) chromate	Tl_2CrO_4	9.8×10^{-13}
Thallium(I) iodide	TlI	6.5×10^{-8}
Thallium(I) phosphate	Tl_3PO_4	6.7×10^{-8}
Tin(II) hydroxide	$Sn(OH)_2$	1.6×10^{-27}
Tin(II) sulfide	SnS	1.3×10^{-27}
Tin(IV) hydroxide	$Sn(OH)_4$	1.0×10^{-57}
Zinc arsenate	$Zn_3(AsO_4)_2$	1.1×10^{-27}
Zinc carbonate	$ZnCO_3$	2.1×10^{-11}
Zinc hydroxide	$Zn(OH)_2$	7.0×10^{-18}
Zinc iodate	$Zn(IO_3)_2$	2.0×10^{-8}
Zinc oxalate	ZnC_2O_4	2.5×10^{-9}
Zinc sulfide (βeta form)	ZnS	3.0×10^{-22}

Table III

Formation Constants of Metal Complexes at 25° C

Metal Ion	log k_1	log k_2	log k_3	log k_4	log k_5	log k_6
			Ammonia			
Ag^+	3.32	3.92	-	-	-	-
Cd^{++}	2.51	1.96	1.30	0.79	-	-
Co^{++}	2.11	1.63	1.05	0.76	0.18	-0.62
Cu^{++}	3.99	3.34	2.73	1.97	-	-
Hg^{++}	8.8	8.7	1.00	0.78	-	-
Ni^{++}	2.67	2.12	1.61	1.07	0.63	-0.09
Zn^{++}	2.18	2.25	2.31	1.96	-	-
			Bromide			
Ag^+	4.38	2.96	0.66	0.73		
Bi^{+++}	2.26	2.19	1.88	1.51	1.58	0.1
Cd^{++}	2.23	0.77	-0.17	0.10		
Co^{++}	-2.30					

Table III – Continued

Formation Constants of Metal Complexes at 25° C

Metal Ion	log k_1	log k_2	log k_3	log k_4	log k_5	log k_6
Cu^{++}	-0.03					
Fe^{+++}	0.49					
Hg^{++}	9.05	8.28	2.41	1.26		
Pb^{++}	1.77	0.15	1.4	-0.3		
Sn^{++}	1.11	0.70	-0.35			
Zn^{++}	-0.60	-0.37	-0.73	0.44		
			Chloride			
Ag^{+}	3.04	2.00	0.00	0.26		
Bi^{+++}	2.43	2.3	0.3	0.6	0.5	0.3
Cd^{++}	2.00	0.70	-0.59			
Co^{++}	-2.4					
Cu^{++}	0.0	-0.7	-1.5	-2.3		
Fe^{+++}	1.48	0.65	-1.0			

Hg^{++}	6.74	6.48	0.95	1.05		
Pb^{++}	1.10	1.16	-0.40	-1.05		
Sn^{++}	1.51	0.73	-0.21	-0.55		
Zn^{++}	-0.50	-0.50	1.00	-1.00		
			Cyanide			
Ag^{+}	($\log k_1k_2 = 19.9$)		0.70	-1.13	-	-
Cd^{++}	5.18	4.42	4.32	3.19	-	-
Cu^{+}	($\log k_1k_2 = 23.8$)		4.61	2.12	-	-
Fe^{++}		($\log k_1k_2k_3k_4k_5k_6 = 24$)				
Fe^{+++}		($\log k_1k_2k_3k_4k_5k_6 = 31$)				
Hg^{++}	18.00	16.70	3.83	2.98	-	-
Ni^{++}		($\log k_1k_2k_3k_4 = 22$)				
Zn^{++}		($\log k_1k_2k_3k_4 = 17$)				

Table III — Continued

Formation Constants of Metal Complexes at 25° C

Metal Ion	log k_1	log k_2	log k_3	log k_4	log k_5	log k_6
			Fluoride			
Ag^{+}	0.36	-	-	-	-	-
Al^{+3}	6.13	5.02	3.85	2.74	1.63	0.47
Cd^{++}	0.46	0.07				
Cr^{+++}	4.36	3.34	2.48			
Cu^{++}	1.23					
Fe^{+++}	5.21	3.95	2.70			
Hg^{++}	1.56					
Mg^{++}	1.82					
Zn^{++}	1.26					

Hydroxide

Ion		
Ag^{+}	2.3	1.9
Al^{+++}	9.15	
Ba^{++}	0.64	
Bi^{+++}	12.15	
Ca^{++}	1.30	
Cd^{++}	6.38	3.09
Co^{++}	1.80	
Cr^{+++}	11.1	7.77
Cu^{++}	6.03	
Fe^{++}	8.08	
Fe^{+++}	11.54	9.30
Hg^{++}	11.51	11.15
Hg_2^{++}	9.7	
Mg^{++}	2.58	
Mn^{++}	3.4	

Table III — Continued

Formation Constants of Metal Complexes at 25° C

Metal Ion	log k_1	log k_2	log k_3	log k_4	log k_5	log k_6
Ni^{++}	3.36					
Pb^{++}	7.51					
Sr^{++}	0.85					
Tl^{+}	0.49					
Zn^{++}	4.36	(log $k_1k_2k_3k_4$ = 15.5)				
Iodide						
Ag^{+}	8.13					
Bi^{+++}						
Cd^{++}	2.28	1.64	1.08	1.10		
Hg^{++}	12.87	10.95	3.78	2.23		
Pb^{++}	1.26	1.54	0.62	0.50		

Thiocyanates

Ag^{+}	4.75	3.48	1.22	0.22		
Bi^{+++}	1.15	1.11	$\log k_3k_4 = 1.15$; $\log k_5k_6 = 0.82$			
Cd^{++}	1.04	0.71	-0.97	1.00		
Co^{++}	3.0	0.0	-0.7	-0.04		
Cr^{+++}	3.08	1.8	1.0	0.3	-0.7	-1.6
Cu^{++}			$(\log k_1k_2k_3 = 5.19)(\log k_1k_2k_3k_4 = 6.52)$			
Fe^{+++}	2.94	1.19	0			
Hg^{++}	$(\log k_1k_2 = 17.47)$		1.68	0.62		
Ni^{++}	1.18	0.46	0.17			
Pb^{++}	1.09	1.43				
Zn^{++}	1.7					

Table III — Continued

Formation Constants of Metal Complexes at 25° C

Metal Ion	log k_1	log k_2	log k_3	log k_4	log k_5	log k_6
			Thiosulfate			
Ag^+	log k_1k_2 = 13.38		0.55			
Cu^{++}	log k_1k_2 = 12.29					
Fe^{+3}	3.25					
Hg^{++}	log k_1k_2 = 29.86		2.4	1.4		
Pb^{++}	log k_1k_2 = 5.13		1.22	0.8		

Table IV

Formation Constants of Metal Complexes of Selected Organic Ligands at 25° C

Metal Ion	log k_1	log k_2	log k_3	log k_1k_2	log $k_1k_2k_3$
			Acetylacetone		
Cd^{++}	3.8	2.8			
Co^{++}	5.4	4.1			
Cu^{++}	8.2	6.7			
Fe^{++}	5.1	3.6			
Fe^{+++}	9.8	9.0	7.4		
Mg^{++}	3.6	2.5			
Mn^{++}	4.2	3.1			
Ni^{++}	5.9	4.5	2.1		
Zn^{++}	5.0	3.8			
			2,2'-Dipyridyl		
Ag^{+}	-	-	-	6.8	-
Cd^{++}	4.5	3.5	2.5	-	-
Cu^{++}	-	-	-	-	17.9
Fe^{++}	4.4	-	-	-	16.4
Mn^{++}	-	-	-	-	6.3
Zn^{++}	5.4	4.4	3.7	-	-
			Dithizone		
Co^{++}	5.3				
Ni^{++}	4.8				
Zn^{++}	5.2				

Table IV — Continued

Formation Constants of Metal Complexes of

Selected Organic Ligands at 25° C

Metal Ion	log k_1	log k_2	log k_3	log k_1k_2	log $k_1k_2k_3$
			EDTA		
Ag^+	7.3				
Ba^{++}	7.8				
Ca^{++}	10.7				
Cd^{++}	16.5				
Co^{++}	16.3				
Cu^{++}	18.8				
Fe^{++}	14.3				
Fe^{+++}	25.1				
Hg^{++}	21.8				
Mg^{++}	8.7				
Mn^{++}	14.0				
Ni^{++}	18.6				
Pb^{++}	18.0				
Sr^{++}	8.6				
Zn^{++}	16.5				

Table IV — Continued

Formation Constants of Metal Complexes of Selected Organic Ligands at 25° C

Metal Ion	log k_1	log k_2	log k_3	log k_1k_2	log $k_1k_2k_3$
		Ethylenediamine			
Ag^+	4.7	3.0			
Cd^{++}	5.5	4.6			
Co^{++}	5.9	4.7			
Cu^{++}	10.6	9.0			
Fe^{++}	4.3	3.3			
Mn^{++}	2.7	2.1			
Ni^{++}	7.5	6.2			
Zn^{++}	5.7	4.7			
		8-Hydroxyquinoline			
Ba^{++}	2.1				
Ca^{++}	3.3				
Cd^{++}	7.8	(6.8)			
Co^{++}	9.1	8.1			
Cu^{++}	12.2	11.2			
Fe^{++}	8.0	7.0			
Fe^{+++}	12.3	11.3			
Mg^{++}	4.7	(3.7)			
Mn^{++}	6.8	5.8			
Ni^{++}	9.9	8.8			
Pb^{++}	9.0	(8.0)			
Sr^{++}	2.6				
Zn^{++}	8.5	(7.5)			

Table IV — Continued

Formation Constants of Metal Complexes of

Selected Organic Ligands at 25° C

Metal Ion	$\log k_1$	$\log k_2$	$\log k_3$	$\log k_1k_2$	$\log k_1k_2k_3$
			Oxalate		
Co^{++}	4.7	2.0	3.0		
Fe^{+++}	9.4	6.8	4.0		
Mg^{++}	2.6	1.8			
Mn^{++}	3.8	1.5			
Ni^{++}	$\log k_1k_2$ =	6.5			
Pb^{++}	$\log k_1k_2$ =	6.5			
Zn^{++}	5.0	2.4	0.7		
			1,10 Phenanthroline		
Cd^{++}	6.4	5.2	4.2	-	-
Cu^{++}	6.3	6.2	5.5	-	-
Fe^{++}	-	-	-	-	21.3
Fe^{+++}	-	-	-	-	14.1
Mn^{++}	-	-	-	-	7.4
Ni^{++}	-	-	-	-	18.3
Zn^{++}	6.4	5.7	4.9	-	-

Table IV — Continued

Formation Constants of Metal Complexes of Selected Organic Ligands at 25° C

Metal Ion	log k_1	log k_2	log k_3	log k_1k_2	log $k_1k_2k_3$
	Tren(2,2',2"-triaminotriethylamine)				
Ag^+	7.8				
Cd^{++}	12.3				
Co^{++}	12.8				
Cu^{++}	18.8				
Fe^{++}	8.8				
Hg^{++}	25.8				
Mn^{++}	5.8				
Ni^{++}	14.8				
Zn^{++}	14.7				
	Trien[N,N'-di(2-aminoethyl)ethylenediamine]				
Ag^+	7.7				
Cd^{++}	10.8				
Co^{++}	11.0				
Cu^{++}	20.4				
Fe^{++}	7.8				
Hg^{++}	25.3				
Mn^{++}	4.9				
Ni^{++}	14.0				
Zn^{++}	12.1				

Table V

Standard and Formal Reduction Potentials at 25° C

Half-Reaction	Standard Potential	Formal Potential
$Ag^+ + e \rightleftarrows Ag$	+ 0.799	
$AgBr + e \rightleftarrows Ag + Br^-$	+ 0.071	
$AgCl + e \rightleftarrows Ag + Cl^-$	+ 0.222	
$Ag_2CrO_4 + 2e^- \rightleftarrows 2Ag + CrO_4^=$	+ 0.45	
$Ag(CN)_2^- + e^- \rightleftarrows Ag + 2CN^-$	- 0.31	
$AgI + e \rightleftarrows Ag + I^-$	- 0.152	
$Ag(NH_3)_2^+ + e \rightleftarrows Ag + 2NH_3$	+ 0.37	
$Ag_2O + H_2O + 2e \rightleftarrows 2Ag + 2OH^-$	+ 0.342	
$Ag_2S + 2e \rightleftarrows 2\ Ag + S^=$	- 0.71	
$Ag(S_2O_3)_2^{3-} + e \rightleftarrows Ag + 2S_2O_3^=$	+ 0.01	
$Al^{+++} + 3e \rightleftarrows Al$	- 1.66	
$Al(OH)_4^- + 3e \rightleftarrows Al + 4OH^-$	- 2.35	
$As + 3H^+ + 3e \rightleftarrows AsH_3$	- 0.60	
$As_2O_3 + 6H^+ + 6e \rightleftarrows 2As + 3H_2O$	+ 0.234	
$H_3AsO_4 + 2H^+ + 2e \rightleftarrows HAsO_2 + 2H_2O$	+ 0.559	+ 0.577 in 1 M HCl
$Ba^{++} + 2e \rightleftarrows Ba$	- 2.90	
$BiO^+ + 2H^+ + 3e \rightleftarrows Bi + H_2O$	+ 0.32	
$BiOCl + 2H^+ + 3e \rightleftarrows Bi + H_2O + Cl^-$	+ 0.16	
$Bi_2O_3 + 3H_2O + 6e \rightleftarrows 2Bi + 6OH^-$	- 0.46	
$Br_2 + 2e \rightleftarrows 2Br^-$	+ 1.087	
$2HOBr + 2H^+ + 2e \rightleftarrows Br_2 + 2H_2O$	+ 1.6	
$2BrO_3^- + 12H^+ + 10e \rightleftarrows Br_2 + 6H_2O$	+ 1.52	

Half-Reaction	Standard Potential	Formal Potential
$C_2N_2 + 2H^+ + 2e \rightleftharpoons 2\ HCN$	+ 0.37	
$Ca^{++} + 2e \rightleftharpoons Ca$	- 2.87	
$Cd^{++} + 2e \rightleftharpoons Cd$	- 0.402	
$Cd(CN)_4^{=} + 2e \rightleftharpoons Cd + 4CN^-$	- 1.03	
$Cd(NH_3)_4^{++} \rightleftharpoons Cd + 4\ NH_3$	- 0.597	
$Cd(OH)_2 + 2e \rightleftharpoons Cd + 2OH^-$	- 0.809	
$CdS + 2e \rightleftharpoons Cd + S^{=}$	-1.2	
$Ce(IV) + e \rightleftharpoons Ce(III)$		+ 0.06 in 2.5 M K_2CO_3
		+ 1.28 in 1 M HCl
		+ 1.70 in 1 M $HClO_4$
		+ 1.60 in 1 M HNO_3
		+ 1.44 in 1 M H_2SO_4
$Cl_2 + 2e \rightleftharpoons 2Cl^-$	+ 1.359	
$2HOCl + 2H^+ + 2e \rightleftharpoons Cl_2 + 2H_2O$	+ 1.63	
$ClO_3^- + 2H^+ + e \rightleftharpoons ClO_2 + H_2O$	+ 1.15	
$ClO_4^- + 2H^+ + 2e \rightleftharpoons ClO_3^- + H_2O$	+ 1.19	
$Co^{++} + 2e \rightleftharpoons Co$	- 0.28	
$Co^{+++} + e \rightleftharpoons Co^{++}$		+ 1.85 in 4 M HNO_3
		+ 1.82 in 8 M H_2SO_4
$Co(NH_3)_6^{+++} + e \rightleftharpoons Co(NH_3)_6^{++}$	+ 0.1	
$Co(OH)_3 + e \rightleftharpoons Co(OH)_2 + OH^-$	+ 0.17	
$Cr^{++} + 2e \rightleftharpoons Cr$	- 0.56	
$Cr(III) + e \rightleftharpoons Cr(II)$	- 0.41	- 0.37 in 0.5 M H_2SO_4
		- 0.40 in 5 M HCl

Half-Reaction	Standard Potential	Formal Potential
$Cr(CN)_6^{-3} + e \rightleftarrows Cr(CN)_6^{-4}$		- 1.13 in 1 M KCN
$Cr(OH)_4^- + 3e \rightleftarrows Cr + 4\ OH^-$	- 1.2	
$CrO_4^= + 2H_2O + 3e \rightleftarrows CrO_2^- + 4\ OH$		- 1.2 in 1 M NaOH
$Cr_2O_7^= + 14\ H^+ + 6e \rightleftarrows 2\ Cr^{+++} + 7\ H_2O$	+ 1.33	+ 1.00 in 1 M HCl
		+ 0.92 in 0.1 M H_2SO_4
		+ 1.15 in 4 M H_2SO_4
$Cs^+ + e \rightleftarrows Cs$	- 2.92	
$Cu^{++} + 2e \rightleftarrows Cu$	+ 0.337	
$Cu(II) + e \rightleftarrows Cu(I)$	+ 0.153	+ 0.01 in 1 M NH_3 + 1 M NH_4^+
$Cu^+ + e \rightleftarrows Cu$	- 0.52	
$Cu(CN)_3^= + e \rightleftarrows Cu + 3CN^-$		- 1.0 in 7 M KCN
$2\ Cu^{++} + 2I^- + 2e \rightleftarrows Cu_2I_2$	+ 0.86	
$CuCl + e \rightleftarrows Cu + Cl^-$	+ 0.137	
$CuI + e \rightleftarrows Cu + I^-$	- 0.185	
$Cu^{++} + Cl^- + e \rightleftarrows CuCl$	+ 0.538	
$F_2 + 2e \rightleftarrows 2F^-$	+ 2.65	
$Fe^{++} + 2e \rightleftarrows Fe$	- 0.440	
$Fe(III) + e \rightleftarrows Fe(II)$	+ 0.771	+ 0.64 in 5 M HCl
		+ 0.735 in 1 M $HClO_4$
		+ 0.46 in 2 M H_3PO_4
		+ 0.68 in 1 M H_2SO_4
$Fe(CN)_6^{-3} + e \rightleftarrows Fe(CN)_6^{-4}$	+ 0.356	+ 0.71 in 1 M HCl
		+ 0.72 in 1 M $HClO_4$

Half-Reaction	Standard Potential	Formal Potential
$Fe(EDTA)^- + e \rightleftarrows Fe(EDTA)^=$		+ 0.12 (in 0.1 M EDTA) (pH 4-6)
$Fe(OH)_3^- + e \rightleftarrows Fe(OH)_2 + OH^-$	- 0.56	
$2H^+ + 2e \rightleftarrows H_2$	0	
$Hg_2^{++} + 2e \rightleftarrows 2\ Hg$	+ 0.792	
$2\ Hg^{++} + 2e \rightleftarrows Hg_2^{++}$	+ 0.907	
$Hg_2Br_2 + 2e \rightleftarrows 2Hg + 2Br^-$	+ 0.139	
$Hg_2Cl_2 + 2e \rightleftarrows 2Hg + 2Cl^-$	+ 0.268	
$Hg_2I_2 + 2e \rightleftarrows 2Hg + 2I^-$	- 0.040	
$HgS + 2e^- \rightleftarrows Hg + S^=$	- 0.72	
$I_2 + 2e \rightleftarrows 2I^-$	+ 0.536	
$I_3^- + 2e \rightleftarrows 3I^-$		+ 0.545 in .5M H_2SO_4
$HOI + H^+ + 2e \rightleftarrows I^- + H_2O$	+ 0.99	
$2IO_3^- + 12H^+ + 10e \rightleftarrows I_2 + 6H_2O$	+ 1.19	
$K^+ + e \rightleftarrows K$	- 2.925	
$Li^+ + e \rightleftarrows Li$	- 3.01	
$Mg^{++} + 2e \rightleftarrows Mg$	- 2.37	
$Mn^{++} + 2e \rightleftarrows Mn$	- 1.19	
$Mn(III) + e \rightleftarrows Mn(II)$		+ 1.5 in 7.5 M H_2SO_4
$MnO_2 + 4H^+ + 2e \rightleftarrows Mn^{++} + 2H_2O$	+ 1.23	
$MnO_4^- + 8H^+ + 5e \rightleftarrows Mn^{++} + 4H_2O$	+ 1.51	
$MnO_4^- + 4H^+ + 3e \rightleftarrows MnO_2 + 2H_2O$	+ 1.69	
$MnO_4^- + e \rightleftarrows MnO_4^=$	+ 0.56	
$Mn(OH)_2 + 2e \rightleftarrows Mn + 2OH^-$	- 1.55	

Half-Reaction	Standard Potential	Formal Potential
$MnO_2 + 2H_2O + 2e \rightleftarrows Mn(OH)_2 + 2OH^-$	- 0.05	
$MnO_4^- + 2H_2O + 3e \rightleftarrows MnO_2 + 4OH^-$	+ 0.59	
$Mo(IV) + e \rightleftarrows Mo(III)$		+ 0.1 in 4.5 M H_2SO_4
$Mo(V) + 2e \rightleftarrows Mo(III)$ (green)		- 0.25 in 2 M HCl
$Mo(V) + 2e \rightleftarrows Mo(III)$(red)		+ 0.11 in 2 M HCl
$Mo(VI) + e \rightleftarrows Mo(V)$	+ 0.45	+ 0.53 in 2 M HCl
$NO_3^- + 3H^+ + 2e \rightleftarrows HNO_2 + H_2O$	+ 0.94	
$2NO_3^- + 4H^+ + 2e \rightleftarrows N_2O_4 + 2H_2O$	+ 0.80	
$NO_3^- + 4H^+ + 3e \rightleftarrows NO + 2H_2O$	+ 0.96	
$HNO_2 + H^+ + e \rightleftarrows NO + H_2O$	+ 1.00	
$NO_3^- + H_2O + 2e \rightleftarrows NO_2^- + 2OH^-$	+ 0.01	
$Na^+ + e \rightleftarrows Na$	- 2.713	
$Ni^{++} + 2e \rightleftarrows Ni$	- 0.23	
$Ni(OH)_2 + 2e \rightleftarrows Ni + 2OH^-$	- 0.72	
$H_2O_2 + 2H^+ + 2e \rightleftarrows 2H_2O$	+ 1.77	
$2H_2O + 2e \rightleftarrows H_2 + 2OH^-$	- 0.828	
$O_2 + 4H^+ + 4e \rightleftarrows 2H_2O$	+ 1.229	
$O_2 + 2H_2O + 4e \rightleftarrows 4OH^-$	+ 0.401	
$O_2^= + 2H_2O + 2e \rightleftarrows 4\ OH^-$	+ 0.88	
$H_3PO_3 + 2H^+ + 2e \rightleftarrows H_3PO_2 + H_2O$	- 0.50	
$H_3PO_4 + 2H^+ + 2e \rightleftarrows H_3PO_3 + H_2O$	- 0.276	
$Pb^{++} + 2e \rightleftarrows Pb$	- 0.126	
$PbO_2 + H_2O + 2e \rightleftarrows PbO + 2OH^-$	+ 0.28	
$PbO_2 + 4H^+ + 2e \rightleftarrows Pb^{++} + 2H_2O$	+ 1.456	

Half-Reaction	Standard Potential	Formal Potential
$Pb(OH)_3^- + 2e \rightleftarrows Pb + 3OH^-$	- 0.54	
$PbO_2 + SO_4^= + 4H^+ + 2e \rightleftarrows PbSO_4 + 2H_2O$	+ 1.685	
$PbCl_2 + 2e \rightleftarrows Pb + 2Cl^-$	- 0.268	
$PbI_2 + 2e \rightleftarrows Pb + 2I^-$	- 0.365	
$PbSO_4 + 2e \rightleftarrows Pb + SO_4^=$	- 0.356	
$Pd^{++} + 2e \rightleftarrows Pd$		+ 0.987 in 4 M $HClO_4$
$Pt^{++} + 2e \rightleftarrows Pt$	+ 1.2	
$PtCl_6^= + 2e \rightleftarrows PtCl_4^= + 2Cl^-$		+ 0.720 in 1 M NaCl
$Rb^+ + e \rightleftarrows Rb$	- 2.92	
$S + 2e \rightleftarrows S^=$	- 0.48	
$S + 2H^+ + 2e \rightleftarrows H_2S$	+ 0.14	
$2SO_3^= + 2H_2O + 2e \rightleftarrows S_2O_4^= + 4OH^-$	- 1.12	
$S_4O_6^= + 2e \rightleftarrows 2S_2O_3^=$	+ 0.09	
$SO_4^= + 4H^+ + 2e \rightleftarrows SO_2 + 2H_2O$	+ 0.17	+ .07 in 1 M H_2SO_4
$2\ H_2SO_3 + 2H^+ + 4e \rightleftarrows S_2O_3^= + 2H_2O$	+ 0.40	
$H_2SO_3 + 4H^+ + 4e \rightleftarrows S + 3H_2O$	+ 0.45	
$SO_4^= + H_2O + 2e \rightleftarrows SO_3^= + 2OH^-$	- 0.93	
$(SCN)_2 + 2e \rightleftarrows 2\ SCN^-$	+ 0.77	
$Sb + 3H^+ + 3e \rightleftarrows SbH_3$	- 0.51	
$Sb_2O_3 + 6H^+ + 6e \rightleftarrows 2\ Sb + 3H_2O$	+ 0.152	
$SbO_2^- + 2H_2O + 3e \rightleftarrows Sb + 4OH^-$		+ 0.675 in 10 M KOH
$SbO^+ + 2H^+ + 3e \rightleftarrows Sb + H_2O$	+ 0.212	
$Sb_2O_5 + 6H^+ + 4e \rightleftarrows 2\ SbO^+ + 3H_2O$	+ 0.58	
$Sb(V) + 2e \rightleftarrows Sb(III)$		+ 0.75 in 3.5 M HCl

Half-Reaction	Standard Potential	Formal Potential
$SbO_3^- + H_2O + 2e \rightleftarrows SbO_2^- + 2OH^-$		- 0.589 in 10 M NaOH
$Sn^{++} + 2e \rightleftarrows Sn$	- 0.140	
$Sn(IV) + 2e \rightleftarrows Sn(II)$	+ 0.154	+ 0.14 in 1 M HCl
$SnCl_6^= + 2e \rightleftarrows Sn^{++} + 6Cl^-$	+ 0.15	
$Sn(OH)_3^- + 2e \rightleftarrows Sn + 3OH^-$	- 0.91	
$Sn(OH)_6^= + 2e \rightleftarrows Sn(OH)_3^- + 3OH^-$	- 0.90	
$Sr^{++} + 2e \rightleftarrows Sr$	- 2.89	
$Ti^{+++} + e \rightleftarrows Ti^{++}$	- 0.37	
$Ti(IV) + e \rightleftarrows Ti(III)$		- 0.05 in 1 M H_3PO_4
		- .01 in .2M H_2SO_4
		+ 0.12 in 2M H_2SO_4
		+ 0.20 in 4 M H_2SO_4
$Tl^+ + e \rightleftarrows Tl$	- 0.336	
$Tl^{+++} + 2e \rightleftarrows Tl^+$	+ 0.128	
$Tl(III) + 2e \rightleftarrows Tl^+$		+ 0.78 in 1 M HCl
$U(IV) + e \rightleftarrows U(III)$		- 0.64 in 1 M HCl
$UO_2^{++} + 4H^+ + 2e \rightleftarrows U(IV) + 2H_2O$		+ 0.41 in .5M H_2SO_4
$Zn^{++} + 2e \rightleftarrows Zn$	- 0.763	
$ZnS + 2e \rightleftarrows Zn + S^=$	- 1.44	
$Zn(CN)_4^= + 2e^- \rightleftarrows Zn + 4CN^-$	- 1.26	
$Zn(OH)_4^= + 2e \rightleftarrows Zn + 4OH^-$	- 1.22	
$Zn(NH_3)_4^{++} + 2e \rightleftarrows Zn + 4\ NH_3$	- 1.03	

Table VI

Acid-Base Indicators

Indicator	pK_a at 25° C	Acid Color	Base Color
Thymol blue	1.65	red	yellow
Quinaldin red	2.60	colorless	red
Methyl orange	3.39	red	orange
2,6-Dinitrophenol	3.67	colorless	yellow
Bromophenol blue	4.10	yellow	blue
Bromocresol green	4.90	yellow	blue
Methyl red	4.97	red	yellow
Chlorophenol red	6.22	yellow	red
Bromocresol purple	6.38	yellow	purple
p-Nitrophenol	6.95	colorless	yellow
Phenol red	7.96	yellow	red
m-Nitrophenol	8.30	colorless	yellow
Thymol blue	9.20	yellow	blue
Phenolphthalein	~9.5	colorless	red
Thymolphthalein	~10.0	colorless	blue

Table VII

Redox Indicators

Indicator	Color of Oxidized Form	Color of Reduced Form	Formal Potential at 25°C in 1 M H_2SO_4 (volts)
5-Nitroferroin (ferrous complex of 5-nitro-1,10-phenanthroline)	Blue	Purple	+ 1.25
Ferroin	Blue	Red	+ 1.11
5-Methylferroin	Blue-green	Red	+ 1.02
Erioglaucine A	Blue-red	Yellow-green	+ 0.98
Diphenylamine	Violet	Colorless	+ 0.76
Methylene blue	Blue	Colorless	+ 0.53
Indigotetrasulfonate	Blue	Colorless	+ 0.36
Diphenylaminesulfonic Acid	Purple	Colorless	+ 0.34
Phenosafranine	Red	Colorless	+ 0.28
m-Bromophenol-indophenol	Red	Colorless	+ 0.25

Table VIII

Metallochromic Indicators for Complexometric Titrations

Indicator	Acid Dissociation Constants of Indicator at 25° C: pK_1	pK_2	Formation Constants of Metal-Indicator Complexes at 25° C
Eriochrome Black T	6.3	11.6	$\log k_{CaIn} = 5.4$
			$\log k_{MgIn} = 7.0$
			$\log k_{ZnIn} = 12.9$
Murexide	9.2	10.5	$\log k_{CaIn} = 5.0$
			$\log k_{NiIn} = 11.3$
			$\log k_{CuIn} = 17.9$
5-Sulfosalicylic acid	2.7	11.7	$\log k_{MgIn} = 4.4$
			$\log k_{FeIn} = 14.4$
			$\log k_{NiIn} = 9.4$
			$\log k_{CuIn} = 16.5$
			$\log k_{PbIn} = 13.3$
1-(2-Pyridylazo)-2-naphthol (PAN)	<2	12.3	$\log k_{NiIn} = 12.7$
			$\log k_{ZnIn} = 11.2$
			$\log k_{MnIn} = 8.5$
Dithizone	4.6	-	$\log k_{ZnIn} = 5.2$
			$\log k_{NiIn} = 4.8$
			$\log k_{CoIn} = 5.3$
Metalphthalein	11.4	12.0	$\log k_{MgIn} = 8.9$
			$\log k_{CaIn} = 7.8$
			$\log k_{BaIn} = 6.2$

Index

Chapter I

(1) 11.3 ml; (2) 22.4% Ni; 5.75% Pb; (3) 0.23M; (4) 3.12g; (5) 5.84% Na; (6) 34.4 ml; (7) 23.7 ml; (8) 2.67% K; 37.40% Na; (9) (a) 3.32g; (b) $[K^+]$ = 0.250M; $[NO_3^-]$ = 0.125M; (10) 258 ml; (11) 900g; (12) 17.5M; (13) 14.8M; 22.9 molal; (14) mole fraction of CH_3OH;.049, C_2H_5OH;0.043 acetone; 0.32, H_2O:0.876; (15) 0.0263; (16) 2.8 molal; (17) 10.6M; (18) 9 : 4; (19) 60g 80% acid and 40g 75% acid, 66g 85% acid and 34g 80% acid; (20) 40.6 molal; (21) 46.2g.

Chapter II

(2) different K's but same equilibrium composition; (3) shift occurs to (a) left; (b) right; (c) left; (d) left; (e) left; (5) heat is absorbed; (6) exothermic; (7) +16 Kcal/mole; (8) -1.5 Kcal/mole; (9) (a) 0.69 Kcal/mole; (b) 0.22 atm; (10) greater; (11) -15.6 Kcal/mole.

Chapter III

(1) (a) 0.05; (b) 0.06; (c) 0.19; (d) 0.116; (e) 0.0; (f) 0.0013; (3) using Eq. III-8; (a) 0.90; (b) 0.82; (c) 0.79; (4) a_{H^+} = 8.5×10^{-3}, 6.3% 0.029; (5) γ_{Cl^-} = 0.80; a_{Cl^-} = 8.0×10^{-3}, γ_{K^+} = 0.80, a_{K^+} = 4.0×10^{-2}, $\gamma_{ClO_4^-}$ = 0.80, $a_{ClO_4^-}$ = 4.0×10^{-2}; (6) H^+, Li^+, Ce^{+3}; (9) pK = 13.76, $[H^+]$ = 1.32×10^{-7} M, neutral; (10) a_{H^+} = 7.2×10^{-3}.

Chapter IV

(4) (a) pH = 3.55; (b) pH = 2.40; (c) pH = 1.01; (d) pH = -.70; (5) $pH_{25°C}$ = 7.00; $pH_{0°C}$ = 7.48, $pH_{60°C}$ = 6.51; (6) (a) pH = 2.70; (b) pH = 6.51; (c) pH = 11.26; (d) pH = 7.09; (7) (a) $[H^+]$ = 1.22×10^{-7} M; (b) $[H^+]$ = 1.00×10^{-7} M; (c) $[H^+]$ = 3.86×10^{-3} M; (8) (a) $[OH^-]$ = 5.32×10^{-3} M; (b) $[OH^-]$ = 1.61×10^{-7} M; (c) $[OH^-]$ = 1.00×10^{-7} M.

Chapter V

(1) (a) 0.364; (b) 3.64×10^{-4}; (c) 13.5%; (2) (a) 2.87; (b) 3.59; (c) 5.97; (d) 6.44; (e) 6.93; (3) 1.47×10^{-4}; (4) (a) 5.12; (b) 8.72; (c) 7.57; (d) 10.76; (e) 11.48; (5) (a) 4.4×10^{-3}; (b) 4.9×10^{-3}; (c) 1.2×10^{-4}; (d) 4.7×10^{-11}; (6) 1.95%; (7) (a) $[H^+] = 10^{-10.58}$ M, $[OH^-] = [HCN] = 10^{-3.30}$ M, $[K^+] = 0.01$M, $[CN^-] = 9.5 \times 10^{-3}$ M; (b) $[H^+] = 10^{-10.85}$, $[OH^-] = [NH_4^+] = 10^{-3.12}$ M, $[NH_3] = 0.03$M; (c) $[H^+] = [NH_3] = 10^{-6.12}$ M, $[OH^-] = 10^{-7.85}$ M, $[NH_4^+] = [Cl^-] = 10^{-3.00}$ M; (d) $[H^+] = [OH^-] = 10^{-6.85}$ M, $[K^+] = [Cl^-] = 0.2$M; (8) (a) 3.30; (b) 8.80; (c) 3.17; (d) 7.05; (e) 6.79; (9) (a) 2.9×10^{-8} M; (b) 1.06×10^{-6} M; (10) 0.30; (11) 4.8g; (12) 0.92g; (13) 79.4 ml; (14) (a) 4.22; (b) 3.15; (c) 9.54; (18) (a) For solutions in 14 (a) and (b), pH increased by 0.008; (b) for solution in 14 (c) pH increases by 0.013; (19) (a) In 100 ml of buffer, $[NH_4^+] = 2.9 \times 10^{-4}$ M and $[NH_3] = 1.1 \times 10^{-4}$ M; (b) in 100 ml of buffer, $[NH_3] = 3.1 \times 10^{-4}$ M and $[NH_4^+] = 8.6 \times 10^{-4}$ M; (20) (a) For the addition of NaOH, $[HOAc] = 1.00 \times 10^{-4}$ M and $[OAc^-] = 3.6 \times 10^{-4}$M; (21) (a) $[H^+] = 1^{-10.10}$ M $[OH^-] = [HCN] = 10^{-3.73}$ M; (b) $[H^+] = 10^{-8.30}$ M $[OH] = 10^{-5.68}$ M $[K^+] = [CN^-] = 10^{-3.60}$ M $[HCN] = 10^{-2.52}$ M.

Chapter VI

(1) (a) 4.35; (b) 1.8; (c) 2.02; (d) 4.03; (2) (a) 1.8×10^{-3} M; (b) $10^{-3.89}$; (3) 2.54; (4) (a) 12.21; (b) $10^{-10.9}$; (c) $10^{-11.15}$; (d) $10^{-10.23}$; (e) $10^{-10.38}$; (5) (a) 2.24×10^{-4} M; (b) 2.57×10^{-5}; (c) 3.24×10^{-6}; (6) 4.02; (7) 6.6×10^{-3} M; (8) $Ka_1 = 1.6 \times 10^{-9}$;

(10)

pH	3.0	6.0	9.0
$[S^{-2}]$	$10^{-15.2}$	$10^{-9.3}$	$10^{-5.4}$
$[PO_4^{-3}]$	$10^{-14.4}$	$10^{-8.5}$	$10^{-4.7}$
$[Y^{-4}]$	$10^{-12.7}$	$10^{-6.8}$	$10^{-3.4}$

Chapter VI (Cont'd)

(11) (a) pH = 10.59; (b) pH = 9.65; (c) pH = 2.98; (d) pH = 2.73; (13) (a) $[H^+]$ = 1.1 x 10^{-6} M; (b) $[H^+]$ = 4.1 x 10^{-9} M; (c) $[H^+]$ = 5.0 x 10^{-9} M; (d) $[H^+]$ = 3.4 x 10^{-5} M; (e) $[H^+]$ = 5.0 x 10^{-10}M; (14) pH = 6.5; (15) 0.014M; (16) $[H_2PO_4^-]$ = .0253M, $[HPO_4^{-2}]$ = 0.0822M.

Chapter VII

(1) (a) pH = 2.64; (b) pH = 4.86; (c) pH = 5.31; (d) pH = 5.43; (e) pH = 3.42; (2) pH = 3.37; (3) pH = 2.35; (4) pH = 1.61; (5) pH = 10.38; (6) (a) pH = 1.60; (b) pH = 0.64; (c) pH = 5.42; (d) pH = 11.36; (7) (a) pH = 8.59 using pKw = 13.80; (b) pH = 8.55 using pKw = 13.82; (c) pH = 6.91; (d) pH = 4.56; (8) (a) pH = 6.97; (b) pH = 9.22; (c) pH = 6.91; (d) pH = 4.64; (9) pH = 9.24; $[NH_3]$ = $[NH_4^+]$ = $[SH^-]$ = 0.02M:$[S^=]$ = $10^{-5.2}$ M; (10) pH = 9.1; (11) (a) pH = 9.32; (b) pH = 8.88; (c) pH = 12.1; (d) pH = 6.41.

Chapter VIII

(1) (a) 2.17 x 10^{-10} M; (b) 4.58 x 10^{-5} M; (c) 6.30 x 10^{-5} M; (d) 6.88 x 10^{-7} M; (e) 7.46 x 10^{-7} M; (3) 4.0 x 10^{-14}; (4) Thallous phosphate: 6.8 x 10^{-8}, lead phosphate; 1.7 x 10^{-32}, ferric phosphate; 4.4 x 10^{-5}; (5) (a) 2.83 x 10^{-18} M; (b) 1.72 x 10^{-26} M; (c) 2.83 x 10^{-14} M; (d) 1.09 x 10^{-5} M; (e) 2.67 x 10^{-2} M; (f) 2.24 x 10^{-3} M; (g) 1.41 x 10^{-8} M; (6) (a) 2.19 x 10^{-8} M; (b) 3.8 x 10^{-6} M; (c) 4.47 x 10^{-8} M; (8) (a) 4.1 x 10^{-7} M; (b) 4.1 x 10^{-7} M; (9) (a) 7.4 x 10^{-10}; (b) 6.9 x 10^{-7}; (10) AgCl precipitates first, $[K^+]$ = 1.5 x 10^{-3}, $[CrO_4^=]$ = 5.0 x 10^{-4} M, $[Ag^+]$ = 8.1 x 10^{-5} M, $[Cl^-]$ = 2.5 x 10^{-6} M; (11) (a) 1.7 x 10^{-4} M; (b) 0.1710M; (c) $[Cl^-]$ = 0.0855M; $[Br^-]$ = 2.35 x 10^{-2} M; (d) $10^{2.53}$; (e) $10^{2.53}$; (12) (a) Using pK_{sp} of AgI = 15.61, $\frac{[I^-]}{[CrO_4^-]}$ = $10^{-8.21}$; (b) Using pK_{sp} of AgI = 15.50 $\frac{[I^-]}{[CrO_4^=]}$ = $10^{-5.97}$; (13) 0.14%; (14) 1.2 x 10^{-6} M; (15) 0.51 M.

Chapter IX

(1) $10^{-10.16}$ M; (2) (a) (1) pAg = 7.01; (a) (2) pAg = 8.52; (a) (3) pAg = 9.19; (3) (a) (1) pCd = 16.06; (a) (2) pCd = 18.02; (a) (3) pCd = 18.86; (a) (4) pCd = 19.66; (4) (a) pNi = 14.11; (6) (a) $10^{-10.75}$ M; (b) $10^{-20.05}$ M;

(7)

	$[I^-]$	$[Hg^{2+}]$	$[HgI^+]$	$[HgI_2]$	$[HgI_3^-]$	$[HgI_4^{2-}]$
a	$10^{-11.91}$	$10^{-4.05}$	$10^{-3.09}$	$10^{-4.05}$	$10^{-12.18}$	$10^{-21.86}$
b	$10^{-7.37}$	$10^{-12.09}$	$10^{-6.59}$	$10^{-3.00}$	$10^{-6.59}$	$10^{-11.73}$
c	$10^{-3.00}$	$10^{-21.75}$	$10^{-11.88}$	$10^{-3.93}$	$10^{-3.16}$	$10^{-3.91}$
d	$10^{-2.33}$	$10^{-23.83}$	$10^{-13.29}$	$10^{-4.67}$	$10^{-3.22}$	$10^{-3.40}$

(9) $10^{-6.86}$; $10^{-11.01}$; $10^{-14.13}$; $10^{-16.17}$ M;

(10)

pH	$[Zn^{++}]$	$[Znen^{++}]$	$[Znen_2^{++}]$	$[ZnY^{-2}]$	K_f'
7	$10^{-13.87}$	$10^{-12.57}$	$10^{-12.19}$	10^{-3}	$10^{+11.34}$
8	$10^{-14.91}$	$10^{-12.15}$	$10^{-10.31}$	10^{-3}	$10^{+9.61}$
9	$10^{-15.91}$	$10^{-12.05}$	$10^{-9.11}$	10^{-3}	$10^{+8.41}$
10	$10^{-16.74}$	$10^{-12.12}$	$10^{-8.42}$	10^{-3}	$10^{+7.72}$

Chapter X

(1) Between pH 1.13 and 3.03, using a pH of 2.08 (mid range), $[HSO_4^-]$ = 1.28 x 10^{-2} M, $[SO_4^{-2}]$ = 8.0 x 10^{-2} M; (2) 0.048 moles; (3) $10^{-23.4}$; (4) 5.4 x 10^{-3} M; (5) (a) 10^{-3} M or if the CO_2 is boiled out, 5 x 10^{-4} M; (b) 1.1 x 10^{-4} M; (7) (a) 3.70 if pK_{sp} for $Ca_3(PO_4)_2$ is 27.25; (b) 1.5 x 10^{-2} M.

TABLE OF ATOMIC WEIGHTS—1961

(Based on Carbon-12)

Element	Symbol	Atomic Number	Atomic Weight	Element	Symbol	Atomic Number	Atomic Weight
Actinium	Ac	89		Mercury	Hg	80	200.59
Aluminum	Al	13	26.9815	Molybdenum	Mo	42	95.94
Americium	Am	95		Neodymium	Nd	60	144.24
Antimony	Sb	51	121.75	Neon	Ne	10	20.183
Argon	Ar	18	39.948	Neptunium	Np	93	
Arsenic	As	33	74.9216	Nickel	Ni	28	58.71
Astatine	At	85		Niobium	Nb	41	92.906
Barium	Ba	56	137.34	Nitrogen	N	7	14.0067
Berkelium	Bk	97		Nobelium	No	102	
Beryllium	Be	4	9.0122	Osmium	Os	76	190.2
Bismuth	Bi	83	208.980	Oxygen	O	8	15.9994[a]
Boron	B	5	10.811[a]	Palladium	Pd	46	106.4
Bromine	Br	35	79.909[b]	Phosphorus	P	15	30.9738
Cadmium	Cd	48	112.40	Platinum	Pt	78	195.09
Calcium	Ca	20	40.08	Plutonium	Pu	94	
Californium	Cf	98		Polonium	Po	84	
Carbon	C	6	12.01115[a]	Potassium	K	19	39.102
Cerium	Ce	58	140.12	Praseodymium	Pr	59	140.907
Cesium	Cs	55	132.905	Promethium	Pm	61	
Chlorine	Cl	17	35.453[b]	Protactinium	Pa	91	
Chromium	Cr	24	51.996[b]	Radium	Ra	88	
Cobalt	Co	27	58.9332	Radon	Rn	86	
Copper	Cu	29	63.54	Rhenium	Re	75	186.2
Curium	Cm	96		Rhodium	Rh	45	102.905
Dysprosium	Dy	66	162.50	Rubidium	Rb	37	85.47
Einsteinium	Es	99		Ruthenium	Ru	44	101.07
Erbium	Er	68	167.26	Samarium	Sm	62	150.35
Europium	Eu	63	151.96	Scandium	Sc	21	44.956
Fermium	Fm	100		Selenium	Se	34	78.96
Fluorine	F	9	18.9984	Silicon	Si	14	28.086[a]
Francium	Fr	87		Silver	Ag	47	107.870[b]
Gadolinium	Gd	64	157.25	Sodium	Na	11	22.9898
Gallium	Ga	31	69.72	Strontium	Sr	38	87.62
Germanium	Ge	32	72.59	Sulfur	S	16	32.064[a]
Gold	Au	79	196.967	Tantalum	Ta	73	180.948
Hafnium	Hf	72	178.49	Technetium	Tc	43	
Helium	He	2	4.0026	Tellurium	Te	52	127.60
Holmium	Ho	67	164.930	Terbium	Tb	65	158.924
Hydrogen	H	1	1.00797[a]	Thallium	Tl	81	204.37
Indium	In	49	114.82	Thorium	Th	90	232.038
Iodine	I	53	126.9044	Thulium	Tm	69	168.934
Iridium	Ir	77	192.2	Tin	Sn	50	118.69
Iron	Fe	26	55.847[b]	Titanium	Ti	22	47.90
Krypton	Kr	36	83.80	Tungsten	W	74	183.85
Lanthanum	La	57	138.91	Uranium	U	92	238.03
Lead	Pb	82	207.19	Vanadium	V	23	50.942
Lithium	Li	3	6.939	Xenon	Xe	54	131.30
Lutetium	Lu	71	174.97	Ytterbium	Yb	70	173.04
Magnesium	Mg	12	24.312	Yttrium	Y	39	88.905
Manganese	Mn	25	54.9381	Zinc	Zn	30	65.37
Mendelevium	Md	101		Zirconium	Zr	40	91.22

[a] The atomic weight varies because of natural variations in the isotopic composition of the element. The observed ranges are boron, ± 0.003; carbon, ± 0.00005; hydrogen, ± 0.00001; oxygen, ± 0.0001; silicon, ± 0.001; sulfur, ± 0.003.

[b] The atomic weight is believed to have an experimental uncertainty of the following magnitude: bromine, ± 0.002; chlorine, ± 0.001; chromium, ± 0.001; iron, ± 0.003; silver, ± 0.003. For other elements the last digit given is believed to be reliable to ± 0.5.

FOUR-PLACE LOGARITHMS

	0	1	2	3	4	5	6	7	8	9	Proportional Parts								
											1	2	3	4	5	6	7	8	9
10	0000	0043	0086	0128	0170	0212	0253	0294	0334	0374	4	8	12	17	21	25	29	33	37
11	0414	0453	0492	0531	0569	0607	0645	0682	0719	0755	4	8	11	15	19	23	26	30	34
12	0792	0828	0864	0899	0934	0969	1004	1038	1072	1106	3	7	10	14	17	21	24	28	31
13	1139	1173	1206	1239	1271	1303	1335	1367	1399	1430	3	6	10	13	16	19	23	26	29
14	1461	1492	1523	1553	1584	1614	1644	1673	1703	1732	3	6	9	12	15	18	21	24	27
15	1761	1790	1818	1847	1875	1903	1931	1959	1987	2014	3	6	8	11	14	17	20	22	25
16	2041	2068	2095	2122	2148	2175	2201	2227	2253	2279	3	5	8	11	13	16	18	21	24
17	2304	2330	2355	2380	2405	2430	2455	2480	2504	2529	2	5	7	10	12	15	17	20	22
18	2553	2577	2601	2625	2648	2672	2695	2718	2742	2765	2	5	7	9	12	14	16	19	21
19	2788	2810	2833	2856	2878	2900	2923	2945	2967	2989	2	4	7	9	11	13	16	18	20
20	3010	3032	3054	3075	3096	3118	3139	3160	3181	3201	2	4	6	8	11	13	15	17	19
21	3222	3243	3263	3284	3304	3324	3345	3365	3385	3404	2	4	6	8	10	12	14	16	18
22	3424	3444	3464	3483	3502	3522	3541	3560	3579	3598	2	4	6	8	10	12	14	15	17
23	3617	3636	3655	3674	3692	3711	3729	3747	3766	3784	2	4	6	7	9	11	13	15	17
24	3802	3820	3838	3856	3874	3892	3909	3927	3945	3962	2	4	5	7	9	11	12	14	16
25	3979	3997	4014	4031	4048	4065	4082	4099	4116	4133	2	3	5	7	9	10	12	14	15
26	4150	4166	4183	4200	4216	4232	4249	4265	4281	4298	2	3	5	7	8	10	11	13	15
27	4314	4330	4346	4362	4378	4393	4409	4425	4440	4456	2	3	5	6	8	9	11	13	14
28	4472	4487	4502	4518	4533	4548	4564	4579	4594	4609	2	3	5	6	8	9	11	12	14
29	4624	4639	4654	4669	4683	4698	4713	4728	4742	4757	1	3	4	6	7	9	10	12	13
30	4771	4786	4800	4814	4829	4843	4857	4871	4886	4900	1	3	4	6	7	9	10	11	13
31	4914	4928	4942	4955	4969	4983	4997	5011	5024	5038	1	3	4	6	7	8	10	11	12
32	5051	5065	5079	5092	5105	5119	5132	5145	5159	5172	1	3	4	5	7	8	9	11	12
33	5185	5198	5211	5224	5237	5250	5263	5276	5289	5302	1	3	4	5	6	8	9	10	12
34	5315	5328	5340	5353	5366	5378	5391	5403	5416	5428	1	3	4	5	6	8	9	10	11
35	5441	5453	5465	5478	5490	5502	5514	5527	5539	5551	1	2	4	5	6	7	9	10	11
36	5563	5575	5587	5599	5611	5623	5635	5647	5658	5670	1	2	4	5	6	7	8	10	11
37	5682	5694	5705	5717	5729	5740	5752	5763	5775	5786	1	2	3	5	6	7	8	9	10
38	5798	5809	5821	5832	5843	5855	5866	5877	5888	5899	1	2	3	5	6	7	8	9	10
39	5911	5922	5933	5944	5955	5966	5977	5988	5999	6010	1	2	3	4	5	7	8	9	10
40	6021	6031	6042	6053	6064	6075	6085	6096	6107	6117	1	2	3	4	5	6	8	9	10
41	6128	6138	6149	6160	6170	6180	6191	6201	6212	6222	1	2	3	4	5	6	7	8	9
42	6232	6243	6253	6263	6274	6284	6294	6304	6314	6325	1	2	3	4	5	6	7	8	9
43	6335	6345	6355	6365	6375	6385	6395	6405	6415	6425	1	2	3	4	5	6	7	8	9
44	6435	6444	6454	6464	6474	6484	6493	6503	6513	6522	1	2	3	4	5	6	7	8	9
45	6532	6542	6551	6561	6571	6580	6590	6599	6609	6618	1	2	3	4	5	6	7	8	9
46	6628	6637	6646	6656	6665	6675	6684	6693	6702	6712	1	2	3	4	5	6	7	7	8
47	6721	6730	6739	6749	6758	6767	6776	6785	6794	6803	1	2	3	4	5	5	6	7	8
48	6812	6821	6830	6839	6848	6857	6866	6875	6884	6893	1	2	3	4	4	5	6	7	8
49	6902	6911	6920	6928	6937	6946	6955	6964	6972	6981	1	2	3	4	4	5	6	7	8
50	6990	6998	7007	7016	7024	7033	7042	7050	7059	7067	1	2	3	3	4	5	6	7	8
51	7076	7084	7093	7101	7110	7118	7126	7135	7143	7152	1	2	3	3	4	5	6	7	8
52	7160	7168	7177	7185	7193	7202	7210	7218	7226	7235	1	2	2	3	4	5	6	7	7
53	7243	7251	7259	7267	7275	7284	7292	7300	7308	7316	1	2	2	3	4	5	6	6	7
54	7324	7332	7340	7348	7356	7364	7372	7380	7388	7396	1	2	2	3	4	5	6	6	7
	0	1	2	3	4	5	6	7	8	9	1	2	3	4	5	6	7	8	9

FOUR-PLACE LOGARITHMS

	0	1	2	3	4	5	6	7	8	9	Proportional Parts 1	2	3	4	5	6	7	8	9
55	7404	7412	7419	7427	7435	7443	7451	7459	7466	7474	1	2	2	3	4	5	5	6	7
56	7482	7490	7497	7505	7513	7520	7528	7536	7543	7551	1	2	2	3	4	5	5	6	7
57	7559	7566	7574	7582	7589	7597	7604	7612	7619	7627	1	2	2	3	4	5	5	6	7
58	7634	7642	7649	7657	7664	7672	7679	7686	7694	7701	1	1	2	3	4	4	5	6	7
59	7709	7716	7723	7731	7738	7745	7752	7760	7767	7774	1	1	2	3	4	4	5	6	7
60	7782	7789	7796	7803	7810	7818	7825	7832	7839	7846	1	1	2	3	4	4	5	6	6
61	7853	7860	7868	7875	7882	7889	7896	7903	7910	7917	1	1	2	3	4	4	5	6	6
62	7924	7931	7938	7945	7952	7959	7966	7973	7980	7987	1	1	2	3	3	4	5	6	6
63	7993	8000	8007	8014	8021	8028	8035	8041	8048	8055	1	1	2	3	3	4	5	5	6
64	8062	8069	8075	8082	8089	8096	8102	8109	8116	8122	1	1	2	3	3	4	5	5	6
65	8129	8136	8142	8149	8156	8162	8169	8176	8182	8189	1	1	2	3	3	4	5	5	6
66	8195	8202	8209	8215	8222	8228	8235	8241	8248	8254	1	1	2	3	3	4	5	5	6
67	8261	8267	8274	8280	8287	8293	8299	8306	8312	8319	1	1	2	3	3	4	5	5	6
68	8325	8331	8338	8344	8351	8357	8363	8370	8376	8382	1	1	2	3	3	4	4	5	6
69	8388	8395	8401	8407	8414	8420	8426	8432	8439	8445	1	1	2	2	3	4	4	5	6
70	8451	8457	8463	8470	8476	8482	8488	8494	8500	8506	1	1	2	2	3	4	4	5	6
71	8513	8519	8525	8531	8537	8543	8549	8555	8561	8567	1	1	2	2	3	4	4	5	5
72	8573	8579	8585	8591	8597	8603	8609	8615	8621	8627	1	1	2	2	3	4	4	5	5
73	8633	8639	8645	8651	8657	8663	8669	8675	8681	8686	1	1	2	2	3	4	4	5	5
74	8692	8698	8704	8710	8716	8722	8727	8733	8739	8745	1	1	2	2	3	4	4	5	5
75	8751	8756	8762	8768	8774	8779	8785	8791	8797	8802	1	1	2	2	3	3	4	5	5
76	8808	8814	8820	8825	8831	8837	8842	8848	8854	8859	1	1	2	2	3	3	4	5	5
77	8865	8871	8876	8882	8887	8893	8899	8904	8910	8915	1	1	2	2	3	3	4	4	5
78	8921	8927	8932	8938	8943	8949	8954	8960	8965	8971	1	1	2	2	3	3	4	4	5
79	8976	8982	8987	8993	8998	9004	9009	9015	9020	9025	1	1	2	2	3	3	4	4	5
80	9031	9036	9042	9047	9053	9058	9063	9069	9074	9079	1	1	2	2	3	3	4	4	5
81	9085	9090	9096	9101	9106	9112	9117	9122	9128	9133	1	1	2	2	3	3	4	4	5
82	9138	9143	9149	9154	9159	9165	9170	9175	9180	9186	1	1	2	2	3	3	4	4	5
83	9191	9196	9201	9206	9212	9217	9222	9227	9232	9238	1	1	2	2	3	3	4	4	5
84	9243	9248	9253	9258	9263	9269	9274	9279	9284	9289	1	1	2	2	3	3	4	4	5
85	9294	9299	9304	9309	9315	9320	9325	9330	9335	9340	1	1	2	2	3	3	4	4	5
86	9345	9350	9355	9360	9365	9370	9375	9380	9385	9390	1	1	2	2	3	3	4	4	5
87	9395	9400	9405	9410	9415	9420	9425	9430	9435	9440	0	1	1	2	2	3	3	4	4
88	9445	9450	9455	9460	9465	9469	9474	9479	9484	9489	0	1	1	2	2	3	3	4	4
89	9494	9499	9504	9509	9513	9518	9523	9528	9533	9538	0	1	1	2	2	3	3	4	4
90	9542	9547	9552	9557	9562	9566	9571	9576	9581	9586	0	1	1	2	2	3	3	4	4
91	9590	9595	9600	9605	9609	9614	9619	9624	9628	9633	0	1	1	2	2	3	3	4	4
92	9638	9643	9647	9652	9657	9661	9666	9671	9675	9680	0	1	1	2	2	3	3	4	4
93	9685	9689	9694	9699	9703	9708	9713	9717	9722	9727	0	1	1	2	2	3	3	4	4
94	9731	9736	9741	9745	9750	9754	9759	9763	9768	9773	0	1	1	2	2	3	3	4	4
95	9777	9782	9786	9791	9795	9800	9805	9809	9814	9818	0	1	1	2	2	3	3	4	4
96	9823	9827	9832	9836	9841	9845	9850	9854	9859	9863	0	1	1	2	2	3	3	4	4
97	9868	9872	9877	9881	9886	9890	9894	9899	9903	9908	0	1	1	2	2	3	3	4	4
98	9912	9917	9921	9926	9930	9934	9939	9943	9948	9952	0	1	1	2	2	3	3	4	4
99	9956	9961	9965	9969	9974	9978	9983	9987	9991	9996	0	1	1	2	2	3	3	3	4
	0	1	2	3	4	5	6	7	8	9	1	2	3	4	5	6	7	8	9